# Timing Jitter in Time-of-Flight Range Imaging Cameras

Gehan Anthonys

# Timing Jitter in Time-of-Flight Range Imaging Cameras

Springer

Gehan Anthonys
Department of Electrical and Computer Engineering
Open University of Sri Lanka
Nawala, Sri Lanka

ISBN 978-3-030-94158-1 ISBN 978-3-030-94159-8 (eBook)
https://doi.org/10.1007/978-3-030-94159-8



This Springer imprint is published by the registered company Springer Nature Switzerland AG
The registered company address is: Gewerbestrasse 11, 6330 Cham, Switzerland

*To my family and all the teachers.*

# Preface

This book continues from fundamental principles in signal processing with time-of-flight (ToF) range imaging (depth) cameras and is suitable for self-study. The reader is assumed to be comfortable with basic signal processing, elementary statistical analysis including the idea of the Gaussian distribution and the concepts of ToF principle; otherwise, the text is self-supportive. This book is focused on ToF cameras that are based on the amplitude modulated continuous wave (AMCW) lidar techniques which measure the phase difference between the emitted and reflected light signals.

The AMCW depth cameras measure the time taken from the modulation light source to the scene and back to the camera, and it is this principle that is used to perform depth measurements. Due to their portable size, feasible design, low weight and less energy consumption, these cameras have high demand in many applications. Commercially available AMCW ToF cameras have relatively high noise levels due to electronic sources such as shot noise, reset noise, amplifier noise, crosstalk, analogue to digital converters quantization and multipath light interference. Many noise sources in these cameras such as harmonic contamination, non-linearity, multipath interferences and light scattering are well investigated. In contrast, the effect of electronic jitter as a noise source in ranging cameras is barely studied.

Jitter is defined to be any timing movement with respect to an ideal (reference) signal. An investigation of the effect of jitter on range imaging is important because timing errors potentially could cause errors in measuring phase, thus in range. This book is organized into three parts: the first is devoted to the operational principles and measurement error of ToF range imaging cameras, the second to the jitter extraction and measurement in ToF depth cameras through proper signal processing techniques with an emerging technique in telecommunication community, and the third to the investigation of jitter influence range measurements.

Those three parts consist of Chaps. 2–10 of this book. In Chap. 1, an understanding of the characteristics of current ToF range imaging systems based on their modulation illumination source, other depth measurement techniques and applications of ToF range imaging cameras are briefly reviewed. The first part, consisting of Chaps. 2–4, provides in more detail an overview and error sources in commercial AMCW ToF depth cameras. Chapters 5–8 present the jitter extraction and measurement in ToF range imaging cameras. The third part of the book addresses the investigation of the influence of the two types of jitter: random and periodic, in the illumination sources of ToF range imaging cameras in Chaps. 9 and 10. Finally, Chap. 11 concludes the significant contributions of this book, summarizes the findings in each chapter and suggests some potential future investigation.

Before this book, there was a huge gap in the analysis of jitter present in the AMCW ToF ranging systems, namely the jitter extraction, the effect of jitter, and what types and what amounts of jitter have more effect on range measurements related to the AMCW ToF ranging. Most of these were largely uninvestigated in the literature. Because of this, analysing, experimenting and designing theoretical models that have not been performed previously became the objectives of this book. Based on this undertaking, enhancements have been developed and successfully incorporated to significantly improve the quality of the achievements. Thus, this book has contributed to minimizing the above-mentioned gaps as much as possible.

This book will be useful to university students, researchers, scientists, engineers and those interested in pursuing research in this area. Not only that, the recommended safe values for the specific types of jitter on the light source signals of the AMCW ToF cameras given in the book are beneficial to the depth sensor manufacturing industries as well as the future industrial applications because the precision and accuracy are important for many applications. However, we hope the content of this book may contribute to stimulate the readers to gain sufficient knowledge to pursue research in this interesting area.

Colombo, Sri Lanka
November 2021

Gehan Anthonys

# Acknowledgements

I wish to express my gratitude to Associate Professor Michael J. Cree, attached to the School of Engineering at the University of Waikato, New Zealand, for his great, fruitful and stimulating discussions in the early stages of this investigation. Then, to Dr Lee Streeter for his valuable suggestions and support given to access the Time of Flight Sensing laboratory in the same department. Next, to a colleague, Carl Lickfold for sharing the valuable ideas.

Also, special thanks for the New Zealand eScience Infrastructure (NeSI) for providing high-performance computing facilities (https://www.nesi.org.nz.) to obtain simulation results in Chaps. 9 and 10 of this book.

# Contents

# Acronyms

| | |
|---|---|
| a.u. | Arbitrary unit |
| w.r.t. | With respect to |
| | |
| AM | Amplitude modulation |
| AMCW | Amplitude modulated continuous wave |
| ADC | Analogue to digital converter |
| | |
| BER | Bit error ratio |
| BERT | Bit error ratio tester |
| BNC | Bayonet Neill-Concelman |
| BUJ | Bounded uncorrelated jitter |
| | |
| CCD | Charged coupled device |
| CMOS | Complementary metal oxide semiconductor |
| CNN | Convolutional neural network |
| CW | Continuous wave |
| | |
| DDJ | Data dependent jitter |
| DJ | Deterministic jitter |
| DUT | Device under test |
| DDC | Digital down converter |
| DAC | Digital to analogue converter |
| DSP | Digital signal processing |
| DC | Direct current |
| DFT | Discrete Fourier transform |
| DCD | Duty cycle distortion |
| | |
| FFT | Fast Fourier transform |
| FM | Frequency modulation |

| | |
|---|---|
| FPN | Fixed pattern noise |
| FT | Fourier transform |
| GPIB | General purpose interface bus |
| GCD | Greatest common divisor |
| IF | Intermediate frequency |
| ISI | Inter symbol interference |
| IFFT | Inverse fast Fourier transform |
| IQR | Interquartile range |
| KDE | Kernel density estimation |
| LED | Light emitting diode |
| LiDAR | Light detection and ranging |
| LO | Local oscillator |
| LOTUS | Law of the unconscious statistician |
| MCS | Monte Carlo simulation |
| MLE | Maximum likelihood estimation |
| MGF | Moment generating function |
| MPI | Multipath interference |
| NA | Not available/applicable |
| PB | Pulse based |
| PJ | Periodic jitter |
| PLL | Phase locked loop |
| PM | Phase modulation |
| PMD | Photonic mixing device |
| PN | Pseudo-noise |
| PSD | Power spectrum density |
| PDF | Probability density function |
| RJ | Random jitter |
| RF | Radio frequency |
| RMS | Root mean square |
| RMSE | Root mean square error |
| SDR | Software defined radio |
| SE | Standard error |
| SG | Savitzky-Golay algorithm |

| | |
|---|---|
| SMA | SubMiniature version A |
| SMB | SubMiniature version B |
| SNR | Signal to noise ratio |
| SRA | Sparse reflections analysis |
| STD | Standard deviation |
| TIE | Time interval error |
| TLC | Time lag correlation |
| ToF | Time-of-flight |
| USB | Universal serial bus |
| UI | Unit interval |

# Chapter 1
# Introduction

**Abstract** This chapter provides an introduction of what is the time-of-flight (ToF) range imaging cameras and the different types of ToF systems based on their modulation light sources. Then, types of depth measurement techniques are briefly reviewed. Next, a sufficient number of applications of ToF range imaging cameras are described. Finally, the overview of this book is presented as a summary and is illustrated as a flowchart.

**Keywords** Time-of-flight (ToF) · Continuous wave · Pulse-based · Pseudo-noise · Range resolution · Modulation frequency · Amplitude modulation · Frequency modulation · Stereoscopic vision · Structured-light · Infrared thermography · Phase shift · Camera triangulation · Applications of ToF · Commercial ToF cameras

## 1.1 Time-of-Flight Range Imaging Systems

The time-of-flight (ToF) principle is measuring the time taken from the modulation light source to the scene and back to the sensor and it is this principle that is used in range imaging systems to perform depth measurements. Sonar and Radar are two widely known ToF systems. Sonar transmits acoustic waves while radar sends out electromagnetic waves. Before the twenty-first century, mostly sound waves have been used for the measurement of short distances since the speed of sound is much slower than the electromagnetic waves and thus the timing does not have to be as accurate [1]. Now, it is possible to use electromagnetic waves for short distance ToF applications with the increasing speed of electronic components. With narrow laser beams possible over long distances, infrared or visible light is desirable since it is relatively easier to consider than sound [2, 3].

From the twenty-first century, depth (range) sensors have been improving in the camera manufacturing industries [4, 5]. Based on the modulation light source, ToF range imaging cameras are categorized as (1) continuous wave (CW) modulation; (2) pulse-based (PB) modulation; and (3) pseudo-noise (PN) modulation [6].



G. Anthonys, *Timing Jitter in Time-of-Flight Range Imaging Cameras*,
https://doi.org/10.1007/978-3-030-94159-8_1

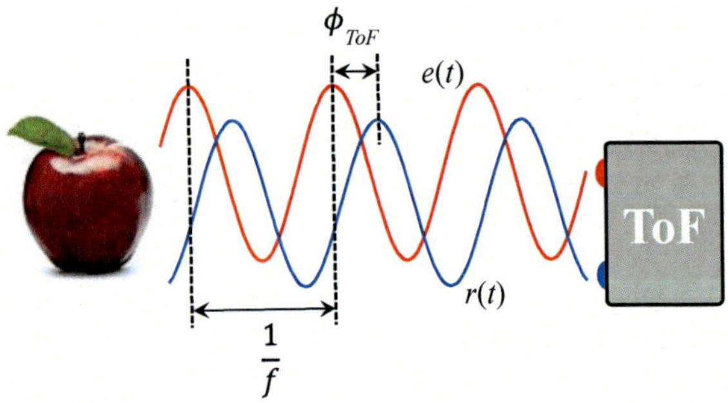


**Fig. 1.1** Continuous wave modulated operation in time-of-flight camera

### *1.1.1 Continuous Wave Modulation*

Continuous wave ToF cameras send a (usually infrared) optical signal e(t) modulated by a sinusoidal or square wave. Signal $e(t)$ is reflected back by the scene surface and travels back towards a receiver co-positioned with the emitter. The signal $r(t)$ is delayed due to the propagation distance thus a phase delay is induced on the received signal [7]. CW ToF cameras measure the phase difference $\phi_{ToF}$ between the emitted signal e(t) and received signal r(t) as shown in Fig. 1.1. The cross-correlation between the two signals allows phase estimation which is directly related to the distance between the camera and scene [8]. The range resolution of this method is inversely proportional to the modulation frequency $f$ of the light source. This type of cameras applies to different modulation techniques such as amplitude (AM), phase (PM), and frequency (FM) modulations. Some advantages of these cameras are the acquisition range and amplitude images and the availability of different light sources (e.g., LED, laser). On the other hand, in order to reduce noise, it needs integration over time that causes the motion blur in the system and the limitation of frame rates are some disadvantages [9].

### *1.1.2 Pulse-Based Modulation*

Figure 1.2 shows the timing diagram for the PB modulation technique which measures the absolute time $t_{ToF}$ (direct ToF) of the light pulse that needs to travel from the source into a scene and back to the sensor of the camera. The sensor contains two shutters $s_1, s_2$ for each pixel. A pixel is accumulating the light in a short and a long interval. When the first shutter $s_1$ is accumulating with the light in the sensor then the second one $s_2$ is active until the whole pulse is returned. The two intervals are used to obtain the corresponding distance calculation [10]. The range resolution of this method is proportional to the pulse width $t_p$ of the light source. These cameras allow long-distance measurements. Thus, the camera needs very short light pulses with high optical power lasers which reduce the influence of background illumination [11]. The arrival time must be accurately

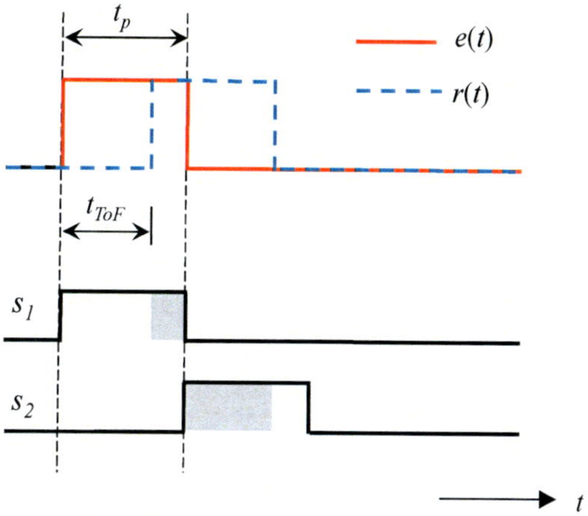


**Fig. 1.2** Timing diagram for pulse-based modulated operation in ToF camera. The shaded areas represent the time period that received the pulse light

detected. However, these cameras typically have low repetition rates for pulses and the measurement of light pulse return may not be accurate due to light scattering.

Sarbolandi et al. [12] investigated some error sources such as temperature drift, multipath effects, motion artifacts, linearity error, and depth inhomogeneity in a pulse-based ToF ranging camera prototype based on the Hamamatsu area sensor S11963-01CR [13]. A quantitative partial comparison for the above intrinsic parameters with their previous study (the Microsoft Kinect v1 based on structured light and Microsoft Kinect v2 based on the ToF principle of the range sensor) is also included [14].

### *1.1.3 Pseudo-Noise Modulation ToF Depth Cameras*

PN modulation uses a repeating pattern (sequence) of binary coding of the data signal [15]. At the receiver, this sequence demodulates the signal to extract the original data as shown in Fig. 1.3. By using minimal interference between the different light sources, the multiple range imaging cameras can operate within the same scene, which is the main advantage of this technique [16]. In addition, from this approach, the objects which occur beyond the system working distance can be identified and removed from the data, but this technique does not allow extended distances to be measured. However, this type of modulation has relatively lower precision than the other two types and it is proportional to the modulation frequency of the light source [17].

A detailed overview of the operational principle of CW, PB, and PN modulation ToF sensors is investigated by Jin and Zeng [9], Pietro Zanuttigh [11], and Consultative Committee for Space Data Systems [18], respectively. The technology and performance of depth cameras are reviewed by Langmann et al. [19].

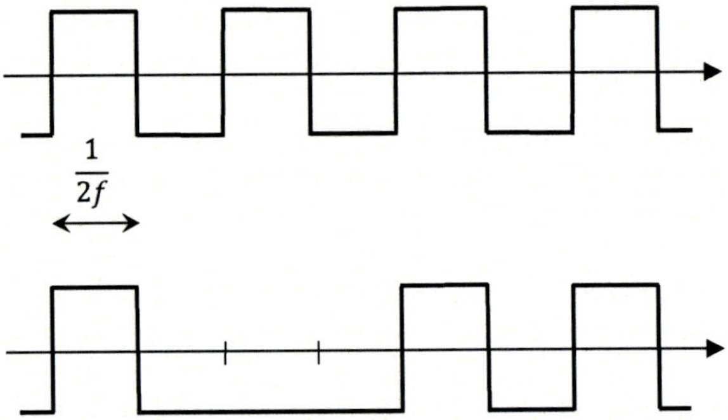


**Fig. 1.3** A portion of (top) the reference range-clock for a square spectrum and (bottom) the ranging sequence spectrum for the binary pattern $\cdots +1-1-1-1+1-1+1\ldots,$ for pseudo-noise modulated operation in ToF camera

## 1.2 Different Depth Measurement Techniques

The ToF principle (this will be separately described in Sect. 2.2 of Chap. 2) is not only the method to find the measurement of depths or location. There are standard well-known techniques such as *stereoscopic vision, structured light*, and *infrared thermography*. Let us consider these techniques in briefly.

### *1.2.1 Stereoscopic Vision*

The stereoscopic vision uses a fundamental concept similar to the human eyes to determine the location of 3D points in the scene that gives their projection into two (or more) images with known calibration [20]. This technique is also known as *camera triangulation* and is required for the intersection of two rays (projected lines) in space. Figure 1.4 shows a point $\mathrm{P} \equiv (x, y, z)$ in the scene that is observed at points $\mathrm{P}_1$ and $\mathrm{P}_2$ in the image planes from Camera—1 and Camera—2 (with lens centers at $\mathrm{C}_1$ and $\mathrm{C}_2$), respectively. Assume the simplest model with two identical cameras (with focal length $f_c$ each) separated only in the $x$ direction by a baseline gap $g$. Let us assume that the origin of the coordinate system coincides with the center $\mathrm{C}_1$. Comparing the similar triangles $\mathrm{C}_1\mathrm{D}_1\mathrm{P}$ and $\mathrm{C}_1\mathrm{O}_1\mathrm{P}_1$, we have

$$\frac{x}{z} = \frac{\mathrm{O}_1\mathrm{P}_1}{\mathrm{O}_1\mathrm{C}_1}. \tag{1.1}$$

Similarly, from the similar triangles $\mathrm{C}_2\mathrm{D}_2P$ and $\mathrm{C}_2\mathrm{O}_2\mathrm{P}_2$, we have

$$\frac{g-x}{z} = \frac{\mathrm{O}_2\mathrm{P}_2}{\mathrm{O}_2\mathrm{C}_2}. \tag{1.2}$$

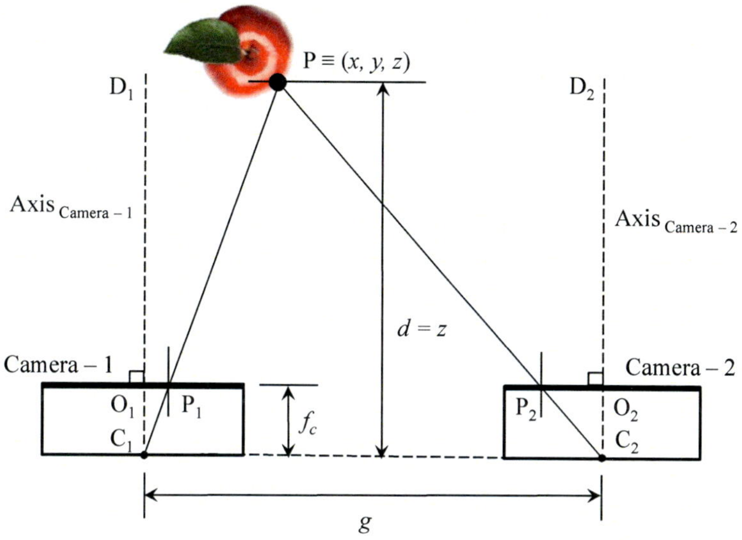


**Fig. 1.4** Any point P in the scene that is visible in both cameras will be projected to a pair of image points in the two images (from the top view, $xz$ plane)

Combining (1.1) and (1.2) with $O_1C_1 = O_2C_2 = f_c$, we get

$$d = z = \frac{gf_c}{O_1P_1 + O_2P_2}. \tag{1.3}$$

Thus, the depth at various points in the scene may be recovered by knowing the disparities of corresponding image points.

In the absence of noise and no errors in calibration, this calculation is straight forward otherwise the two projected lines will not generally intersect each other, in which case it is necessary to find the best point of intersection. Then some optimized techniques (e.g., maximum likelihood and weighted midpoint methods) need to be used which take high computational cost [21]. Therefore, this is a disadvantage, in addition to the requiring of at least two cameras. On the other hand, no need for special conditions in terms of scene illumination and the high-resolution images are some advantages of this technique.

### *1.2.2 Structured Light*

The structured light method captures the 3D surface imaging using a specific pattern of light such as single dot, slit or grating stripes, and a 2D imaging camera [22]. A spatially varying 2D structured illumination is generated by a special projector or a light source modulated by a spatial light modulator. For example, an imaging sensor in a video camera is used to acquire a 2D image of the scene under structured light illumination. Structured light encoding depends on the projector's resolution. The information is encoded into a sequence of patterns performed in the temporal domain [23, 24].

However, the geometric shape of the surface distorts the projected structured light pattern when the surface in the scene is non-planar [25]. Various structured light principles and algorithms are used to obtain accurate 3D surface profiles of objects in the scene [26–28]. The main advantage of structured light is that it can achieve relatively high spatial resolution and is better for indoor applications. Also, this method has become popular, in part due to the price and availability of the sensors in comparison to other sensor technologies. However, problems with transparent and reflective surfaces, the superposition of the light pattern with reflections are some disadvantages [29].

### *1.2.3 Infrared Thermography*

The process to detect radiation from an object by using a thermal imager is called infrared thermography. This radiation heat is converted to temperature and displays an image of the temperature distribution. The infrared radiation is invisible to the human eye. Since every object above absolute zero gives off thermal infrared energy, thermal imagers can easily detect and display infrared wavelengths regardless of ambient light. This method was applied to recognize human gait patterns [30, 31]. Some advantages of this method are it is fast, reliable, and accurate output; a large surface area can be scanned at the same time. On the other hand, the cost of the instrument is relatively high; it is unable to detect the inside temperature if the medium is separated by glass/polythene hence difficult to measure the distance, are some of the disadvantages of this method [29].

## 1.3 Applications of ToF Range Imaging Cameras

The ToF technology applies to a wide range of applications in day-to-day life across consumer, automotive, entertainment, and health science to many engineering fields. Advancement in semiconductor miniaturization has led to compact solutions at smaller form factors that make integrated ToF sensors almost available in mobile phones today. On the other hand, these ToF range imaging (depth) cameras have interesting properties such as portable size, feasible design, low weight, and low energy consumption [32] cause to have high demand in many areas such as mobile robotics [33, 34], medical imaging [35–38], 3D reconstruction [39], human identification [40, 41], human motion capturing [42], underwater 3D scanning [43–48], computer graphics [49], and plant industry [50–53] are some of them. Among these, let us briefly consider a few of them as follows.

*Robotics* is one of the major areas that use depth sensors vastly. For avoiding collision, navigation, mapping, and localization are the main functions handled in this area. When modern robots learn to see and in order to be able to keep to given cycle times, speed is just as important as accuracy. In addition, autonomous vehicle manufactures are used depth sensor systems with a computer vision system for daily routine use without increasing the cost too dramatically. Many warehouses are already taking advantage of the usage of fully autonomous vehicles for the day-to-day transport of their products.

The ToF cameras are also used in *medical imaging* such as in a patient positioning which detects the respiratory motion of humans in real-time [37, 38]. In addition, operating room registration systems employing ToF have been used for matching it with 3D endoscopy images in laboratory setup [54], or in an intra-modality ToF-to-ToF approach [55].

Earlier, most *human identification* systems are used 2D cameras to capture a photo to decide the person's identity using some special algorithms which have significant drawbacks. But in the present, depth cameras are used to see the real 3D face of humans. For example, present smartphones have depth sensors for facial recognition with more features of the face for more accurate recognition.

*Underwater 3D scanning* is being increasingly used in marine biology and oceanology. This process is very useful when scientists need to examine objects on the seafloor over time. Active optical 3D scanners are commonly used due to the domain-specific challenges of underwater imaging. For example, the light is strongly attenuated by water and this depends on the wavelength, also flickering and higher back-scatter due to the presence of suspended particles [47]. Also, preliminary results of using a commercial ToF depth camera, Microsoft Kinect™ v2, for 3D scanning of underwater objects have been investigated in [46].

The plant industry is another growing area that uses ToF depth cameras for various stages of their products such as monitoring of seedling and plant growth [56, 57], structures of plants [58], crop protection, weed volume estimation [59, 60], etc. Syed et al. [57] investigated different growth stages of cucumber, pepper, tomato, and lettuce seedlings by using RGB image acquisitions taken by Intel® RealSense™ SR300 depth sensor [61]. An adaptive methodology using a Microsoft Kinect™ v2 for point cloud rigid registration and stitching for maize plants investigated by Vázquez-Arellano et al. [53]. To control the environment in a greenhouse while continuously monitoring plant growth, the sunlit leaf area within a plant canopy was non-destructively measured and verified using the Kinect 3D depth sensor in [56]. Paulus et al. [62] measured the volumetric shape of sugar beet taproots, the leaves of sugar beets, and the shape of wheat ears by comparing the Microsoft Kinect Device with a laser scanner. Furthermore, Chéné et al. [63] have shown that the leaves with a high rate of success can be identified from 3D imagining taken by Kinect. Many applications have arisen using Kinect sensors, from robotics to human applications. Condotta et al. [64] have concluded that the ToF technology is the best to be used for indoor applications. Indoor and outdoor depth imaging of leaves with time-of-flight and stereo vision sensors are analyzed and compared by Kazmi et al. [65].

Besides, ToF cameras can also be able to use in low light conditions since the sensor uses infrared light to discover the distance to the scene; it could help with applications even in a pitch-black environment. Furthermore, some versions of the amplitude modulated continuous wave (AMCW) ToF cameras such as MESA Imaging SwissRanger 4000 and 4500 allow multiple cameras to operate simultaneously [66]. In the last few years, the ToF cameras have become commercially available at reasonable prices (i.e., a couple of hundred US dollars). Some applications are necessary to have high accuracy with quick measurements, therefore in those areas, expensive depth cameras are used.

## 1.4 Book Overview

This introduction provides an understanding of the characteristics of current ToF range imaging systems based on their modulation illumination source. Apart from the time-of-flight principle other depth measurement techniques and applications of ToF depth cameras are briefly reviewed. This book is mainly focused on the range imaging cameras with the ToF depth measurement technique. As summarized in Fig. 1.5, the book is structured across eleven chapters with three parts, namely **ToF range imaging cameras and measurement error, jitter extraction and measurement**, and **jitter influence range measurements**. The major contributions in each chapter are pointed out with the five appendices in the figure. In addition to that, a brief description of each chapter is also presented as follows.

The first part of this book, consisting of the next three chapters, describes in more detail an overview and error sources in commercial AMCW ToF range imaging cameras.

Chapter 2 contributes an overview of range imaging systems including the ToF depth camera components, the operational principles of ToF ranging systems in homodyne and heterodyne operations, the concepts of signal's aliasing and unambiguous range, and the details of two AMCW ToF range imaging cameras which are used in the experimental chapters of this book are pointed out. Then, the summarized specifications of selected commercial ToF range imaging cameras in tabular form which will be beneficial to the reader are included in the same chapter. Next, Chap. 3 mainly focuses on measurement error in ToF depth cameras, including the definitions of error sources: noise and jitter with noise sources in range imaging cameras are analyzed. Furthermore, noise mitigating techniques in ToF range imaging cameras in the literature are briefly described. Chapter 4 contributes to the investigation of electronic jitter in signals including the types of electronic jitter, the random and periodic jitter sources in ranging systems, jitter analysis techniques in signals: using graphical methods, in time and frequency domains, with the statistical and signal processing approaches. Finally, the impact of jitter on range imaging reported in the literature is described.

The next four chapters in Part II comprise jitter extraction and measurement in ToF range imaging cameras.

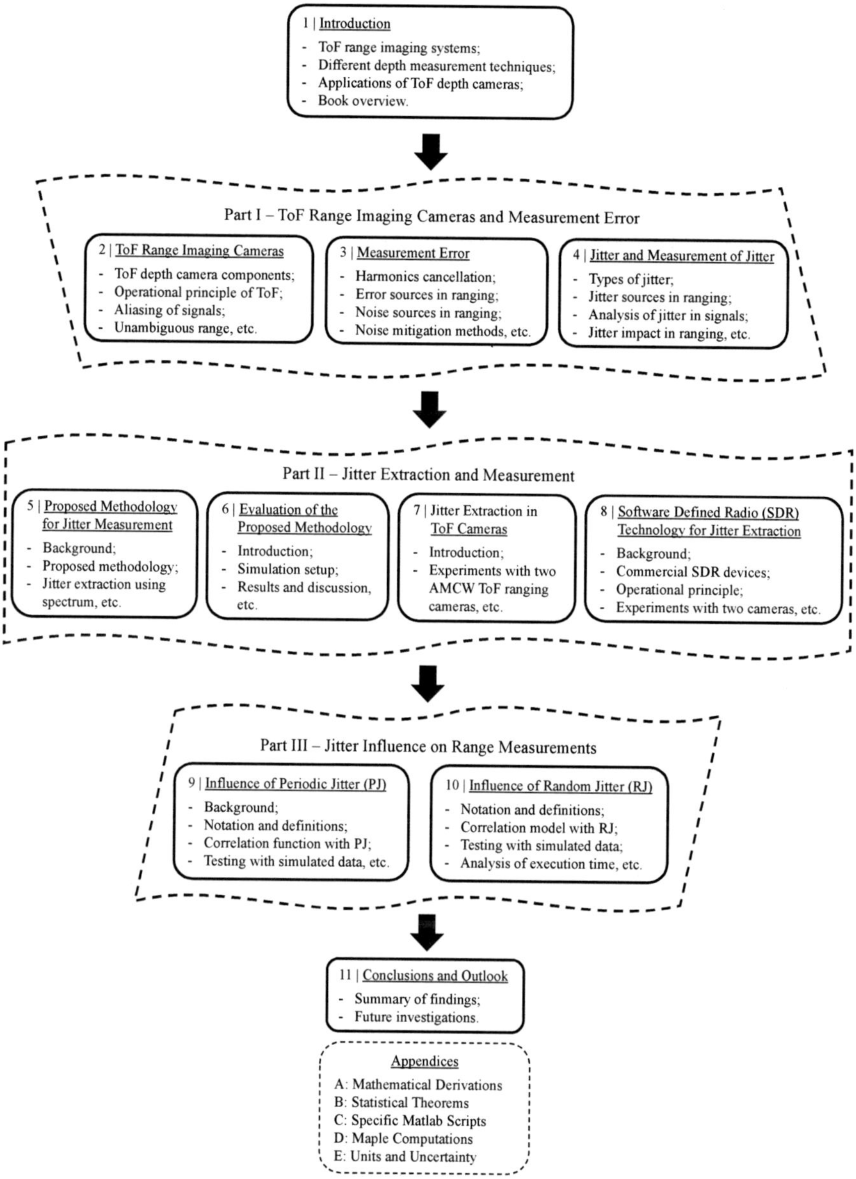


**Fig. 1.5** Summary of the chapters of the book

Chapter 5 describes a proposing novel algorithm to extract the periodic and random jitter present in the light source of the AMCW ToF ranging cameras. The most important advantage of this method is that it does not require a reference or sampling clock signal to trigger the time measurement unit. Instead, signal processing techniques with Fourier analysis is used. This algorithm is camera independent and can be used in real-time jitter investigation in most of the ToF systems where it is impossible to access their clock signals. Chapter 6 verifies the proposed algorithm by testing with the generated simulation data for various simulation setups such as only random jitter, only periodic jitter, and the combination of random and periodic jitter, separately. The approach is validated by several simulation setups with numerous values for the parameters of the light source signal and jitter. In addition, three types of standard geometrical shapes of signals: sinusoidal, triangular, and rectangular (square) are considered for each type of simulation setup, separately. The last two chapters in Part II are included experimental results. The measurement of jitter present in two commercial AMCW ToF cameras is reported in Chap. 7. All experiments are measured with a medium speed (6 GHz) real-time digital oscilloscope. First, the detailed summary of the sub-steps of the proposed algorithm is figured out as a flow chart. Then, the experimental setup with the corresponding results for each camera is described. Evidence that random jitter and periodic jitter at multiple frequencies are present in both cameras is presented. The oscilloscope for the experiments used in Chap. 7 is an expensive instrument. Can one replace the oscilloscope with a much cheaper instrument? Chapter 8 is an investigation of the possibility of re-purposing a vastly cheaper software defined radio (SDR) USB dongle for jitter extraction in signals. The operational principle of the receiver side of the SDR technology is described and the results of experiments on the same two AMCW ToF cameras are presented. Comparison is made with the results in Chap. 7, and the limitations of the SDR technique are described.

The third part of the book addresses the investigation of the influence of the two types of jitter: random and periodic, in the illumination sources of ToF range imaging cameras in two chapters.

Chapter 9 is an investigation of the influence of the periodic jitter on range measurements. The primary contribution is the derivation of an analytical model in the Fourier domain. The heterodyne operation principle of the correlation model of the AMCW ToF ranging systems is obtained in the Fourier domain. Then, a novel analytical model for the correlation function with periodic jitter for both heterodyne and homodyne operations in ToF systems are investigated. In order to verify the results, two common numerical approaches are described. The obtained model is tested by the simulation data with various parameters such as modulation frequency, the amplitude and frequency of the periodic jitter, the phase errors due to the periodic jitter and the integration period in the camera. Lastly, the remarkable findings of this novel investigation are included at the end of this chapter. Chapter 10 is an investigation of the effect of random jitter on range measurements. Using Fourier analysis and standard statistical theorems, benchmark stochastic models for the influence of random jitter in the heterodyne and homodyne operations of the AMCW ToF range imaging are constructed. The model is tested with the Monte Carlo method through the simulated data with various parameters such as modulation frequency and the RMS of random jitter in the model. Two non-parametric estimations are performed to find the probability density function for the standard deviation of the phase error which is the characteristic parameter for random jitter in the correlation model. Finally, interesting and useful findings for present and predictions for future AMCW ToF cameras on range measurements due to the random jitter are figured out.

Chapter 11 concludes the significant contributions of this book, summarizes the findings in each chapter, and suggests some potential future investigation.

In addition, five appendices are included.

**Appendix A** Mathematical Derivations: This appendix contains some of the advanced mathematical concepts and derivations used in the book.

**Appendix B** Statistical Theorems: The relevant statistical theorems used in this book are stated in this appendix. Proofing of some of them is also included.

**Appendix C** Specific Matlab Scripts: The algorithms in Chaps. 6–10 are implemented using Matlab and among them; specific scripts are included in this appendix.

**Appendix D** Maple Computations: In this appendix, some expressions obtained in Chap. 9 are verified by using the mathematical tool, Maple.

**Appendix E** Units and Uncertainty: This appendix contains the units for the standard measurements and prefixes that are used with these units. In addition, the uncertainty derivations for measurements that are used throughout the book are included.

## 1.5 Further Reading

For an overview of ToF range imaging cameras that cover topics including, 3D camera systems, accuracy and performance evaluations, geometric calibration, indoor and robotics navigation, applications, etc., refer [67, 68]. Furthermore, for the evaluation of the depth sensors and scanners, see [4, 69]. The book by Pietro Zanuttigh [11] is devoted to the topics of ToF and structured light depth cameras. In addition, optimal patterns of algorithms and principles for the applications with structured light are investigated in [70–72].

## References

1. Chevalier François Le. *Principles of radar and sonar signal processing*. Artech House, 2002.
2. Daniele Marioli, Claudio Narduzzi, Carlo Offelli, Dario Petri, Emilio Sardini, and Andrea Taroni. Digital time-of-flight measurement for ultrasonic sensors. *IEEE Transactions on Instrumentation and Measurement*, 41(1):93–97, 1992.
3. Michael D Bedford and Gareth A Kennedy. Evaluation of ZigBee (IEEE 802.15. 4) time-of-flight-based distance measurement for application in emergency underground navigation. *IEEE Transactions on Antennas and Propagation*, 60(5):2502–2510, 2012.
4. François Blais. Review of 20 years of range sensor development. *Journal of Electronic Imaging*, 13(1):231–244, 2004.
5. Larry Li. Time-of-flight camera - an introduction. Technical report, Texas Instruments - Technical White Paper, Dallas, Texas 75265, U.S.A., 2014.

6. Bernd Jähne. *Computer vision and applications: a guide for students and practitioners.* Elsevier, 2000.
7. Carlo Dal Mutto, Pietro Zanuttigh, and Guido M Cortelazzo. *Time-of-flight cameras and Microsoft Kinect™*. Springer Science & Business Media, 2012.
8. Glenn Beheim and Klaus Fritsch. Range finding using frequency-modulated laser diode. *Applied Optics*, 25(9):1439–1442, 1986.
9. Xiang-Liang Jin and Shao-Qing Zeng. Design of 3D TOF camera system based on CW modulation technique. In *2016 13th IEEE International Conference on Solid-State and Integrated Circuit Technology (ICSICT)*, pages 373–375, Oct 2016. https://doi.org/10.1109/ICSICT.2016.7998925.
10. I Moring, T Heikkinen, R Myllyla, and A Kilpela. Acquisition of three-dimensional image data by a scanning laser range finder. *Optical engineering*, 28(8):288897, 1989.
11. Carlo Dal Mutto Pietro Zanuttigh, Giulio Marin. *Time-of-Flight and Structured Light Depth Cameras: Technology and Applications*. Springer International Publishing, 1 edition, 2016. ISBN 978-3-319-30971-2, 978-3-319-30973-6.
12. Hamed Sarbolandi, Markus Plack, and Andreas Kolb. Pulse based time-of-flight range sensing. *Sensors*, 18(6), 2018. ISSN 1424-8220. https://doi.org/10.3390/s18061679. URL http://www.mdpi.com/1424-8220/18/6/1679.
13. *Distance area image sensor – S11963-01CR*. Hamamatsu Photonics K.K., Solid State Division, August 2020. URL https://www.hamamatsu.com/resources/pdf/ssd/s11963-01cr_kmpd1142e.pdf.
14. Hamed Sarbolandi, Damien Lefloch, and Andreas Kolb. Kinect range sensing: Structured-light versus Time-of-Flight Kinect. *Computer vision and image understanding*, 139:1–20, 2015.
15. Bernhard Büttgen, Felix Lustenberger, Peter Seitz, et al. Pseudonoise optical modulation for real-time 3-d imaging with minimum interference. *IEEE Transactions on Circuits and Systems I: Regular Papers*, 54(10):2109–2119, 2007.
16. B. Büttgen and P. Seitz. Robust optical time-of-flight range imaging based on smart pixel structures. *IEEE Transactions on Circuits and Systems I: Regular Papers*, 55(6):1512–1525, July 2008. ISSN 1549-8328. https://doi.org/10.1109/TCSI.2008.916679.
17. Refael Z Whyte, Andrew D Payne, Adrian A Dorrington, and Michael J Cree. Multiple range imaging camera operation with minimal performance impact. In *Image Processing: Machine Vision Applications III*, volume 7538, page 75380I. International Society for Optics and Photonics, 2010.
18. Consultative Committee for Space Data Systems. Pseudo-Noise (PN) Ranging Systems, 2014.
19. Benjamin Langmann, Klaus Hartmann, and Otmar Loffeld. Depth camera technology comparison and performance evaluation. In *ICPRAM (2)*, pages 438–444, 2012.
20. Ramesh Jain, Rangachar Kasturi, and Brian G Schunck. *Machine vision*, volume 5. McGraw-hill New York, 1995.
21. William H Press, H William, Saul A Teukolsky, A Saul, William T Vetterling, and Brian P Flannery. *Numerical recipes 3rd edition: The art of scientific computing*. Cambridge university press, 2007.
22. Paul J Besl. Active optical range imaging sensors. In *Advances in machine vision*, pages 1–63. Springer, 1989.
23. Van-Luan Tran and Huei-Yung Lin. Accurate RGB-D camera based on structured light techniques. In *2017 International Conference on System Science and Engineering (ICSSE)*, pages 235–238. IEEE, 2017.
24. Van Luan Tran and Huei-Yung Lin. A structured light RGB-D camera system for accurate depth measurement. *International Journal of Optics*, 2018, 2018.
25. Jason Geng. Structured-light 3d surface imaging: a tutorial. *Advances in Optics and Photonics*, 3(2):128–160, 2011.
26. Eli Horn and Nahum Kiryati. Toward optimal structured light patterns. *Image and Vision Computing*, 17(2):87–97, 1999.

27. Li Zhang, Brian Curless, and Steven M Seitz. Rapid shape acquisition using color structured light and multi-pass dynamic programming. In *Proceedings. First International Symposium on 3D Data Processing Visualization and Transmission*, pages 24–36. IEEE, 2002.
28. Srinivasa G Narasimhan and Shree K Nayar. Structured light methods for underwater imaging: light stripe scanning and photometric stereo. In *Proceedings of OCEANS 2005 MTS/IEEE*, pages 2610–2617. IEEE, 2005.
29. Alvaro Muro-De-La-Herran, Begonya Garcia-Zapirain, and Amaia Mendez-Zorrilla. Gait analysis methods: an overview of wearable and non-wearable systems, highlighting clinical applications. *Sensors*, 14(2):3362–3394, 2014.
30. Zhaojun Xue, Dong Ming, Wei Song, Baikun Wan, and Shijiu Jin. Infrared gait recognition based on wavelet transform and support vector machine. *Pattern recognition*, 43(8):2904–2910, 2010.
31. W Samson, A Van Hamme, S Sanchez, L Chèze, Serge Van Sint Jan, and Véronique Feipel. Dynamic footprint analysis by time-of-flight camera. *Computer methods in biomechanics and biomedical engineering*, 15(sup1):180–182, 2012.
32. Ying He, Bin Liang, Yu Zou, Jin He, and Jun Yang. Depth errors analysis and correction for time-of-flight (tof) cameras. *Sensors*, 17(1):92, 2017.
33. Stefan May, Bjorn Werner, Hartmut Surmann, and Kai Pervolz. 3d time-of-flight cameras for mobile robotics. In *Intelligent Robots and Systems, 2006 IEEE/RSJ International Conference on*, pages 790–795. IEEE, 2006.
34. Thomas Schamm, Marcus Strand, Thomas Gumpp, Ralf Kohlhaas, J Marius Zollner, and Rudiger Dillmann. Vision and tof-based driving assistance for a personal transporter. In *Advanced Robotics, 2009. ICAR 2009. International Conference on*, pages 1–6. IEEE, 2009.
35. Joel S Karp, Suleman Surti, Margaret E Daube-Witherspoon, and Gerd Muehllehner. The benefit of time-of-flight in pet imaging: experimental and clinical results. *Journal of nuclear medicine: official publication, Society of Nuclear Medicine*, 49(3):462, 2008.
36. Stefan Soutschek, Jochen Penne, Joachim Hornegger, and Johannes Kornhuber. 3-d gesture-based scene navigation in medical imaging applications using time-of-flight cameras. pages 1–6, 2008.
37. Jochen Penne, Christian Schaller, Joachim Hornegger, and Torsten Kuwert. Robust real-time 3D respiratory motion detection using time-of-flight cameras. *International Journal of Computer Assisted Radiology and Surgery*, 3(5):427–431, 2008.
38. Christian Schaller, Andre Adelt, Jochen Penne, and Joachim Hornegger. Time-of-flight sensor for patient positioning. In *Medical Imaging 2009: Visualization, Image-Guided Procedures, and Modeling*, volume 7261, page 726110. International Society for Optics and Photonics, 2009.
39. Peter Henry, Michael Krainin, Evan Herbst, Xiaofeng Ren, and Dieter Fox. Rgb-d mapping: Using depth cameras for dense 3d modeling of indoor environments. In *12th International Symposium on Experimental Robotics (ISER*. Citeseer, 2010.
40. Xia Liu and Kikuo Fujimura. Hand gesture recognition using depth data. In *6th International Conference on Automatic Face and Gesture Recognition*, page 529. IEEE, 2004.
41. Jing Tong, Jin Zhou, Ligang Liu, Zhigeng Pan, and Hao Yan. Scanning 3d full human bodies using kinects. *IEEE transactions on visualization and computer graphics*, 18(4):643–650, 2012.
42. Varun Ganapathi, Christian Plagemann, Daphne Koller, and Sebastian Thrun. Real time motion capture using a single time-of-flight camera. In *2010 IEEE Computer Society Conference on Computer Vision and Pattern Recognition*, pages 755–762. IEEE, 2010.
43. Chi-Leung Tsui, David Schipf, Keng-Ren Lin, Jonathan Leang, Feng-Ju Hsieh, and Wei-Chih Wang. Using a Time of Flight method for underwater 3-dimensional depth measurements and point cloud imaging. In *OCEANS 2014-TAIPEI*, pages 1–6. IEEE, 2014.
44. Miquel Massot-Campos and Gabriel Oliver-Codina. Optical sensors and methods for underwater 3D reconstruction. *Sensors*, 15(12):31525–31557, 2015.

45. Sundara Tejaswi Digumarti, Gaurav Chaurasia, Aparna Taneja, Roland Siegwart, Amber Thomas, and Paul Beardsley. Underwater 3D capture using a low-cost commercial depth camera. In *2016 IEEE Winter Conference on Applications of Computer Vision (WACV)*, pages 1–9. IEEE, 2016.
46. Atif Anwer, Syed Saad Azhar Ali, Amjad Khan, and Fabrice Mériaudeau. Underwater 3D scanning using Kinect v2 time of flight camera. In *Thirteenth International Conference on Quality Control by Artificial Vision 2017*, volume 10338, page 103380C. International Society for Optics and Photonics, 2017.
47. Miguel Castillón, Albert Palomer, Josep Forest, and Pere Ridao. State of the art of underwater active optical 3D scanners. *Sensors*, 19(23):5161, 2019.
48. Patrizio Mariani, Iñaki Quincoces, Karl H Haugholt, Yves Chardard, Andre W Visser, Chris Yates, Giuliano Piccinno, Giancarlo Reali, Petter Risholm, and Jens T Thielemann. Range-gated imaging system for underwater monitoring in ocean environment. *Sustainability*, 11(1): 162, 2019.
49. Andreas Kolb, Erhardt Barth, Reinhard Koch, and Rasmus Larsen. Time-of-flight cameras in computer graphics. In *Computer Graphics Forum*, volume 29, pages 141–159. Wiley Online Library, 2010.
50. Ralph Klose, Jaime Penlington, and Arno Ruckelshausen. Usability study of 3D time-of-flight cameras for automatic plant phenotyping. *Bornimer Agrartechnische Berichte*, 69(93-105):12, 2009.
51. A Ruckelshausen, L Busemeyer, R Klose, A Linz, K Moeller, M Thiel, K Alheit, F Rahe, D Trautz, and U Weiss. Sensor and system technology for individual plant crop scouting. In *The 10th International Conference on Precision Agriculture Denver*, 2010.
52. Lucas Busemeyer, Daniel Mentrup, Kim Möller, Erik Wunder, Katharina Alheit, Volker Hahn, Hans Maurer, Jochen Reif, Tobias Würschum, Joachim Müller, et al. BreedVision — A multi-sensor platform for non-destructive field-based phenotyping in plant breeding. *Sensors*, 13(3): 2830–2847, 2013.
53. Manuel Vázquez-Arellano, David Reiser, Dimitris S Paraforos, Miguel Garrido-Izard, Marlowe Edgar C Burce, and Hans W Griepentrog. 3-D reconstruction of maize plants using a time-of-flight camera. *Computers and Electronics in Agriculture*, 145:235–247, 2018.
54. Sven Haase, Christoph Forman, Thomas Kilgus, Roland Bammer, Lena Maier-Hein, and Joachim Hornegger. ToF/RGB sensor fusion for augmented 3D endoscopy using a fully automatic calibration scheme. In *Bildverarbeitung für die Medizin 2012*, pages 111–116. Springer, 2012.
55. Simon Placht, Joseph Stancanello, Christian Schaller, Michael Balda, and Elli Angelopoulou. Fast time-of-flight camera based surface registration for radiotherapy patient positioning. *Medical physics*, 39(1):4–17, 2012.
56. H Umeda, Y Mochizuki, T Saito, T Higashide, and Y Iwasaki. Diagnosing method for plant growth using a 3d depth sensor. In *International Symposium on New Technologies for Environment Control, Energy-Saving and Crop Production in Greenhouse and Plant 1227*, pages 631–636, 2017.
57. Tabinda Naz Syed, Liu Jizhan, Zhou Xin, Zhao Shengyi, Yuan Yan, Sami Hassan Ahmed Mohamed, and Imran Ali Lakhiar. Seedling-lump integrated non-destructive monitoring for automatic transplanting with Intel RealSense depth camera. *Artificial Intelligence in Agriculture*, 3:18–32, 2019.
58. George Azzari, Michael L Goulden, and Radu B Rusu. Rapid characterization of vegetation structure with a Microsoft Kinect sensor. *Sensors*, 13(2):2384–2398, 2013.
59. Dionisio Andújar, José Dorado, César Fernández-Quintanilla, and Angela Ribeiro. An approach to the use of depth cameras for weed volume estimation. *Sensors*, 16(7):972, 2016.
60. Dionisio Andujar, Angela Ribeiro, César Fernández-Quintanilla, and Jose Dorado. Using depth cameras to extract structural parameters to assess the growth state and yield of cauliflower crops. *Computers and Electronics in Agriculture*, 122:67–73, 2016.
61. Intel® RealSense™ camera sr300. https://ark.intel.com/content/www/us/en/ark/products/92329/intel-realsense-camera-sr300.html, 2021. [Accessed April 2021].

62. Stefan Paulus, Jan Behmann, Anne-Katrin Mahlein, Lutz Plümer, and Heiner Kuhlmann. Low-cost 3D systems: suitable tools for plant phenotyping. *Sensors*, 14(2):3001–3018, 2014.
63. Yann Chéné, David Rousseau, Philippe Lucidarme, Jessica Bertheloot, Valérie Caffier, Philippe Morel, Étienne Belin, and François Chapeau-Blondeau. On the use of depth camera for 3D phenotyping of entire plants. *Computers and Electronics in Agriculture*, 82:122–127, 2012.
64. Isabella CFS Condotta, Tami M Brown-Brandl, Santosh K Pitla, John P Stinn, and Késia O Silva-Miranda. Evaluation of low-cost depth cameras for agricultural applications. *Computers and Electronics in Agriculture*, 173:105394, 2020.
65. Wajahat Kazmi, Sergi Foix, Guillem Alenyà, and Hans Jørgen Andersen. Indoor and outdoor depth imaging of leaves with time-of-flight and stereo vision sensors: analysis and comparison. *ISPRS journal of photogrammetry and remote sensing*, 88:128–146, 2014.
66. *SR4000/SR4500 User Manual*. MESA Imaging AG. URL https://acroname.com/sites/default/files/assets/sr4000_sr4500_manual.pdf.
67. Fabio Remondino and David Stoppa. *TOF Range-Imaging Cameras*. Springer Publishing Company, Incorporated, 2013. ISBN 3642275222, 9783642275227.
68. Marcin Grzegorzek, Christian Theobalt, Reinhard Koch, and Andreas Kolb. *Time-of-Flight and Depth Imaging. Sensors, Algorithms and Applications: Dagstuhl Seminar 2012 and GCPR Workshop on Imaging New Modalities*, volume 8200. Springer, 2013.
69. Radu Horaud, Miles Hansard, Georgios Evangelidis, and Clément Ménier. An overview of depth cameras and range scanners based on time-of-flight technologies. *Machine vision and applications*, 27(7):1005–1020, 2016.
70. JJ Le Moigne and Allen Mark Waxman. Structured light patterns for robot mobility. *IEEE Journal on Robotics and Automation*, 4(5):541–548, 1988.
71. TP Monks and John N Carter. Improved stripe matching for colour encoded structured light. In *International Conference on Computer Analysis of Images and Patterns*, pages 476–485. Springer, 1993.
72. Nelson G Durdle, Jaishankar Thayyoor, and VJ Raso. An improved structured light technique for surface reconstruction of the human trunk. In *Conference Proceedings. IEEE Canadian Conference on Electrical and Computer Engineering (Cat. No. 98TH8341)*, volume 2, pages 874–877. IEEE, 1998.

# Part I
# ToF Range Imaging Cameras and Measurement Error

This part consists of three chapters that focus on the overview and error sources in commercial AMCW ToF range imaging cameras. Chapter 2 contributes an overview of range imaging systems including the ToF depth camera components, operational principle of ToF, unambiguous range, commercial ToF range imaging cameras, etc. Next, noise sources in ranging cameras and noise mitigating techniques are briefly described in Chap. 3, while jitter sources and measurement of jitter in range imaging, and impact of jitter on range imaging are investigated in Chap. 4.

# Chapter 2
# ToF Range Imaging Cameras

**Abstract** In this chapter, first, a development of the ToF camera with a typical model is briefly reviewed. Next, the three types of approaches: shuttered light pulse, direct ToF, and indirect ToF, based on the ToF principle are described. Among them, the theoretical framework for homodyne and heterodyne operations of indirect ToF range imaging cameras, i.e., AMCW systems, are investigated. Then, the concepts of aliasing and unambiguous range are briefly stated. Configurations of two commercial AMCW ToF cameras that will be used in experimental chapters are also described. Finally, the specifications of the commercial ToF range imaging cameras are pointed out in the tabular form.

**Keywords** Light detection and ranging (LiDAR) · Shuttered light pulse · Direct ToF · Indirect ToF · Reflected light · Emitted shutter light · Correlation function · Integration period · Homodyne operation · Heterodyne operation · Aliasing of signals · Unambiguous range · Frame rate · Infrared LEDs · Infrared Laser light

## 2.1 ToF Depth Camera Components

In 1977, the first ToF ranging cameras for scene analysis were invented by Stanford Research Institute (presently SRI International) [1]. In the 1990s, the invention of the lock-in charge-coupled device (CCD) technique enabled fast sampling of received light [2]. This technique was used by Schwarte [3] to investigate a method of measuring the phase and magnitudes of electromagnetic waves. As a result of that, in 1999, his team invented the first CCD-based ToF camera prototype [4]. Following that, ToF cameras were rapidly improved over the last couple of decades [5].

A typical building block of a ToF camera is illustrated in Fig. 2.1. These cameras have three main components: a modulation source (an emitter), a lens (optics), and a depth sensor. Typically, the modulation source is based on a solid-state laser or light emitting diode (LED) to produce an unobtrusive infrared that drives an illumination source. This emitter should be capable of switching at high frequencies and provides

 
G. Anthonys, *Timing Jitter in Time-of-Flight Range Imaging Cameras*,
https://doi.org/10.1007/978-3-030-94159-8_2

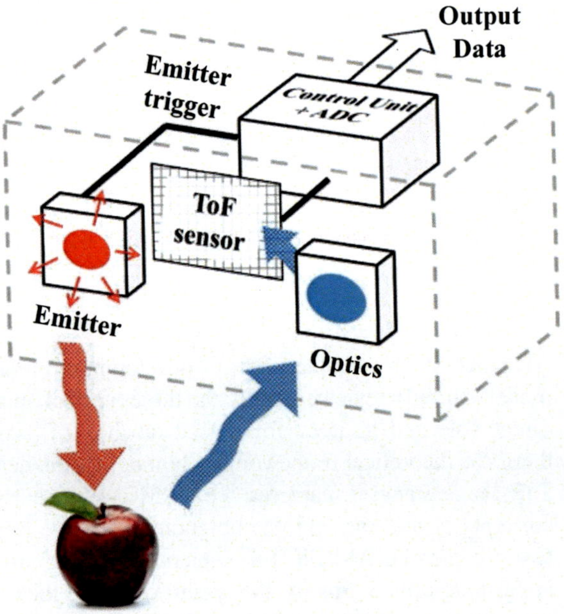


**Fig. 2.1** A typical block diagram of a ToF range imaging camera

modulated light to illuminate the scene. The lens collects the light reflected back from the target by using an intensity based detector that measures the returning light. A depth sensor is composed of an array of shuttered photo-detectors or CCDs (pixels) that are capable of distance measurements. In general, the sensor array and imaging optics are used to provide a full field of view. The shutter is driven by the modulation source usually at the same frequency as the light source. Then, either indirect time measurement through the phase difference of the modulation signal or direct time measurement with the laser pulse is used to interpret the distance to the scene. Finally, the sensor interface communicates with the external world such as a PC. In addition, there are other components such as the power supply, corresponding biasing current or voltage signals, control unit, and reading out from the sensor data stream [6].

## 2.2 Operational Principle in ToF Systems

The ToF has become a key depth sensing technique in a wide range of light detection and ranging (LiDAR) applications. ToF sensors are becoming increasingly relevant with the increasing range of applications. Now, let us consider the operational behavior in these systems. As stated in Sect. 1.1, the ToF principle is used to measure the distance $d$ to an object and can be expressed as

$$d = \frac{1}{2} v \, \delta t_{ToF}, \tag{2.1}$$

where $v$ is the signal propagation velocity and $\delta t_{ToF}$ is the time taken for the signal to return to the sensor. Due to the increase in the speed of electronic components, the timing accuracy in the ToF systems is much higher. Several approaches can be used to measure depth using the ToF principle and the most common methods: shuttered light pulse, direct ToF, and indirect ToF are considered.

### 2.2.1 *Shuttered Light Pulse Principle*

This is one of the promising technique for range measurement using the ToF principle. The sensors based on shutter light pulse project a pulse of light with known dimensions (say, width $t_w$) to an object and observe the reflected illumination with a shuttered camera. The reflected is observed only a portion of the reflected light by using a gating mechanism [7]. Figure 2.2 shows the light is reflected from closer scene returns to the camera before the light reflected by a far-distant scene. That is, from nearby scenes camera is captured more light than far-distant scenes. Then, the depth of objects is proportional to the amount of integrated light, i.e., the shaded part of the pulse (see Fig. 2.2).

Generally, the optical properties of the scene are variable, therefore, another calibration measurement is required. During this measurement, the gate is kept open to acquire the full reflected pulse from the scene. Then, the measured intensities from both cases can be used to compute the distance to the scene as [7]

$$d = \frac{c \, t_w}{2} \left( 1 - \frac{I_1}{I_2} \right), \tag{2.2}$$

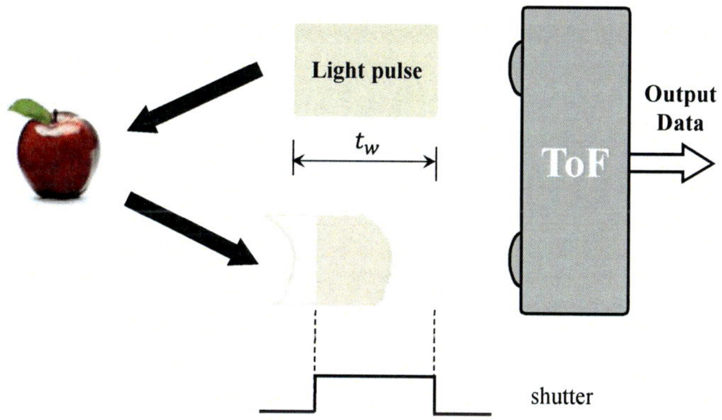


**Fig. 2.2** Shuttered light pulse ToF principle; A light pulse of duration $t_w$ radiates the scene

where $c$ is the speed of light (the exact value is defined as $2.99792458 \times 10^8$ m/s). $I_1$ and $I_2$ are the intensity measurements of the range and calibration frames, respectively. Furthermore, this method can be used to compute the depth under ambient illumination conditions [8, 9].

### 2.2.2 *Direct ToF Principle*

Figure 2.3 shows a simple block outlining the operating principles of the direct ToF system. As indicated in the name, the direct ToF (dToF) method directly measures the time difference between the emitted pulse, $t_{start}$, and the received pulse, $t_{stop}$. The measured timing information is then combined with the speed of light, $c$, to estimate the distance, $d$, of the target from the sensor as

$$d = \frac{1}{2} c \left( t_{stop} - t_{start} \right). \tag{2.3}$$

The dToF sensors use a diode or laser to emit a pulse of light and require accurate and robust timing references for precise depth measurements over a large range of distances [10, 11]. Various approaches can be used to implement a dToF senor, including, single-photon avalanche diode (SPAD) [12], single-shot with silicon photomultipliers (SiPM) [13], time-gated SPAD quanta image sensors [14], time-correlated single-photon counting [15] and mutually coupled time-to-digital converters [16].

The dToF systems are available in a single point, line and full-field varieties and they are utilized in harsh environments such as heavy rain, fog or snow and high background light conditions. Besides night vision applications, requiring very high sensitivity, these SPAD-based systems find applications for complex imaging tasks such as looking around corners [17].

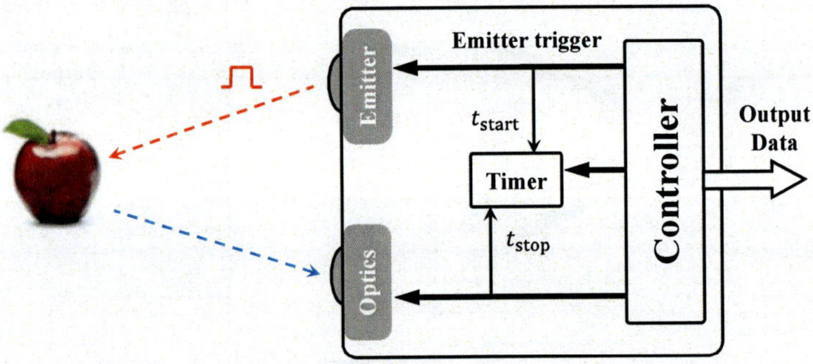


**Fig. 2.3** Principle of direct ToF measurement

### 2.2.3 Indirect ToF Principle

In indirect ToF systems, the sensors send out continuous modulated light and measure the phase of the reflected light to calculate the distance to an object. Therefore, these are called continuous wave ToF ranging imaging systems (especially AMCW type). Figure 2.4 illustrates the operation of an indirect ToF camera system. The camera illuminates the scene with a modulated light source (using laser or a LED) and captures (typically, using an infrared CCD/CMOS sensor) the reflected light via a shuttered pixelated sensor in the camera resulting in distance information and a picture of the complete scene simultaneously.

The received reflected light back from the scene with a modulation frequency of the emitted light source (illumination) and camera sensor $f$ is described by,

$$l(t, \phi) = A_l \cos(2\pi f t - \phi) + B_l, \tag{2.4}$$

where $A_l$ is the amplitude of the reflected light which depends on the scene's reflectivity and the sensor's sensitivity, $B_l$ is the offset or intensity which is the average of the received signal due to the background illumination, and the phase shift $\phi$ is directly related to the time of flight. Even though both $A_l$ and $B_l$ are amplitudes of infrared radiation (measured in volts), generally, they are called amplitude and intensity, respectively [18]. From Fig. 2.4, the shutter signal $s(t)$ of amplitude $a$ with background offset $b$ is given by

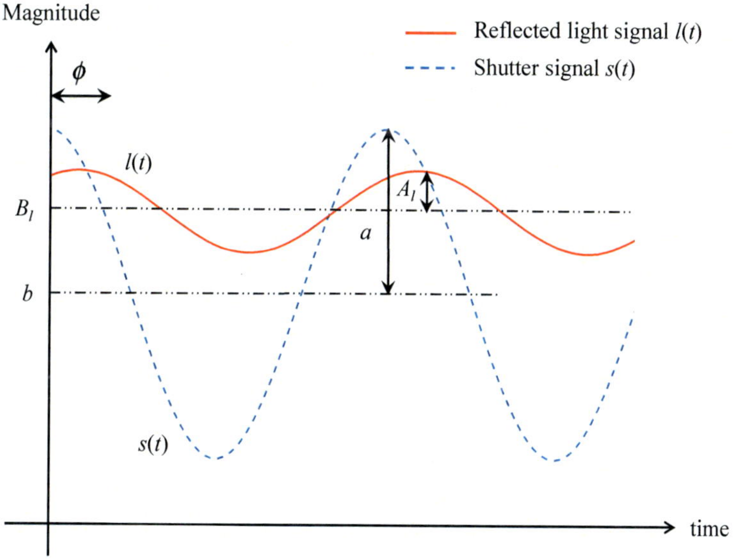


**Fig. 2.4** Light and shutter signals relationship in a ToF camera

$$s(t) = a\cos(2\pi f t) + b. \tag{2.5}$$

A practical implementation using light imposes, it should be satisfied $l(t) \geq 0$ thus $B_l \geq A_l$ and also $s(t) \geq 0$ (see Fig. 2.4). In practice, the phase shift cannot be measured directly. Therefore, the phase shift $\phi$ due to the time of flight between the illumination and reflection is measured by correlating the two signals. As there are three unknown variables, at least three measurements of the scene must be made. This is achieved by including an extra added phase delay $\theta_k$ for the $k$th phase acquisition, which the camera controls, gives

$$s(t, \theta_k) = a\cos(2\pi f t + \theta_k) + b. \tag{2.6}$$

Then, the correction function is given by

$$C(\theta_k) = s(t, \theta_k) \otimes l(t, \phi) = \lim_{T\to\infty} \frac{1}{T} \int_{-T/2}^{T/2} s(t, \theta_k)\, l(t, \phi)\, dt, \tag{2.7}$$

where $T$ is the integration period and the product of the integration in (2.7) can be expressed as

$$\begin{aligned} s(t, \theta_k)\, l(t, \phi) &= \Big(a\cos(2\pi f t + \theta_k) + b\Big)\Big(A_l \cos(2\pi f t - \phi) + B_l\Big) \\ &= aA_l \cos(2\pi f t + \theta_k)\cos(2\pi f t - \phi) + aB_l \cos(2\pi f t + \theta_k) \\ &\quad + bA_l \cos(2\pi f t - \phi) + bB_l \\ &= \frac{aA_l}{2}\Big(\cos(\phi + \theta_k) + \cos(4\pi f t - \phi + \theta_k)\Big) + aB_l \cos(2\pi f t + \theta_k) \\ &\quad + bA_l \cos(2\pi f t - \phi) + bB_l. \end{aligned} \tag{2.8}$$

Substituting (2.8) into (2.7) and performing the integration give

$$\begin{aligned} C(\theta_k) = \lim_{T\to\infty} \frac{1}{T} \Bigg\{ \frac{aA_l}{2}\bigg(t\, \cos(\phi + \theta_k) + \frac{\sin(4\pi f t - \phi + \theta_k)}{4\pi f}\bigg) \\ + aB_l \frac{\sin(2\pi f t + \theta_k)}{2\pi f} + bA_l \frac{\sin(2\pi f t - \phi)}{2\pi f} + bB_l\, t \Bigg\}_{-T/2}^{T/2} \end{aligned} \tag{2.9}$$

Now apply the limits and by simplification, (2.9) leads to

$$C(\theta_k) = \frac{A}{2}\cos(\phi + \theta_k) + B, \tag{2.10}$$

where $A = aA_l$ and $B = bB_l$. Note that all $\sin(\cdot)$ terms in (2.9) are vanished. For convenient performance of the phase shift $\phi$ in (2.10), the phase step $\theta_k$ should be spaced over $2\pi$ radians [19]. For simplicity, if the samples are evenly spaced, the phase step is given by

$$\theta_k = \frac{2\pi k}{K}, \quad k = 0, 1, \ldots, K - 1. \tag{2.11}$$

Then, this correlation measurement in (2.10) is repeated $K$ times with a phase step of $\theta_k$, introduced to the camera sensor or light source.

The relation between the phase difference $\phi$ due to the time of flight between the illumination and the reflection is measured and from that, the radial distance to the scene $d$ can be expressed as [20]

$$d = \frac{c\phi}{4\pi f}. \tag{2.12}$$

According to (2.12), it is clear that the depth $d$ depends on $\phi$ and $f$. However, the phase difference $\phi$ is unknown. Now, let us see how to make a measurement of $\phi$ (hence $d$) in the two types of operations of the camera as follows.

#### 2.2.3.1 Homodyne Operation

In homodyne operation the modulation frequency of the emitted light source and sensor signals are identical. In this mode, the phase shifted illumination is mixed with the camera shutter to produce a constant intensity level. Even though three measurements of phase frames are sufficient to perform a single calculation of range, traditionally four samples in the period of the reflected signal are sequentially acquired for each pixel of the sensor since the multiple of $\pi/2$ radians phase frames can be easily generated electronically, and the calculation is straightforward [20] as

$$I_k = C(\theta_k) \quad \text{for} \quad \theta_k = \pi k/2, \ k = 0, \ldots, 3, \tag{2.13}$$

giving,

$$\begin{aligned} I_0 &= 0.5A\cos\phi + B, \\ I_1 &= -0.5A\sin\phi + B, \\ I_2 &= -0.5A\cos\phi + B, \\ I_3 &= 0.5A\sin\phi + B, \end{aligned} \tag{2.14}$$

where $I_0, \ldots, I_3$ are the raw intensities as shown in Fig. 2.5.

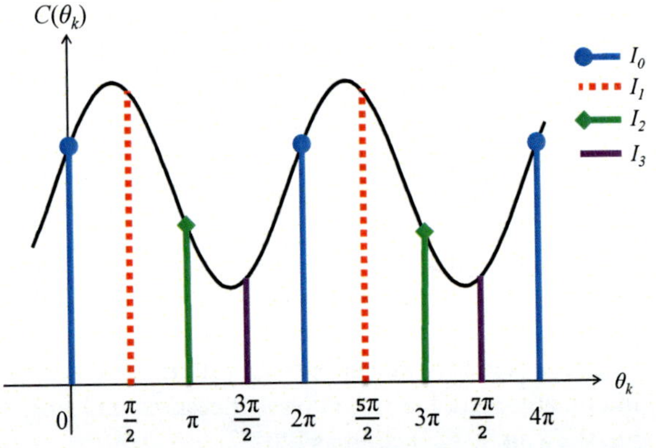


**Fig. 2.5** Estimation of four samples of the correlation function in (2.10)

The solutions for the three unknowns in (2.14) can be obtained as

$$\phi = \tan^{-1}\left(\frac{I_3 - I_1}{I_0 - I_2}\right), \tag{2.15a}$$

$$A = \frac{1}{2}\sqrt{(I_0 - I_2)^2 + (I_3 - I_1)^2}, \tag{2.15b}$$

$$B = \frac{1}{4}\sum_{k=0}^{3} I_k. \tag{2.15c}$$

Alternatively, the discrete Fourier transform (DFT) over the integration period $T$ can be used to calculate the three unknowns. The phase and amplitude of the first frequency bin contain the $\phi$ and $A$, respectively, while the DC component of the spectrum gives $B$ [21].

#### 2.2.3.2 Heterodyne Operation

When the modulation frequencies of the two signals are different, the camera is in heterodyne operation. In this mode, the phase difference of the illumination signal is maintained in the low frequency beat and the camera can be performed by taking at least four synchronous samples per cycle for each pixel of the sensor. The phase shift can be determined through analysis of the time varying intensities for each pixel using a more general form than in the homodyne case [22]. Then, from DFT, the unknowns $\phi$, $A$, and $B$ can be found from the phase and amplitude of the beat frequency bin, and the DC component of the spectrum, respectively, [21] as

$$\phi = \tan^{-1}\left(\frac{\sum_k I_k \cos\theta_k}{\sum_k I_k \sin\theta_k}\right), \tag{2.16a}$$

$$A = \frac{1}{2}\sqrt{\left(\sum_k I_k \cos\theta_k\right)^2 + \left(\sum_k I_k \sin\theta_k\right)^2}, \tag{2.16b}$$

$$B = \frac{1}{K}\sum_k I_k, \qquad k = 0, 1, \ldots, K-1. \tag{2.16c}$$

A continuous phase shift in heterodyne operation is an effective result rather than the discrete phase steps between samples of the homodyne method. However, the signal amplitude is reduced by the phase shift during the sensor integration period, in particular attenuating higher frequency harmonic components which can contaminate the phase measurement [23]. By changing the beat frequency, the ability of the system to easily alter the value of $K$ from three to several hundred is another advantage. This allows the user to select between high-speed, high-precision measurements with $K$ large (many frames) or reduced-precision measurements with $K$ small (few frames).

Once the phase shift $\phi$ is known, the depth measurements (the radial distance to the scene $d$) can be obtained via (2.12). The parameters $\phi$ and $A$ in both (2.15) and (2.16) depend on the assumption that the light source or camera sensor amplitude modulation signals follow a sinusoidal shape, and the correlation function, thus, the function $C(\theta_k)$ does not contain higher-order harmonics [24]. Typically, the light source in most current ToF cameras consists of a bank of infrared LEDs that are amplitude modulated using square-shaped currents to minimize the LED rise and fall times [25].

## 2.3 Aliasing of Signals

Physically, the signals of different frequencies become indistinguishable from each other, hence the term *aliasing*. When an analog signal is digitized, any component of the signal that is above one-half the digitizing or sampling frequency will be aliased. This limit of the frequency is known as the Nyquist frequency [26]. The high frequency components of the signal spectrum take the identity of the lower frequencies in the spectrum of the sampled signal. Thus, the aliasing can introduce low frequency errors in the acceleration signal whether it is collected experimentally or through simulation. These errors due to the aliasing will cause the integrated signals to drift.

The aliasing is impossible to prevent if a continuous signal is represented by a finite collection of samples. Since all experimentally obtained signals have been discretized by an analog to digital converter with a finite duration, some level of aliased content will be presented in these signals. The aliased content can be minimized but not eliminated by using a proper anti-aliasing filter and choosing a high sample rate than the cut-off frequency of the filter. Analog signals are usually low-pass filtered to remove most or all of the components above the Nyquist frequency in order to avoid aliasing. However, aliased content that has a negligible effect on the magnitude of a signal can have a significant impact on its integration [27].

## 2.4 Unambiguous Range

In CW modulation, the depth measurement $d$ is calculated by the phase $\phi$ which wraps around every $2\pi$. Therefore, the distance will also have an aliasing distance which is called unambiguous distance (range), $d_{\text{unamb}}$, and is equal to half of the maximum distance that the light can reach during a period. It can be defined as

$$d_{\text{unamb}} = \frac{\lambda}{2} = \frac{c}{2f}, \tag{2.17}$$

where $\lambda$ represents the wavelength of the modulated light signal. This means that objects which are located at distances greater than $d_{\text{unamb}}$ will be incorrectly calculated in (2.12). For example, from (2.17) gives the unambiguous range is 5 m when $f = 30\,\text{MHz}$ (note that the speed of light is taken as, $c \approx 3 \times 10^8$ m/s). Then the phase value is given by $\phi = \pi/2$ each for the objects located at 1.25 m and 6.25 m. Both objects will result in a computed range of 1.25 m. Therefore, we are unable to make a difference between a signal coming back from a far distance and another signal coming back from a close distance with the same phase shift.

According to (2.17), the simplest method to avoid ambiguity problems is to reduce the modulation frequency $f$, to extend the measurable distance. But, the frequency reduction has proportionally increased the uncertainty of the measurement [28], hence this method is not suitable for applications that require high precision. However, some advanced ToF systems deploy a multi-frequency method to extend the distance [29]. In this method, true location is the one where the different frequencies agree even each modulation frequency will have a different ambiguity distance.

## 2.5 Commercial ToF Range Imaging Cameras

The technologies used to build depth sensors of AMCW lidar systems have rapidly improved in camera manufacturing industries in recent years [30, 31]. Each pixel of the sensor captures both greyscale image measurement (intensity) and distance to the scene at a frame rate. Both the depth accuracy and the frame rate are limited by the required integration period of the camera. The appropriate tuning of the integration period is very important in ToF measurements. Long integration periods lead to good ToF depth measurements repeatability [32]. During the last decade, significant progress has been made in increasing the resolution and the sensitivity of the camera sensor [33]. At present, the modulation frequencies of the ToF cameras are relatively small (less than 100 MHz) and all these cameras have different accuracy and precision levels [5, 34]. There are a number of commercial ToF cameras available in the market. Most of them (MESA Imaging AG [35–37], SoftKinetic [38], PMD Technologies [39], Panasonic [40], etc.,) use the AMCW phase shift method while others (Advanced Scientific Concepts [41], Hamamatsu Photonics K.K. [42], ams AG [43], etc.,) use direct measurement time of flight (i.e., pulse method) to measure distances.

The price of commercial ToF cameras varies enormously and different technologies are used in their design. ToF cameras with high accuracy in ranging are very expensive while the cameras with substantial inaccuracies in ranging are relatively cheap. Therefore, some ToF cameras are not suitable for applications that need greater accuracy levels. In this book, we are using two AMCW ToF range imaging cameras, namely MESA Imaging SwissRanger 4000 [36] (expensive one, US $10,000) and SoftKinetic DepthSense 325 [44] (relatively less price, US $25) in the experimental set ups which will be used in Chaps. 7 and 8. The modulation frequencies of these cameras are self-generated inside and their configurations are discussed in the rest of this section.

### 2.5.1 MESA Imaging SwissRanger 4000

MESA Imaging SwissRanger 4000 (SR4000) uses infrared LEDs for its illumination source and uses a four-sample homodyne operation. The illumination process includes an optical feedback system. The optical feedback redirects some of the LED illuminations directly back to the sensor region. In SR4000, the LEDs appear relatively bright with a wavelength of 850 nm and affect very low sensitivity to the human eye and the camera has been verified for eye safety. A picture of a MESA SwissRanger 4000 is given in Fig. 2.6.

The SR4000 was manufactured in two versions. From one version, the user can select the modulation frequency among 29, 30, and 31 MHz. For the other version, the modulation frequency can be chosen among 14.5, 15, and 15.5 MHz. The calibrated ranges for 30 and 15 MHz are 0.8−5.0 m and 0.8−8.0 m, respectively. The size of the image captured from this camera is 176×144 pixels and the integration time, $IT$, can be controlled by the user with the formula [45],

$$IT = 0.3\text{ ms} + 0.1 \times intTime\text{ ms}, \tag{2.18}$$

where $intTime$ is a value from 0 to 255 that sets the $IT$ in ms. The frame rate, $FR$, depends on integration time and readout time, $RO$, as [45]

$$FR = \frac{1}{4(IT + RO)}, \tag{2.19}$$

(a) (b)

**Fig. 2.6** The MESA SwissRanger 4000 with power (**a**) off and (**b**) on

where $RO \approx 4.6$ ms and 4 represents since this camera uses four samples to produce the phase measurement, requiring four separate integration periods. By default the parameter $intTime$ is set to 30. This leads to an integration time of $IT = 3.3$ ms and an effective frame rate of $FR \approx 31.5$ fps. This depth camera operates with a maximum frame rate of 50 fps.

Too high integration times lead to saturation while low integration times lead to noise. Higher integration times also cause the higher temperature of the sensor and therefore the heat sink requirements become even more important [45]. In addition, SR4000 allows multiple cameras to operate simultaneously. The MESA SR4000 depth camera is now discontinued in the market.

### 2.5.2 *SoftKinetic DepthSense 325*

SoftKinetic DepthSense 325 (DS325) uses infrared laser (diffused) light for its illumination source and uses the homodyne operation. The camera has been verified for eye safety and has a low intensity signal. DS325 has the capability of a short range distance, 0.15–1 m only. The modulation frequencies for this camera are 50 and 60 MHz with frame rates of 25 and 30 fps, respectively. A picture of a SoftKinetic DepthSense 325 is shown in Fig. 2.7. The size of the image captured from DS325 is $320 \times 240$ pixels and the integration time is fixed. New versions of this camera are available in the market today.[1]

Regardless of whether LEDs or laser diodes are used, the emitted light is modulated with a modulation signal by its specific frequency, whose characteristics

[1] https://www.sony-semicon.co.jp/products_en/index.html.

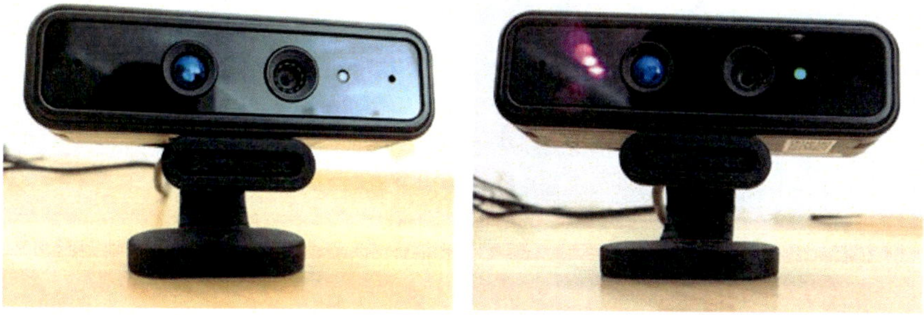

(a) (b)

**Fig. 2.7** The SoftKinetic DepthSense 325 with power (**a**) off and (**b**) on

are used for measuring of the depth. LEDs are cheaper and often smaller than lasers. However, some commercial ToF ranging cameras allow the simultaneous implementation of the ToF principle for several range measurements at the same time. Therefore, each pixel here represents an own range measurement system.

All cameras perform with their own configurations. The relevant camera parameters such as resolution, maximum range, frame rate, modulation frequency, and precision/accuracy, of the selected commercial ToF cameras, are tabulated in Table 2.1.

Since this book focuses only on the AMCW ToF range imaging cameras, from now, the terms "AMCW" and "range imaging" will be omitted for readability. Further references to ToF cameras will normally imply the homodyne operation in AMCW systems unless otherwise stated.

## 2.6 Chapter Remarks and Further Reading

This chapter was mainly focused on the background related to the amplitude modulation continuous wave (AMCW) ToF range imaging cameras in addition to the other types of ToF depth cameras.

Typically, a ToF camera consists mainly of a modulation source, a lens, and a depth sensor. An introductory analysis of the components of ToF range imaging cameras can be found in [6], and the common depth measurement techniques for depth cameras can be found in [18]. Based on the time-of-flight principle, different types of technologies can be found in [31]. Among them, for indirect ToF (i.e., AMCW) systems, the theoretical framework for heterodyne operation is well analyzed in [21]. The issues of signal's aliasing and unambiguous range on ToF systems are well investigated in [56].

The configurations of earlier ToF cameras can be found in Sect. 3.4 of [57] and the latest commercial AMCW ToF cameras can be found in [39, 40, 49, 52] while the pulse-based ToF cameras/sensors can be found in [41–43].

**Table 2.1** Specifications of the selected commercial time-of-flight range imaging cameras

| Manufacturer and ToF camera/sensor | | | Image size [pixel×pixel] | Modulation freq. [MHz] | Maximum range [m] | Max. frame rate [fps] | Precision or accuracy [mm] |
|---|---|---|---|---|---|---|---|
| Amplitude modulated continuous wave ToF cameras/sensors (indirect ToF) | | | | | | | |
| MESA Imaging | SwissRanger [35–37] | SR3000 | 176 × 144 | 20 | 7.5 | 25 | NA |
| | | SR4000 | | 29, 30, 31 | 5 | 54 | ±10 |
| | | | | 14.5, 15, 15.5 | 10 | 50 | ±15 |
| | | SR4500 | | 15–30 | 9 | 10–30 | ±20 for ≤5 m |
| | | | | | | | ±40 for >5 m |
| SoftKinetic | DepthSense [38] | DS311 | 160 × 120 | 20 | 1.5–4.5 | 60 | <300 at 3 m |
| | | DS325 | 320 × 240 | 50, 60 | 1.5–3.0 | 30 | <14 at 1 m |
| | | DS525 | 320 × 240 | Not available | 0.15–1.0 | 25–30, 50–60 | <14 at 1 m |
| | | DS536A | 78 × 58 | NA | 0.1–5.0 | 5–60 | NA |
| Microsoft | Kinect | V1 – (uses structured light) | 640 × 480 | NA | 0.4–4 | 30 | NA |
| | | V2 [46] | 512 × 424 | 16, 80, 120 | 0.5–4.5 | 30 | ±30 at 2 m |
| PMD Technologies [39] | GmbH | PMD19k | 160 × 120 | 20 | 7.5 | 15 | ±30 |
| | | CamCube3.0 | 200 × 200 | 21 | 15 | 15 | ±30 |
| | | CamCube3.0 | 200 × 200 | 21 | 15 | 15 | ±30 |
| | | A2 | 64 × 16 | 16 | 15 | 15 | ±100 for <40 m |
| | CamBoard | pico monstar | 352 × 288 | 100 | 0.5–6.0 | 60 | ≤1% |
| | | pico flexx | 224 × 172 | NA | 0.1–4.0 | 5–45 | ≤1% at 5 fps |
| | | | | | | | ≤2% at 45 fps |

| | | | | | | | |
|---|---|---|---|---|---|---|---|
| Panasonic Corporation | D-Imager [40] | EKL3104 | 160 × 120 | 3 modes—the selection is available using application software. | 9 | | ±40, $\sigma = 30$ (low ambient) |
| | | | | | | | ±40, $\sigma = 140$ (high ambient) |
| | | EKL3105 | | | 1.2–5 | 30 | $\sigma = 20$ (low ambient) |
| | | | | | 1.5–5 | | $\sigma = 50$ (high ambient) |
| | | EKL3106 | | | 1.2–5 | | $\sigma = 30$ (low ambient) |
| | | | | | 1.5–5 | | $\sigma = 140$ (high ambient) |
| | GN1C series [47, 48] | | RGB 320 × 240 | NA | [Indoor 0.1–7,8 | RGB 24 | [1 at 2.5 m, |
| | | | NIR 640 × 480 | | Outdoor 0.1–4,5] | NIR 30 | 4 at 12 m] |
| Intel® RealSense™ Technology [49] | Depth Camera [50] | D415 | Up to | NA | 0.3–10 | 90 | ≤ 2% up to 2 m |
| | | D435/D435i | 1280×720 | NA | 0.2–10 | – do – | ≤ 2% up to 2 m |
| | | D455 | active stereo depth resolution | NA | 0.4–10 | – do – | ≤ 2% up to 4 m |
| | | LiDAR Camera [51] | | | | | |
| | | L515 | 1024 × 768 | NA | 0.25–9 | 30 | <5 at 1 m |
| | | | | | | 30 | <14 at 9 m |

(continued)

**Table 2.1** (continued)

| Manufacturer and ToF camera/sensor | | | Image size [pixel×pixel] | Modulation freq. [MHz] | Maximum range [m] | Max. frame rate [fps] | Precision or accuracy [mm] |
|---|---|---|---|---|---|---|---|
| Amplitude modulated continuous wave ToF cameras/sensors (indirect ToF) | | | | | | | |
| Brookman Technology [52] | | | 320 × 240 | NA | 0.5–10 | 60 | ≤2.5% at 8 m |
| BT008D CMOS– | | ToF 3 tap sensor [53] | | | | | |
| Pulse modulation ToF cameras/sensors (direct ToF) | | | | | | | |
| Hamamatsu Photonics K.K. [42] | | | 160 × 120 | NA | Depends on integration time (IT) and readout time (RT), then frame rate = 1/(IT + RT) | | NA |
| Distance area image sensor—S11963-01CR | | | | | | | |
| Advanced Scientific Concepts Inc. [41] | 3D Flash LIDAR | Portable | 128 × 128 | NA | 70–1100 | 15 | NA |
| | | TigerEye | | | 60–1100 | 30 | ±40 at 60 m |
| | | TigerCub | | | NA | 30 | NA |
| | | DragonEye | | | 1500 | 10 | <±100, $3\sigma = \pm 150$ |
| | | GoldenEye | | | 3000 | 10 | <±100, $3\sigma = \pm 150$ |
| | | Peregrine | 128 × 32 | | Lens depend | 20 | NA |
| ams AG [43] | | ToF sensor | NA | NA | | NA | NA |
| | | TMF8701 [54] | | | 0.1–0.6 | | |
| | | TMF8801 [55] | | | 0.02–2.5 | | |

NA ≡ Not available

# References


1. David Nitzan, Alfred E Brain, and Richard O Duda. The measurement and use of registered reflectance and range data in scene analysis. *Proceedings of the IEEE*, 65(2):206–220, 1977.
2. Thomas Spirig, Peter Seitz, Oliver Vietze, and Friedrich Heitger. The lock-in CCD-two-dimensional synchronous detection of light. *IEEE Journal of quantum electronics*, 31(9): 1705–1708, 1995.
3. R Schwarte. Method and apparatus for determining the phases and/or amplitude information of an electromagnetic wave, 1997.
4. Robert Lange, Peter Seitz, Alice Biber, and Rudolf Schwarte. Time-of-flight range imaging with a custom solid state image sensor. In *Laser Metrology and Inspection*, volume 3823, pages 180–192. International Society for Optics and Photonics, 1999.
5. Ying He, Bin Liang, Yu Zou, Jin He, and Jun Yang. Depth errors analysis and correction for time-of-flight (tof) cameras. *Sensors*, 17(1):92, 2017.
6. Fabio Remondino and David Stoppa. *TOF Range-Imaging Cameras*. Springer Publishing Company, Incorporated, 2013. ISBN 3642275222, 9783642275227.
7. S Christie, SL Hill, B Bury, JO Gray, and KM Booth. Design and development of a multi-detecting two-dimensional ranging sensor. *Measurement Science and Technology*, 6(9):1301, 1995.
8. James Davis and Hector Gonzalez-Banos. Enhanced shape recovery with shuttered pulses of light. In *IEEE Workshop on Projector-Camera Systems*. Citeseer, 2003.
9. Hector Gonzalez-Banos and James Davis. Computing depth under ambient illumination using multi-shuttered light. In *Proceedings of the 2004 IEEE Computer Society Conference on Computer Vision and Pattern Recognition, 2004. CVPR 2004.*, volume 2, pages II–II. IEEE, 2004.
10. Paulo Crespo, Georgy Shakirin, Fine Fiedler, Wolfgang Enghardt, and Andreas Wagner. Direct time-of-flight for quantitative, real-time in-beam PET: a concept and feasibility study. *Physics in Medicine & Biology*, 52(23):6795, 2007.
11. Cristiano Niclass, Mineki Soga, Hiroyuki Matsubara, Satoru Kato, and Manabu Kagami. A 100-m Range 10-Frame\s 340 × 96-Pixel Time-of-Flight Depth Sensor in 0.18-$\mu$m CMOS. *IEEE Journal of Solid-State Circuits*, 48(2):559–572, 2012.
12. Augusto Ronchini Ximenes, Preethi Padmanabhan, Myung-Jae Lee, Yuichiro Yamashita, Dun-Nian Yaung, and Edoardo Charbon. A 256× 256 45/65nm 3D-stacked SPAD-based direct TOF image sensor for LiDAR applications with optical polar modulation for up to 18.6 dB interference suppression. In *2018 IEEE International Solid-State Circuits Conference-(ISSCC)*, pages 96–98. IEEE, 2018.
13. Salvatore Gnecchi and Carl Jackson. A 1× 16 SiPM array for automotive 3D imaging LiDAR systems. In *Proceedings of the 2017 International Image Sensor Workshop (IISW), Hiroshima, Japan*, pages 133–136, 2017.
14. Ximing Ren, Peter WR Connolly, Abderrahim Halimi, Yoann Altmann, Stephen McLaughlin, Istvan Gyongy, Robert K Henderson, and Gerald S Buller. High-resolution depth profiling using a range-gated CMOS SPAD quanta image sensor. *Optics express*, 26(5):5541–5557, 2018.
15. Chockalingam Veerappan, Justin Richardson, Richard Walker, Day-Uey Li, Matthew W Fishburn, Yuki Maruyama, David Stoppa, Fausto Borghetti, Marek Gersbach, Robert K Henderson, et al. A 160× 128 single-photon image sensor with on-pixel 55ps 10b time-to-digital converter. In *2011 IEEE International Solid-State Circuits Conference*, pages 312–314. IEEE, 2011.
16. Augusto Ronchini Ximenes, Preethi Padmanabhan, and Edoardo Charbon. Mutually coupled time-to-digital converters (TDCs) for direct time-of-flight (dTOF) image sensors. *Sensors*, 18 (10):3413, 2018.

17. Andreas Velten, Thomas Willwacher, Otkrist Gupta, Ashok Veeraraghavan, Moungi G Bawendi, and Ramesh Raskar. Recovering three-dimensional shape around a corner using ultrafast time-of-flight imaging. *Nature communications*, 3(1):1–8, 2012.
18. Carlo Dal Mutto, Pietro Zanuttigh, and Guido M Cortelazzo. *Time-of-flight cameras and Microsoft Kinect™*. Springer Science & Business Media, 2012.
19. Todd C Monson, Jeffrey W Grantham, Steve W Childress, John T Sackos, Robert O Nellums, and Steve M Lebien. Characterization of scannerless ladar. In *Laser Radar Technology and Applications IV*, volume 3707, pages 409–421. International Society for Optics and Photonics, 1999.
20. R. Lange and P. Seitz. Solid-state time-of-flight range camera. *IEEE Journal of Quantum Electronics*, 37(3):390–397, Mar 2001. ISSN 0018-9197. https://doi.org/10.1109/3.910448.
21. Andrew D Payne. *Development of a full-field time-of-flight range imaging system*. PhD thesis, The University of Waikato, 2008.
22. Richard M Conroy, Adrian A Dorrington, Rainer Künnemeyer, and Michael J Cree. Range imager performance comparison in homodyne and heterodyne operating modes. In *Three-Dimensional Imaging Metrology, Proc. SPIE*, volume 7239, pages 1–10, 2009.
23. Adrian PP Jongenelen, Andrew D Payne, Dale A Carnegie, Adrian A Dorrington, and Michael J Cree. Development of a real-time full-field range imaging system. In *Recent Advances in Sensing Technology*, pages 113–129. Springer, 2009.
24. Chrysostomos L Nikias and Jerry M Mendel. Signal processing with higher-order spectra. *IEEE Signal processing magazine*, 10(3):10–37, 1993.
25. E Fred Schubert, Thomas Gessmann, and Jong Kyu Kim. Light emitting diodes. *Kirk-Othmer Encyclopedia of Chemical Technology*, 2000.
26. L Eyer and P Bartholdi. Variable stars: which Nyquist frequency? *Astronomy and Astrophysics Supplement Series*, 135(1):1–3, 1999.
27. Timothy S Edwards. Effects of aliasing on numerical integration. *Mechanical systems and signal processing*, 21(1):165–176, 2007.
28. B. Büttgen and P. Seitz. Robust optical time-of-flight range imaging based on smart pixel structures. *IEEE Transactions on Circuits and Systems I: Regular Papers*, 55(6):1512–1525, July 2008. ISSN 1549-8328. https://doi.org/10.1109/TCSI.2008.916679.
29. Adrian P Jongenelen, Donald G Bailey, Andrew D Payne, Adrian A Dorrington, and Dale A Carnegie. Analysis of errors in tof range imaging with dual-frequency modulation. *IEEE transactions on instrumentation and measurement*, 60(5):1861–1868, 2011.
30. Francois Blais, J Angelo Beraldin, and Sabry F El-Hakim. Range error analysis of an integrated time-of-flight, triangulation, and photogrammetric 3d laser scanning system. In *Laser Radar Technology and Applications V*, volume 4035, pages 236–248. International Society for Optics and Photonics, 2000.
31. Radu Horaud, Miles Hansard, Georgios Evangelidis, and Clément Ménier. An overview of depth cameras and range scanners based on time-of-flight technologies. *Machine vision and applications*, 27(7):1005–1020, 2016.
32. Peter Fürsattel, Simon Placht, Michael Balda, Christian Schaller, Hannes Hofmann, Andreas Maier, and Christian Riess. A comparative error analysis of current time-of-flight sensors. *IEEE Transactions on Computational Imaging*, 2(1):27–41, 2016.
33. H Kraft, J Frey, T Moeller, M Albrecht, M Grothof, B Schink, H Hess, and B Buxbaum. 3D-camera of high 3D-frame rate, depth-resolution and background light elimination based on improved PMD (photonic mixer device)-technologies. *OPTO, Nuremberg, May*, 2004.
34. Aongus McCarthy, Nils J Krichel, Nathan R Gemmell, Ximing Ren, Michael G Tanner, Sander N Dorenbos, Val Zwiller, Robert H Hadfield, and Gerald S Buller. Kilometer-range, high resolution depth imaging via 1560 nm wavelength single-photon detection. *Optics express*, 21(7):8904–8915, 2013.
35. *SwissRanger SR-3000 Manual*. MESA Imaging AG, October 2006. URL https://aiweb.techfak.uni-bielefeld.de/files/SR3000_manual_V1.03.pdf.
36. *SR4000 Data Sheet*. MESA Imaging AG, February 2011. URL http://www.adept.net.au/cameras/Mesa/pdf/SR4000.pdf.

37. *SR4500 Data Sheet*. MESA Imaging AG, July 2013. URL http://www.adept.net.au/cameras/Mesa/pdf/SR4500.pdf.
38. DEPTHSENSE SENSORS. https://www.sony-depthsensing.com/\penalty\z@Depthsense/\penalty\z@Depthsense\penalty\z@Sensors, 2021a. [Accessed April 2021].
39. World class 3D depth sensing Development Kits. For everyone. https://pmdtec.com/picofamily/, 2021b. [Accessed April 2021].
40. Panasonic GC1N Series Time Of Flight Camera Module. https://na.industrial.panasonic.com/products/sensors/sensors-\penalty\z@automotive-\penalty\z@industrial-\penalty\z@applications/\penalty\z@lineup/\penalty\z@time-\penalty\z@flight-\penalty\z@tof-\penalty\z@camera-\penalty\z@module/series/142944, 2021c. [Accessed April 2021].
41. Leaders in 3D Global shutter Flash LIDAR™ Cameras and Vision Systems. https://asc3d.com/, 2021d. [Accessed April 2021].
42. Distance & position sensors. https://www.hamamatsu.com/jp/en/product/optical-sensors/distance-position-sensor/index.html, 2021e. [Accessed April 2021].
43. Time-of-Flight Sensor. https://ams.com/time-of-flight, 2021f. [Accessed April 2021].
44. *DS325 Data Sheet*. SoftKinetic, September 2012. URL https://www.sony-depthsensing.com/Portals/0/Download/WEB_20120907_SK_DS325_Datasheet_V2.1.pdf.
45. *SR4000/SR4500 User Manual*. MESA Imaging AG. URL https://acroname.com/sites/default/files/assets/sr4000_sr4500_manual.pdf.
46. Oliver Wasenmüller and Didier Stricker. Comparison of kinect v1 and v2 depth images in terms of accuracy and precision. In *Asian Conference on Computer Vision*, pages 34–45. Springer, 2016.
47. TIME OF FLIGHT (TOF) CAMERA MODULE - PANASONIC DIRECT TECHNOLOGY. https://na.industrial.panasonic.com/whats-new/time-flight-tof-camera-module-panasonic-direct-technology, 2021g. [Accessed April 2021].
48. *Time of Flight Depth-Color Image Camera: GC1N*. Panasonic Corporation of North America, March 2021. URL https://b2b-api.panasonic.eu/file_stream/pids/fileversion/244670.
49. Intel® RealSense™ technology. https://www.intel.com/content/www/us/en/architecture-and-technology/realsense-overview.html, 2021h. [Accessed April 2021].
50. *Intel® RealSense™ Product Family D400 Series*. Intel Corporation, June 2020. URL https://www.intelrealsense.com/wp-content/uploads/2020/06/Intel-RealSense-D400-Series-Datasheet-June-2020.pdf.
51. *Intel® RealSense™ LiDAR Camera L515*. Intel Corporation, January 2021. URL https://www.intelrealsense.com/download/7691/.
52. 3d depth image sensor. https://brookmantech.com/en/products/#sensor, 2021i. [Accessed April 2021].
53. *Time of Flight 3D Depth Sensor Product Catalog*. Brookman Technology, Inc., February 2021. URL https://brookmantech.com/en/mgr/wp-content/uploads/2020/10/BT-CMOS-ToF_Brochure_V20210201.pdf.
54. *TMF8701 Time-of-Flight Sensor*. ams AG, January 2021a. URL https://ams.com/documents/20143/4409395/TMF8801_DS000648_7-00.pdf.
55. *TMF8801 1D Time-of-Flight Sensor*. ams AG, January 2021b. URL https://ams.com/documents/20143/4409395/TMF8801_DS000648_7-00.pdf.
56. S. B. Gokturk, H. Yalcin, and C. Bamji. A Time-Of-Flight Depth Sensor - System Description, Issues and Solutions. In *Computer Vision and Pattern Recognition Workshop, 2004. CVPRW '04. Conference on*, pages 35–35, June 2004. https://doi.org/10.1109/CVPR.2004.17.
57. Carlo Dal Mutto Pietro Zanuttigh, Giulio Marin. *Time-of-Flight and Structured Light Depth Cameras: Technology and Applications*. Springer International Publishing, 1 edition, 2016. ISBN 978-3-319-30971-2, 978-3-319-30973-6.

# Chapter 3
# Measurement Error in Range Imaging Systems

**Abstract** The commercial ToF range imaging cameras have error sources: noise and jitter, in different amounts which potentially affect range measurements. In this chapter, first, the harmonic cancellation on ToF range imaging cameras is briefly investigated. Then, the definitions of the accuracy, precision, noise, and jitter terms are introduced. Next, noise error sources related to the commercial ToF cameras and noise error mitigation techniques in range measurements are briefly described. Finally, as a summary, common noise errors and their extraction techniques in ToF cameras are pointed out in the tabular form.

**Keywords** Harmonic cancellation · Accuracy · Precision · Noise · Jitter · Noise sources · Temperature noise · Depth calibration · Wiggling effect · Fixed pattern noise · Light scattering · Nonlinearity · Multipath interference · Motion blurs · Signal-to-noise ratio

## 3.1 Harmonic Cancellation

The generation of perfect sinusoids of the relevant frequency is impossible. Sampling a received signal takes a finite non-zero time, thus the sampling of the received signal is not ideal. This causes harmonic distortion in the received signal, which negatively impacts the quality of the measured data. Typically a calibration is performed to correct these errors [1, 2]. In addition, the unwanted harmonics in the received signal considerably impacts the depth measurements. By using a lookup table, Kahlmann et al. [1] described a method to calibrate for this kind of error while Lindner and Kolb [2] used a fitted cubic B-spline in their method. However, if the temperature changes in the system, the calibration methods can be invalidated. Therefore, a phase encoding approach that attenuates the harmonics during the sampling process was introduced by Payne et al. [3]. That method is independent of the amplitude modulation waveform shape. Moreover, Streeter and Dorrington [4] demonstrated a simple method to remove the third harmonic that typically causes the main harmonic error in AMCW based ToF cameras. They verified that the standard deviation of the noise results has not been significantly changed after applying



G. Anthonys, *Timing Jitter in Time-of-Flight Range Imaging Cameras*,
https://doi.org/10.1007/978-3-030-94159-8_3

the method and recommended that the method can be applied to the higher-order harmonics as well.

## 3.2 Error Sources in Range Measurement

The accuracy and precision of ToF range imaging cameras are important for many applications, however, these cameras have many noise sources that degrade both accuracy and precision. The commercial ToF cameras have error sources: noise and jitter, in different amounts which potentially affect range measurements.

### *3.2.1 Accuracy and Precision*

Accuracy and precision are both refer to the quality of measurement, but they are two dissimilar indicators of measurement. Accuracy is the closeness of the set of measurements to the actual value, while precision is the closeness of those measurements to each other [5]. In other words, accuracy is the degree of veracity while precision is the degree of reproducibility. For the understanding purpose, the two terms for the parameter, *measured depth* can be illustrated as shown in Fig. 3.1.

Since these two terms are independent of each other, so a specific set of measurements can be called to be either accurate, or precise, or both, or neither. In terms of accuracy and precision, the comparison of the captured depth images from two versions of Microsoft Kinect v1 and v2 is investigated by Wasenmüller and Stricker [6]. Applications of 3D construction and visual odometry are performed for their investigation. They recommended Kinect v2 has a higher accuracy which is difficult to enhance algorithmically and Kinect v1 which makes it suitable for fast prototypes [6]. Furthermore, some investigations on the accuracy and precision of ToF depth cameras are found in [1, 3, 7, 8].

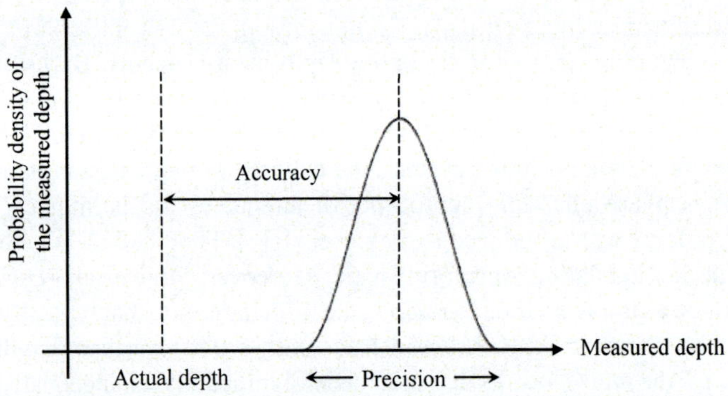


**Fig. 3.1** An illustration of the terms: accuracy and precision of measured depth

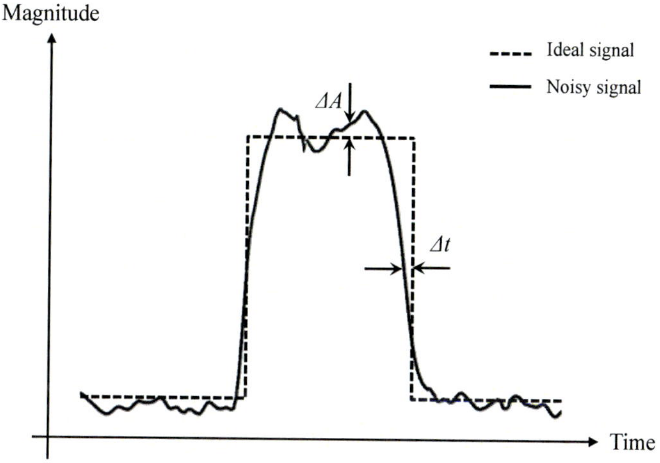


**Fig. 3.2** A noisy signal and its ideal signal

### 3.2.2 *Noise and Jitter*

The raw intensities in either (2.15) or (2.16) that are used to compute $A$, $B$ and $\phi$ are subject to noise. As a result, the modulation amplitude, background offset, and the range information that one gets from the camera signals are also noisy. Any signal can deviate from its ideal occurrence in any magnitude and timing as shown in Fig. 3.2. If the deviation is in magnitude, it is said to be noise (or amplitude noise $\Delta A$), whereas if the variation is in timing, it is said to be jitter (or timing jitter $\Delta t$). These definitions are used throughout this book. Often both noise and jitter occur at the same time. Jitter reveals itself as variations in phase, width, duty cycle, or period [9]. To reduce the uncertainty in depth measurements, it is useful to understand some of the common error sources of noise and jitter in ToF range imaging cameras. There are significant sources that are present in almost every measurement of instrument accuracy and precision, electronic noise and jitter. Therefore, the rest of this chapter focuses on the noise sources in ToF ranging cameras and noise mitigation techniques.

## 3.3 Noise Sources in Range Imaging Cameras

It is clear that the depth measurements depend on the phase angle and modulation frequency in the system. Therefore, it is more important to investigate what factors contribute (for $\phi$ and $f$) directly or indirectly to the precision and accuracy of ToF range measurements, because the noise sources affect the precision while the measurement errors affect the accuracy. In ToF cameras, the noise can be categorized as systematic and non-systematic (random) errors [10, 11]. Systematic errors can be further categorized according to their source in the process of range measurement. Noise sources are due to random pixel noise, temperature noise, built-in pixel-related noise, the amplitude of reflected light, and depth distortion.

### *3.3.1 Integration Time Related Noise*

It is noticed that there is a relationship between the integration time and the depth measurements for the same scene [11]. This affects the range of depths that the camera is sensing with more precision. Too long an integration time leads to saturation while short integration time increases the noise level. The effects of the integration time on the data measurements of various ToF cameras have been investigated in literature [12, 13]. Many range cameras use the four-phase shift algorithm [14] which needs four samples to produce the phase measurement. This requires four separate integration periods. Some cameras use even more integration periods, therefore the total time required to capture a depth image is three (or more) times the integration time, plus three (or more) times the readout time. This is reflected in the achievable frame rate for given integration time. Integration time can typically be set by the user and is adjusted to the observed scene.

### *3.3.2 Temperature Noise*

The inside and outside temperatures of ToF cameras are always unstable, therefore it will take some time to stabilize at some level after turning on. Until the temperature of the camera is stabilized, the depth measurements can drift. The warm up effects on different cameras have been investigated by Pfeifer et al. [13], Chiabrando et al. [15], and Kahlmann and Ingensand [8]. As an example, Pfeifer et al. [13] explored the warm up effect in MESA SwissRanger 4000 and showed that there is a 2 mm shift in measurement over the first hour of operation. After one hour the measurement is much more stable. The inside heating of ToF cameras may be increased if the camera is running for long hours. This causes a change in the behavior of the materials inside the camera. Furthermore, the effect of the warm up of some ToF cameras has been investigated by Kahlmann et al. [1] and Piatti and Rinaudo [12]. As an example, Kahlmann et al. [1] pointed out the influence of the integration time and temperature on measured distance by SwissRanger™.

### *3.3.3 Built-in Pixel-Related Noise*

This kind of error occurs in the sensor array due to properties of different material in the complementary metal-oxide-semiconductor gates or charge-coupled device, and the time delays of capacitor charge during the signal correlation process. As results of these, two neighboring pixels that measure the same real depth may nevertheless provide different depth measurements [1, 10].

### 3.3.4 *Intensity of Reflected Light*

Since the ToF measurements are based on the light reflected from the scene, the accuracy of the measurements is highly related to the amount of incident light. The higher reflected intensity implies higher depth accuracy. On the other hand, saturation of the sensor can occur when the scene is too close to the camera and invalidates the measurement [11, 16].

### 3.3.5 *Depth Distortion*

Depth distortion is referred to as wiggling or circular error [11]. In practice, the emitted infrared light cannot be exactly generated in the modulation process as described in the theory. This causes errors in the depth measurements and depends only on the measured depth for each pixel. This type of error can be addressed either by considering the average values from multiple relative measurements or by comparing the depth measurements with a reference truth distance [17, 18].

On the other hand, several non-systemic errors in ToF cameras can be identified as light scattering, multipath interference, motion blurs, and distortion in signal-to-noise ratios.

### 3.3.6 *Multipath Interference*

Multipath interference (MPI) is due to the interference of multiple light reflections captured at each pixel of the sensor. This is highly scene dependent and is one of the major problems in ToF measurements, because (1) it is ambiguous and difficult to prevent, (2) it can produce large errors in range, and (3) we do not know the scene until we have measured it. MPI is mainly due to the presence of surface edges, reflections of translucent objects (e.g., glass, white sheet), volumetric scattering (e.g., fog, dust), and inter-reflections of objects in the surrounding area [19, 20]. The behavior between phase and amplitude is not linear with the frequency when MPI is present. The impact of multipath interference in ToF systems is well explored in literature [21–24]. The non-linearity and multipath interference in the signal is one of the major problems in AMCW ToF ranging systems.

### 3.3.7 *Motion Blurs*

Motion blurs is another type of error that more adversely affects ranging after the MPI. Due to the movement of the scene or the camera during the integration time, the measurement of the scene is blurred. Also, the motion that occurs between integration periods is a major problem. Lottner et al. [25] have used a combination of a photonic mixing device (PMD) camera and a conventional 2D image sensor to detect lateral motion artifacts. Also, Lindner and Kolb [26] have investigated compensation methods for motion artifacts for ToF cameras. If the motion is restricted to a single direction, then the correction process is relatively faster than the motion in multiple directions [27, 28]. Streeter and Dorrington [29] have investigated both motion blur within each frame and errors arising due to changes between frames.

### 3.3.8 *Signal-to-Noise Ratio*

The signal-to-noise ratio (SNR) compares the levels of signal power to noise power and is most often expressed in decibels (dB). The achievable SNR varies between cameras and scenes since some cameras have a stronger emitted signal than others. When this ratio is increasing with time, it is called that the system is in the better specification, since there is more useful data (the signal) than the unwanted data (the noise) [11].

### 3.3.9 *Random Noise*

The random noise in the radial motion in ToF ranging was investigated by Streeter [30]. Two methods were developed: one was by using the maximum likelihood estimation (MLE) and a new method based on the measured correlation signal and its derivative. The author concluded that the MLE performed better than the other in terms of random noise, but MLE was slower and was not able to provide an estimate in every example used. In contrast, the other method always produces an estimate of radial motion.

## 3.4 Noise Mitigation Techniques

There are different types of de-noising techniques for the error sources in ToF systems. A mathematical model to predict systematic errors and statistical uncertainties of ToF camera was investigated by Rapp et al. [31]. An error propagation model from the measured data is used to analyze them. The pixel base non-uniformity of the depth signal was also examined. The amplitude was used for pixel validation purposes and temperature noise was not considered in their work. Let us briefly consider other corresponding de-noising techniques as follows.

### 3.4.1 Temperature Noise

With increases in temperature, a higher rate of thermally generated electrons is present in range sensors technology. In CCD and CMOS photo sensors increased internal heating causes a larger number of thermal generated electrons [1]. Even though some cameras have a cooling process, it remains a small effect and can be factored in the calibration.

### 3.4.2 Depth Calibration

Due to the systematic appearance of sources of noise errors, there is a deviation of actual depth and depth measured by the ToF camera. Thus, a thorough depth calibration has to be performed to record accurate data. Zhang [32] developed a flexible technique for the calibration of radial distance to the scene which is based on a closed-form solution, followed by a non-linear refinement based on the maximum likelihood criterion. Kahlmann et al. [1] and Kahlmann and Ingensand [8] introduced a calibration pattern consisting of filled white circles on a black background while Fuchs and Hirzinger [18] used a classic checker-board pattern where both methods are suited to accurately determine the intrinsic parameters such as the depth measurement distortion and focal length, and the pose of the camera. However, Kahlmann et al. [1] presented a parameter based calibration approach while Kahlmann and Ingensand [8] used a differential measurement principle setup. Both of them were considered a photogrammetric camera calibration and a distance system calibration with respect to the reflectivity and the distance itself.

Fuchs and Hirzinger [18] introduced a calibration procedure based on the distance and amplitude measurements with respect to an external positioning system. They simultaneously estimate the distance parameters and the extrinsic parameters (e.g., optical tracking system or a robot) with respect to the external positioning system for a ToF camera. Lindner et al. [33] presented a lightweight calibration framework in which intrinsic calibration, wiggling error calibration, and reflectivity related depth error calibration are integrated. The method is based on analysis-by-synthesis, a planar checker-board pattern with different levels of reflectivity. Furthermore, Hansard et al. [34] introduced an automatic detection of calibration grids of ToF images with a checker-board pattern. They used the Hough transform which is based on careful reasoning about the global geometric structure of the board, before and after perspective projection.

### 3.4.3 Camera Distortion

Due to the wiggling effect of the camera, the depth measurements offset can be distorted. To overcome this offset, some researchers choose one integration time and perform the depth distortion calibration process with the same value [16, 35]. On the other hand, high integration time was used to compensate the higher optical output power of their proposed illumination module and implemented the model by using pseudo four-phase shift algorithm [36]. Longer integration times cause a higher temperature of the sensor, therefore some 3D cameras have an auto mode settings for the integration time. Hussmann et al. [37] described a CW modulation method based on sine waves with various higher integration time for their noise distribution model.

### 3.4.4 Built-in Pixel-Related Noise

A simple model which predicts how the distance error at one pixel depends on image intensity at that pixel and on the distance itself is introduced by Falie and Buzuloiu [10]. The model is based on shot noise (correlated with a light source) with additional noise source (due to the presence of dark pixels). They experimented on three features: intensity-dependent variance of the noise, intensity-dependent distance error, and surround-dependent distance errors of ToF cameras. They concluded that due to the non-linearity in the phase shift estimation, the darker objects seem to be either further away or closer depending on the distance. This error can be obtained by comparing the calculated depths with a reference distance, and representing them as a table or matrix, known as a Fixed Pattern Noise (FPN) matrix. This matrix depends on the integration time in the system and has to be determined for every different integration period [1]. Recently, Li et al. [38] have demonstrated a calibration method based on *delay line* to optimize the camera's linearity, accuracy, and pixel-related error.

### 3.4.5 Light Scattering

Jamtsho and Lichti [39] noticed that the light scattering is highly dependent on the surface area and the distance of the scattering object from the camera and the background scene. They concluded that scattering range bias is independent of the integration time while the scattering amplitude bias monotonically increases with the integration time. However, they tested only for MESA SwissRanger 3000 and 4000 ToF ranging cameras. On the other hand, Schäfer et al. [40] presented a new model, based on raw data calibration with only one additional intrinsic camera parameter and have shown that their approach effectively removes the errors of in camera light scattering. They, however, assumed a constant scattering parameter that can be averaged for every pixel in the measurement area.

### *3.4.6 Non-linearity and Multipath Interference*

Godbaz et al. [41] show that the standard amelioration methods such as lookup tables and B-spline models are invalidated due to temperature changes. They consider heterodyning and harmonic cancellation as partial solutions to non-linearity issues in practical systems. Commonly, there are two types of algorithms to describe multipath interference (MPI) in the literature. One type is focused on *diffuse* multipath, arising from Lambertian surfaces (i.e., uniformly all pixels to all pixels) [19, 42] and other the type is focused on *two-path* multipath, which arises from specular surfaces (i.e., per pixel basis) [21, 22].

Jiménez et al. [42] have proposed a radiometric model that represents the working principle of ToF cameras, including the MPI phenomenon. They show the original scene can be recovered from contaminated measurements from the ToF camera, with an iterative method that converges to a depth correction. It reduced highly the MPI influence on the results. Godbaz et al. [22] propose two models; one is based on a Cauchy distribution over the range and uses four measurements at different modulation frequencies to determine the amplitude, phase, and reflectivity distribution of up to two component returns within each pixel. The other method is based on the attenuation ratios to determine the amplitude and phase of up to two component returns within each pixel.

Freedman et al. [43] introduced an algorithm based on sparse reflections analysis. They show the method can be used with many types of multipath interference. The investigation of the properties of a binary sequence as the modulation waveform is another technique available that is commonly used. Moreover, Whyte et al. [44] resolved these errors by using the direct and global separation methods: a technique that is also applicable to normal camera images. It is considered that the multiple paths are a direct return which is the return from the shortest path length, and a global return which includes all other returns. Recently, Marco et al. [45] have proposed a different approach, a convolution neural network (CNN) based on depth reconstruction methods for the correction of MPI by using an unmodified, off-the-shelf camera, with a single frequency in real time. An encoder-decoder (convolutional-deconvolutional) neural network with a two-stage training process is designed using both captured and synthetic data. The network is trained by synthetic depth data with MPI as input and it returned the corrected depth map. This investigation was worked in several real world scenarios with more general reflections and it failed for extremely transparent or glossy surfaces.

### 3.4.7 Motion Blurs

Lottner et al. [25] recommended two possible approaches to minimize the effect due to the motion blurs without discarding the corrupted data. One is to take two-phase depth computation after the phase sampling of the images and the other is to take an average of positionally weighted neighboring pixels. In addition, the usage of coded exposure and optical flow techniques together in AMCW range measurement is used by Streeter and Dorrington [29]. In addition, Jimenez et al. [46] have proposed a method that removes motion artefacts from a single frame captured by a PMD ToF camera without altering current hardware. They recovered the depth of each pixel by exploiting the consistency of the correlation samples and local neighbors of the pixel. Recently, Streeter [47] used the Kalman filter to perform noisy time-series analysis for the problem of transverse motion by raw ToF data.

### 3.4.8 Signal-to-Noise Ratio

To increase the signal-to-noise ratio (SNR), the average value can be computed with a certain accuracy threshold with respect to the pixel's variance or simply removing the corrupted values by using the low amplitude filtering method [10]. Fuchs and May [17] used a procedure to decide the optimal integration time depending on the required areas. Mufti and Mahony [48] proposed SNR estimator for ToF cameras based on statistical modelling and maximum likelihood estimation. They provided noise models for the three parameters of amplitude, phase, and intensity. The model was used for prediction of errors in a ToF camera under various SNR conditions.

Among those error sources, light scattering and depth distortion can be easily mitigated. In addition to that, the effect of SNR can be easily overcome. Table 3.1 is a summary of some selected error reductions related to ToF systems attained by the main compensation approaches in the literature.

## 3.5 Chapter Remarks and Further Reading

For a given frequency, it is impossible to generate perfect sinusoidal signals. AMCW ToF range imaging cameras measure three independent parameters of intensity, amplitude, and phase which are primarily used for range estimation. Most of these cameras are used in the homodyne operation. In this chapter, first, the harmonic cancellation in ToF cameras was briefly described. Then, definitions of accuracy, precision, noise, and jitter were introduced. Next, the main contribution of this chapter was a sufficient investigation of the noise error sources in commercial ToF range imaging cameras with noise mitigation techniques.

**Table 3.1** Noise extraction techniques for time-of-flight range imaging cameras

| Investigation on | Author [Reference] | Methodology/algorithm |
|---|---|---|
| Harmonic cancellation | Kahlmann et al. [1] | Lookup table |
| | Lindner and Kolb [2] | Fitted cubic B-spline |
| | Payne et al. [3] | Phase encoding method |
| | Streeter and Dorrington [4] | Modification on the data acquisition process |
| Wiggling and reflectivity error calibration | Lindner et al. [33] | Checker-board pattern and an Analysis-by-Synthesis method |
| Depth distortion and amplitude | Oprisescu et al. [16] | Spatially and adaptive neighborhood filters |
| | Fuchs and Hirzinger [18] | B-Splines and multiple range amplitudes |
| | Kim et al. [35] | 6-degree polynomial and radial pattern |
| | Hussmann and Edeler [36] | Pseudo four-phase shift algorithm |
| | Hussmann et al. [37] | A CW modulation based on sine waves |
| Photogrammetric and distance calibrations | Kahlmann et al. [1], Kahlmann and Ingensand [8] | A parameter based approach; differential measurement principle |
| Distance calibration | Zhang [32] | Closed-form solution and maximum likelihood estimation (MLE) |
| | Fuchs and Hirzinger [18] | Checker-board pattern and a non-linear optimization algorithm |
| | Hansard et al. [34] | Checker-board pattern and Hough transform |
| Built-in pixel & integration time | Kahlmann et al. [1] | Fixed Pattern Noise (FPN) matrix and lookup table |
| | Lindner and Kolb [2] | FPN matrix with constant integration time |
| | Falie and Buzuloiu [10] | Shot noise and correlated with light source |
| | Fuchs and Hirzinger [18] | Pan and tilt coefficient with different integration time |
| Lateral motion artefacts | Lottner et al. [25] | Classical 2D image edge detector |
| Motion artefacts | Schmidt and Jähne [49] | Analyzing the temporal raw data signal |
| | Jimenez et al. [46] | A single frame (i.e., 4 phase images) |

(continued)

**Table 3.1** (continued)

| Investigation on | Author [Reference] | Methodology/algorithm |
|---|---|---|
| Axial and lateral motions | Lindner and Kolb [26] | A theoretical model |
| Motion blur | Hussmann et al. [27], Hussmann et al. [28] | Optical distance measurement and phase shift algorithm |
| | Lee [50] | Pixel-wise motion blur elimination method |
| Transverse motion | Streeter and Dorrington [29], Streeter [47] | Coded exposure and optical flow techniques; Kalman filter |
| Light scattering | Mure-Dubois and Hügli [51] | Hypotheses of linearity and space invariance |
| | Jamtsho and Lichti [39] | An analytical compensation method |
| | Schäfer et al. [40] | A model, based on raw data calibration and one additional intrinsic camera parameter |
| Non-linearity | Godbaz et al. [41] | Harmonic cancellation and heterodyning |
| Multipath interference | Dorrington et al. [21] | Mixed pixel restoration and multipath separation algorithm |
| | Godbaz et al. [22] | Cauchy distribution over range and attenuation ratios |
| | Fuchs et al. [19] | Photometric calibration and environment integration |
| | Freedman et al. [43] | Sparse Reflections Analysis algorithm |
| | Bhandari et al. [20] | Spectral estimation theory |
| | Jiménez et al. [42] | Radiometric model under some mild conditions to the surfaces and lights in the scene |
| | Whyte et al. [44] | Direct and global separation methods |
| | Marco et al. [45] | Convolutional-deconvolutional neural network |
| Signal-to-noise ratio | Falie and Buzuloiu [10] | Low amplitude filtering method |
| | Gudmundsson et al. [52] | Compare the average of data with a fixed threshold |
| | Fuchs and May [17] | Optimal integration time parametrization |
| | Mufti and Mahony [48] | Rice distribution and MLE |
| Random noise | Streeter [30] | MLE and based on the derivative of correlation signal |

The precision and accuracy of the ToF full-field ranging cameras are important for many applications, however, several noise sources degrade both precision and accuracy. These errors arise primarily due to the factors such as the hardware inadequacies in camera design leading to problems, and light traveling multiple paths in the scene.

The error sources in ToF cameras directly affect the range measurement and many of them are well examined in the literature. Among those error sources, the motion blurs, non-linearity, and multipath interference more adversely affect ranging. Some error sources in these cameras can be mitigated by calibration techniques. Detailed investigations on noise error sources with compensation methods can be found in Sect. 3.3 of [53] and in [11].

## References


1. T. Kahlmann, F. Remondino, and H. Ingensand. Calibration for increased accuracy of the range imaging camera Swissranger$^{TM}$. In *Image Engineering and Vision Metrology*, volume 36 of *3*, pages 136–141, 2006.
2. Marvin Lindner and Andreas Kolb. Lateral and depth calibration of PMD-distance sensors. In *Proceedings of the Second International Conference on Advances in Visual Computing - Volume Part II*, ISVC'06, pages 524–533, Berlin, Heidelberg, 2006. Springer-Verlag. https://doi.org/10.1007/11919629_53.
3. Andrew D. Payne, Adrian A. Dorrington, Michael J. Cree, and Dale A. Carnegie. Improved measurement linearity and precision for AMCW time-of-flight range imaging cameras. *Appl. Opt.*, 49(23):4392–4403, Aug 2010. https://doi.org/10.1364/AO.49.004392.
4. Lee Streeter and Adrian A Dorrington. Simple harmonic error cancellation in time of flight range imaging. *Optics Letters*, 40(22):5391–5394, 2015. https://doi.org/10.1364/OL.40.005391.
5. ISO Accuracy. Accuracy (trueness and precision) of measurement methods and results — Part 1: General principles and definitions. *International Organization for Standardization*, pages 5725–1, 1994.
6. Oliver Wasenmüller and Didier Stricker. Comparison of kinect v1 and v2 depth images in terms of accuracy and precision. In *Asian Conference on Computer Vision*, pages 34–45. Springer, 2016.
7. Timo Kahlmann and Hilmar Ingensand. High-precision investigations of the fast range imaging camera SwissRanger. In *Photonics in the Transportation Industry: Auto to Aerospace*, volume 6758, page 67580J. International Society for Optics and Photonics, 2007.
8. Timo Kahlmann and Hilmar Ingensand. Calibration and development for increased accuracy of 3D range imaging cameras. *Journal of Applied Geodesy*, 2(1):1–11, 2008.
9. Mike Li. *Jitter, Noise, and Signal Integrity at High-speed*. Prentice Hall Press, Upper Saddle River, NJ, USA, first edition, 2007. ISBN 9780132429610.
10. Dragos Falie and Vasile Buzuloiu. Noise characteristics of 3D time-of-flight cameras. In *2007 International Symposium on Signals, Circuits and Systems*, volume 1, pages 1–4. IEEE, 2007.
11. S. Foix, G. Alenya, and C. Torras. Lock-in time-of-flight (tof) cameras: A survey. *IEEE Sensors Journal*, 11(9):1917–1926, Sept 2011. ISSN 1530-437X. https://doi.org/10.1109/JSEN.2010.2101060.
12. Dario Piatti and Fulvio Rinaudo. SR-4000 and CamCube3.0 time of flight (ToF) cameras: Tests and comparison. *Remote Sensing*, 4(4):1069–1089, 2012.
13. Nobert Pfeifer, Derek Lichti, Jan Böhm, and Wilfried Karel. 3D cameras: Errors, calibration and orientation. In *TOF Range-Imaging Cameras*, pages 117–138. Springer, 2013.

14. M. Servin, J. C. Estrada, and J. A. Quiroga. The general theory of phase shifting algorithms. *Opt. Express*, 17(24):21867–21881, Nov 2009. https://doi.org/10.1364/OE.17.021867.
15. Filiberto Chiabrando, Roberto Chiabrando, Dario Piatti, and Fulvio Rinaudo. Sensors for 3D imaging: Metric evaluation and calibration of a CCD/CMOS time-of-flight camera. *Sensors*, 9 (12):10080–10096, 2009.
16. S. Oprisescu, D. Falie, M. Ciuc, and V. Buzuloiu. Measurements with tof cameras and their necessary corrections. In *Signals, Circuits and Systems, 2007. ISSCS 2007. International Symposium on*, volume 1, July 2007. https://doi.org/10.1109/ISSCS.2007.4292691.
17. Stefan Fuchs and Stefan May. Calibration and registration for precise surface reconstruction with time-of-flight cameras. *Int. J. Intell. Syst. Technol. Appl.*, 5(3/4):274–284, 2008. ISSN 1740-8865. https://doi.org/10.1504/IJISTA.2008.021290.
18. S. Fuchs and G. Hirzinger. Extrinsic and depth calibration of tof-cameras. In *Computer Vision and Pattern Recognition, 2008. CVPR 2008. IEEE Conference on*, pages 1–6, June 2008. https://doi.org/10.1109/CVPR.2008.4587828.
19. Stefan Fuchs, Michael Suppa, and Olaf Hellwich. Compensation for multipath in tof camera measurements supported by photometric calibration and environment integration. In *Computer Vision Systems*, pages 31–41. Springer, 2013.
20. Ayush Bhandari, Achuta Kadambi, Refael Whyte, Christopher Barsi, Micha Feigin, Adrian Dorrington, and Ramesh Raskar. Resolving multipath interference in time-of-flight imaging via modulation frequency diversity and sparse regularization. *Optics letters*, 39(6):1705–1708, 2014.
21. AA Dorrington, JP Godbaz, MJ Cree, AD Payne, and LV Streeter. Separating true range measurements from multi-path and scattering interference in commercial range cameras. In *IS&T/SPIE Electronic Imaging, Proc. SPIE*, volume 7864, pages 1–10, 2011.
22. John P Godbaz, Michael J Cree, and Adrian A Dorrington. Closed-form inverses for the mixed pixel/multipath interference problem in AMCW lidar. In *IS&T/SPIE Electronic Imaging*, pages 829618–829618. International Society for Optics and Photonics, 2012.
23. Refael Whyte, Ayush Bhandari, Lee Streeter, Michael J. Cree, and Adrian A. Dorrington. Time frequency duality of time-of-flight range cameras for resolving multi-path interference. In *Proceedings of the 29th International Conference on Image and Vision Computing New Zealand*, pages 247–252, New York, NY, USA, 2014a. ACM. https://doi.org/10.1145/2683405.2683448.
24. R. Whyte, L. Streeter, M.J. Cree, and A.A. Dorrington. Review of methods for resolving multi-path interference in time-of-flight range cameras. In *Sensors Conference, 2014 IEEE*, pages 629–632, Valencia, Spain, Nov 2014b. https://doi.org/10.1109/ICSENS.2014.6985077.
25. O. Lottner, A. Sluiter, K. Hartmann, and W. Weihs. Movement artefacts in range images of time-of-flight cameras. In *2007 International Symposium on Signals, Circuits and Systems*, volume 1, pages 1–4, July 2007. https://doi.org/10.1109/ISSCS.2007.4292665.
26. M. Lindner and A. Kolb. Compensation of Motion Artifacts for Time-of-Flight Cameras. *Lecture Notes in Computer Science*, 5742:16, 2009. https://doi.org/10.1007/978-3-642-03778-8_2.
27. Stephan Hussmann, Alexander Hermanski, and Torsten Edeler. Real-time motion supression in tof range images. In *Instrumentation and Measurement Technology Conference (I2MTC), 2010 IEEE*, pages 697–701. IEEE, 2010.
28. Stephan Hussmann, Alexander Hermanski, and Torsten Edeler. Real-time motion artifact suppression in tof camera systems. *IEEE Transactions on Instrumentation and Measurement*, 60(5):1682–1690, 2011.
29. Lee Streeter and Adrian A. Dorrington. Coded exposure correction of transverse motion in full-field range imaging. *Optical Engineering*, 53(10):102109, 2014. https://doi.org/10.1117/1.OE.53.10.102109.
30. Lee Streeter. Stochastic calculus analysis of optical time-of-flight range imaging and estimation of radial motion. *Journal of the Optical Society of America A*, 34(7):1063–1072, 2017.

31. H. Rapp, M. Frank, F.A. Hamprecht, and B. Jähne. A theoretical and experimental investigation of the systematic errors and statistical uncertainties of time-of-flight-cameras. *Int. J. Inte. Sys. Tech. and Appl.*, 5(3/4), 2008.
32. Zhengyou Zhang. A flexible new technique for camera calibration. *IEEE Transactions on Pattern Analysis and Machine Intelligence*, 22:1330–1334, December 2000.
33. Marvin Lindner, Ingo Schiller, Andreas Kolb, and Reinhard Koch. Time-of-flight sensor calibration for accurate range sensing. *Comput. Vis. Image Underst.*, 114(12):1318–1328, December 2010. ISSN 1077-3142. https://doi.org/10.1016/j.cviu.2009.11.002.
34. Miles Hansard, Seungkyu Lee, Ouk Choi, and Radu Patrice Horaud. *Time-of-flight cameras: principles, methods and applications*. Springer Science & Business Media, 2012.
35. Young Min Kim, D. Chan, C. Theobalt, and S. Thrun. Design and calibration of a multi-view tof sensor fusion system. In *Computer Vision and Pattern Recognition Workshops, 2008. CVPRW '08. IEEE Computer Society Conference on*, pages 1–7, June 2008. https://doi.org/10.1109/CVPRW.2008.4563160.
36. S. Hussmann and T. Edeler. Pseudo-four-phase-shift algorithm for performance enhancement of 3D-TOF vision systems. *IEEE Transactions on Instrumentation and Measurement*, 59(5): 1175–1181, May 2010. ISSN 0018-9456. https://doi.org/10.1109/TIM.2010.2040881.
37. S. Hussmann, F. Knoll, and T. Edeler. Modulation method including noise model for minimizing the wiggling error of tof cameras. *IEEE Transactions on Instrumentation and Measurement*, 63(5):1127–1136, May 2014. ISSN 0018-9456. https://doi.org/10.1109/TIM.2013.2286932.
38. Liwu Li, Hongwei Liu, Yuan Xu, and Yanghao Zheng. Measurement linearity and accuracy optimization for Time-of-Flight range imaging cameras. In *2020 IEEE 4th Information Technology, Networking, Electronic and Automation Control Conference (ITNEC)*, volume 1, pages 520–524. IEEE, 2020.
39. Sonam Jamtsho and Derek D Lichti. Modelling scattering distortion in 3D range camera. *International Archives of Photogrammetry, Remote Sensing and Spatial Information Sciences*, 38(5):299–304, 2010.
40. Henrik Schäfer, Frank Lenzen, and Christoph S Garbe. Model based scattering correction in time-of-flight cameras. *Optics express*, 22(24):29835–29846, 2014.
41. John P Godbaz, Michael J Cree, and Adrian A Dorrington. Understanding and ameliorating non-linear phase and amplitude responses in AMCW lidar. *Remote Sensing*, 4(1):21–42, 2011.
42. David Jiménez, Daniel Pizarro, Manuel Mazo, and Sira Palazuelos. Modeling and correction of multipath interference in time of flight cameras. *Image and Vision Computing*, 32(1):1–13, 2014.
43. Daniel Freedman, Yoni Smolin, Eyal Krupka, Ido Leichter, and Mirko Schmidt. Sra: Fast removal of general multipath for tof sensors. In *Computer Vision–ECCV 2014*, pages 234–249. Springer, 2014.
44. R. Whyte, L. V. Streeter, M. J. Cree, and A. A. Dorrington. Resolving multiple propagation paths in time of flight range cameras using direct and global separation methods. *Optical Engineering*, 54(11):1–9, 2015. https://doi.org/10.1117/1.OE.54.11.113109.
45. Julio Marco, Quercus Hernandez, Adolfo Munoz, Yue Dong, Adrian Jarabo, Min H Kim, Xin Tong, and Diego Gutierrez. DeepToF: off-the-shelf real-time correction of multipath interference in time-of-flight imaging. *ACM Transactions on Graphics (ToG)*, 36(6):219, 2017.
46. David Jimenez, Daniel Pizarro, and Manuel Mazo. Single frame correction of motion artifacts in pmd-based time of flight cameras. *Image and Vision Computing*, 32(12):1127–1143, 2014.
47. Lee Streeter. Time-of-flight range image measurement in the presence of transverse motion using the Kalman filter. *IEEE Transactions on Instrumentation and Measurement*, 67(7):1573–1578, 2018.
48. Faisal Mufti and Robert Mahony. Statistical analysis of signal measurement in time-of-flight cameras. *{ISPRS} Journal of Photogrammetry and Remote Sensing*, 66(5):720 – 731, 2011. ISSN 0924-2716. https://doi.org/10.1016/j.isprsjprs.2011.06.004.

49. Mirko Schmidt and Bernd Jähne. Efficient and robust reduction of motion artifacts for 3d time-of-flight cameras. In *3D Imaging (IC3D), 2011 International Conference on*, pages 1–8. IEEE, 2011.
50. Seungkyu Lee. Time-of-flight depth camera motion blur detection and deblurring. *IEEE Signal Processing Letters*, 21(6):663–666, 2014.
51. James Mure-Dubois and Heinz Hügli. Real-time scattering compensation for time-of-flight camera. In *Proceedings of the ICVS Workshop on Camera Calibration Methods for Computer Vision Systems*, 2007.
52. Sigurjon Ami Gudmundsson, Henrik Aanæs, and Rasmus Larsen. Environmental effects on measurement uncertainties of time-of-flight cameras. In *Signals, Circuits and Systems, 2007. ISSCS 2007. International Symposium on*, volume 1, pages 1–4. IEEE, 2007.
53. Carlo Dal Mutto Pietro Zanuttigh, Giulio Marin. *Time-of-Flight and Structured Light Depth Cameras: Technology and Applications*. Springer International Publishing, 1 edition, 2016. ISBN 978-3-319-30971-2, 978-3-319-30973-6.

# Chapter 4
# Jitter and Measurement of Jitter

**Abstract** The previous chapter focused on the noise analysis in range imaging systems. The jitter also affects the noise, and often both noise and jitter occur at the same time. Therefore, in the literature, it is important to investigate the behavior of jitter in signals of range measurements. This chapter contributes by investigating the jitter in ranging systems. First, mathematical representations for several measurement methods of jitter are stated. Then the different types of jitter in signals and the phase of sinusoidal signals are briefly described. Next, jitter sources in ranging systems and three types of approaches, namely *graphical representation, time and frequency domains*, and *statistical and signal processing domains* for analyzing the jitter in signals are described. Next, the jitter impact on range imaging systems and currently available jitter measurement studies relevant to the ToF range imaging cameras are reviewed. Only a couple of studies of jitter that related to range measurements are found in the literature. Those methods are camera dependent and cannot be used for jitter measurement in all ToF range imaging cameras.

**Keywords** Jitter sources · Period jitter · Cycle-to-cycle jitter · Time-interval error jitter · Phase jitter · Nondeterministic (random) jitter · Deterministic jitter · Eye diagram · Histogram · Bathtub curve · Bit error ratio · Unit interval · Time-lag correlation · Reference clock · Fourier domain

## 4.1 Jitter

Another potential noise source is jitter which arises because of mistiming in electronic signals on the camera sensor such as in the ADC. As stated in Chap. 3, the jitter is variation in timing. Jitter is defined as any timing movement with reference to the timing of the intended (ideal) signal. Figure 4.1 shows a signal with jitter (without noise) and its ideal signal. The dashed and solid lines indicate the ideal signal and jittery signal, respectively. All parameters in the figure are defined and explained in the following sub-sections. The effect of timing jitter as a noise source in ranging cameras is barely recognized or addressed in the literature [1, 2]. But, now, we believe the amount of jitter is also impacting on ranging. Because



G. Anthonys, *Timing Jitter in Time-of-Flight Range Imaging Cameras*,
https://doi.org/10.1007/978-3-030-94159-8_4

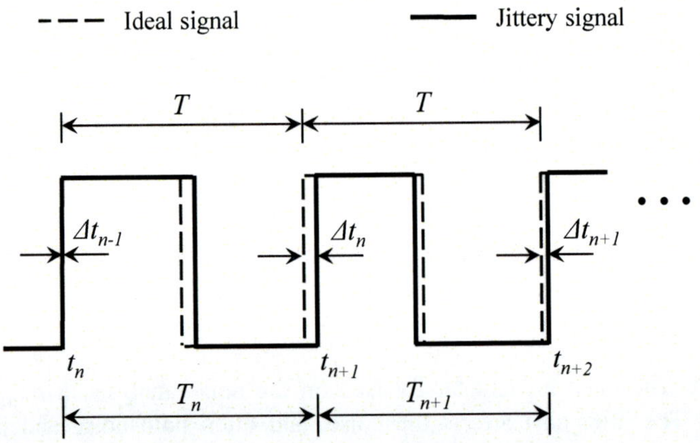


**Fig. 4.1** Graphical representation of clock-like signal with jitter

at present, the depth sensors for the ToF systems have been rapidly improved in camera manufacturing industries by increasing the resolution of the camera range with higher modulation frequencies and higher frame rates. These improvements are potentially more affected by the mistiming of electronic signals and thus, by jitter.

There are several methods by which jitter may be measured on a waveform. These are period jitter, cycle-to-cycle jitter, and time interval error jitter. It is important to understand what these measurements disclose and how they relate to each other.

### 4.1.1 Period Jitter

Period jitter is the time difference between a measured clock cycle period and the ideal clock cycle period. In practice, it is difficult to quantify an ideal clock cycle period, thus an average period is used. Period jitter is typically specified over a set number of clock cycles. For a given number of individual clock periods, we can measure each one and calculate the average clock period. From Fig. 4.1, each individual clock period of the ideal signal is $T$, then the average period of the ideal signal is given by $T$. From these measurements, the standard deviation and the peak-to-peak value can be computed [3]. Mathematically, the period jitter for $n$th cycle, $j_n^{\text{PRD}}$, can be expressed as

$$
\begin{aligned}
j_n^{\text{PRD}} &= T_n - T \\
&= t_{n+1} - t_n - T,
\end{aligned}
\tag{4.1}
$$

where $n$ is the time-domain index and $T_n$ is the $n$th period (cycle) of the signal.

### 4.1.2 Cycle-to-Cycle Jitter

Cycle-to-cycle (CtoC) jitter is defined as the time difference between any consecutive clock periods. This can be found by applying a first order difference operation to the period jitter. It is always specified as an absolute magnitude and is typically reported as a peak value in ps. Sometimes this is expressed as an RMS value in ps. This type of jitter is commonly used to illustrate the stability of spread spectrum clocks because the period jitter is more sensitive to the frequency spreading feature while CtoC jitter is not [4]. CtoC jitter is important for specifying the performance of phase-locked loop (PLL), however, it does not establish a relationship between non-consecutive clocks. Specifically, CtoC measurement does not address frequency drift over time and it shows the instantaneous dynamics of a clock-recovery PLL [5]. Mathematically, CtoC jitter for $n$th cycle, $j_n^{\mathrm{CtoC}}$, can be expressed as

$$\begin{aligned} j_n^{\mathrm{CtoC}} &= |T_{n+1} - T_n| = |t_{n+2} - t_{n+1} - (t_{n+1} - t_n)| \\ &= |t_{n+2} - 2t_{n+1} + t_n| , \end{aligned} \tag{4.2}$$

where $|\cdot|$ represents the absolute value.

### 4.1.3 Time Interval Error Jitter

Time interval error (TIE) is the difference between clock edge time of ideal and jittery signals for each active clock edge present (either rising or falling edge). It is measured by subtracting the actual clock edge from the ideal clock edge [3]. From Fig. 4.1, the deviation of the signal $n$th timing event $t_n$, from its ideal occurrence in time, $\Delta t_n = j_n^{\mathrm{TIE}}$, is given by

$$\begin{aligned} j_n^{\mathrm{TIE}} &= \Delta t_n = T_n + \Delta t_{n-1} - T \\ &= t_{n+1} - t_n + \Delta t_{n-1} - T, \end{aligned} \tag{4.3}$$

where $\Delta t_{n-1}$ is the deviation of the signal $(n-1)$th timing event $t_{n-1}$.

### 4.1.4 Phase Jitter

The term phase jitter has not been standardized, but there seems to be a general consensus on its meaning [6]. Generally, noise and jitter can affect system performance all the time and often both of them occur at the same time. The impacts of jitter and noise are not linear. Let us consider a sinusoidal signal with the phase modulated by phase noise $\theta(t)$ is given by

$$s(t) = \sin(2\pi f t + \theta(t)) \\ = \sin\left(2\pi f\left(t + \frac{\theta(t)}{2\pi f}\right)\right), \tag{4.4}$$

and the phase jitter is [7]

$$j^{\text{PHS}} = \frac{\theta(t)}{2\pi f}. \tag{4.5}$$

From this relationship, it is clear that for a given frequency $f$ of the signal, the phase jitter amount increases (or decreases) when the phase noise $\theta(t)$ increases (or decreases). Thus, the phase jitter in a signal is the impact on the phase noise and vice versa causing the effect of the jitter on other noise sources in range measurements.

## 4.2 Types of Jitter in Signals

The total jitter in a signal is specified by the sum of the components of non-deterministic jitter and deterministic jitter as shown in Fig. 4.2. Non-deterministic jitter is due to the random process and is referred to as random jitter (RJ). In contrast, the deterministic jitter is due to deterministic processes and is linearly additive with a predictable function. It can be characterized by its peak-to-peak (or peak) value. Periodic jitter (PJ), bounded uncorrelated jitter (BUJ), and data dependent jitter (DDJ) which consists of duty cycle distortion (DCD) and inter symbol interference (ISI) are sub-categories of the deterministic jitter [9].

Each type of these can be mathematically represented in three different domains: time, frequency, and statistical. These representations are appropriately used in this book. Jitter and noise are physical quantities, and each has a corresponding physical unit or dimension. Now, let us consider each of these jitter types in briefly.

### *4.2.1 Random Jitter*

The random jitter sources are usually corresponding of shot noise, thermal noise, flicker noise ($1/f$), and other high-order noise processes. This type of jitter often follows a Gaussian distribution and the probability density function (PDF) of the random jitter can be formed as

$$f_{\text{RJ}}(\Delta t) = \frac{1}{\sigma_{\text{RJ}}\sqrt{2\pi}} \exp\left\{ -\frac{1}{2}\left(\frac{\Delta t - \mu_{\text{RJ}}}{\sigma_{\text{RJ}}}\right)^2 \right\}, \tag{4.6}$$

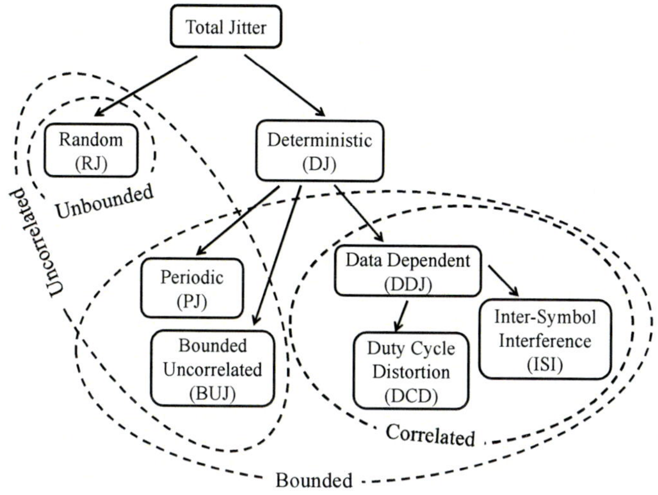


**Fig. 4.2** Classification of jitter

where $\mu_{RJ}$ and ${\sigma_{RJ}}^2$ are the mean and variance of the RJ, respectively. The RJ can be characterized by its root-mean-square (RMS) value or standard deviation (STD) $\sigma$, thus,

$$\Delta t^{RJ}(t) = \epsilon(t), \tag{4.7}$$

where $\epsilon(t)$ is the random process which is a function of time and has the PDF, $f_{RJ}(\Delta t)$. This function is characterized as unbounded because its PDF is not zero unless the jitter $\Delta t$ approaches infinity.

In addition, there are higher-order jitter processes that have a general form of $f^{-\alpha}$ for random jitter processes. When $\alpha = 0$, 1, and 2, it corresponds to white (or Gaussian) jitter, flick jitter, and an integrated white, respectively [9].

## 4.2.2 *Periodic Jitter*

The periodic jitter refers to periodic variations of signal edge positions over time at a certain frequency or period. From a signal perspective, it is the same as any periodic signal in terms of frequency and phase, but its amplitude is jitter in units of timing. Sometimes, this PJ is referred to as sinusoidal jitter. In a signal, there can be multiple PJs of various frequencies and these follow the superposition rule, and each can be characterized by its amplitude and frequency, thus [9]

$$\Delta t^{\mathrm{PJ}}(t) = \sum_{p=1}^{P} A_{\mathrm{PJ}}{}^{(p)} \sin\left(2\pi f_{\mathrm{PJ}}{}^{(p)} t + \phi_{\mathrm{PJ}}^{(p)}\right), \qquad P \in \mathbb{Z}^{+} \tag{4.8}$$

where $A_{\mathrm{PJ}}{}^{(p)}$, $f_{\mathrm{PJ}}{}^{(p)}$, and $\phi_{\mathrm{PJ}}^{(p)}$ are the amplitude, frequency, and phase difference of the $p$th component of the PJ, respectively. In addition, this equation can change appropriately to other PJs with different profiles such as triangle, sawtooth, or trapezoid.

### 4.2.3 *Bounded Uncorrelated Jitter*

The BUJ is typically caused by coupling from adjacent data-carrying links. This type of jitter is bounded due to coupling strength, and uncorrelated because there is no correlation to the channel's data pattern. The models depend on the data pattern, coupling signal, and its mechanism [10]. From a mathematical point of view, the BUJ can be modeled by many independent PJs if it is composed of many independent root sources. Then, its time-domain PDF is a truncated Gaussian, is defined by [9]

$$f_{\mathrm{BUJ}}(\Delta t) = \begin{cases} \frac{\rho_{\mathrm{BUJ}}}{\sigma_{\mathrm{BUJ}}\sqrt{2\pi}} \exp\left\{ -\frac{1}{2}\left(\frac{\Delta t}{\sigma_{\mathrm{BUJ}}}\right)^{2} \right\}, & |\Delta t| \le A_{\mathrm{BUJ}} \\ 0, & |\Delta t| > A_{\mathrm{BUJ}} \end{cases} \tag{4.9}$$

where $A_{\mathrm{BUJ}}$ and $\sigma_{\mathrm{BUJ}}$ are the peak and STD values for the BUJ, respectively. $\rho_{\mathrm{BUJ}}$ is the normalization probability for the PDF. This $\rho_{\mathrm{BUJ}}$ is introduced to ensure that the integral of (4.9) will be unit 1.

### 4.2.4 *Data Dependent Jitter*

The DDJ corresponds to a variable jitter that depends on the bit pattern transmitted on the link under test. Since this type of jitter is closely related to the run length of the data pattern, there are only limited run-length permutations for a given pattern. So, the DDJ value will be finite and discrete. Based on this, general math models of the PDF for the ISI and DCD can be expressed.

#### 4.2.4.1 Duty Cycle Distortion Jitter

The DCD describes a jitter amounting to a signal having unequal pulse widths for high and low logic values, this means that the duty cycle of the high speed clock will be directly translated into the overall DCD. This can be modeled by [9]

$$f_{\mathrm{DCD}}(\Delta t) = \sum_{k=1}^{K} p_{\mathrm{DCD}}^{(k)}\, \delta\left(\Delta t - D_{\mathrm{DCD}}^{(k)}\right), \qquad K \in \mathbb{Z}^{+} \tag{4.10}$$

where $p_{\mathrm{DCD}}^{(k)}$ and $D_{\mathrm{DCD}}^{(k)}$ are the probability and amount for the $k$th component of the DCD jitter, respectively. Term $\delta(\ldots)$ is conventional Dirac delta function. This model should be satisfied the following probability condition

$$\sum_{k=1}^{K} p_{\mathrm{DCD}}^{(k)} = 1. \tag{4.11}$$

#### 4.2.4.2 Inter Symbol Interference Jitter

The ISI refers to the band limited characteristic of the transmitted patterns (i.e., cable, driver, connectors); this effect is proportional to the number of consecutive equal transmitted bits and patterns on the same trace. Various edge patterns have various frequency components. Because of the conductors filtering effects, different patterns propagate at different speeds through the conductors. This difference in propagation speeds causes bits to smear into adjacent bits, resulting in ISI [9, 11]. The same kind of sum of delta functions method in (4.10) can be applied as,

$$f_{\mathrm{ISI}}(\Delta t) = \sum_{k=1}^{K} p_{\mathrm{ISI}}^{(k)}\, \delta\left(\Delta t - D_{\mathrm{ISI}}^{(k)}\right), \qquad K \in \mathbb{Z}^{+} \tag{4.12}$$

where $p_{\mathrm{ISI}}^{(k)}$ and $D_{\mathrm{ISI}}^{(k)}$ are the probability and amount for the $k$th component of the ISI jitter, respectively, and satisfies the condition

$$\sum_{k=1}^{K} p_{\mathrm{ISI}}^{(k)} = 1. \tag{4.13}$$

## 4.3 Phase of Sinusoidal Signals

Throughout this book, we are considering sinusoidal geometrical shape for waveform unless otherwise stated. As stated in Chap. 3, the generation of a sinusoidal signal is impossible for a given frequency. On the other hand, the phase value of the sinusoidal signals is not from a single contribution. In practice, the total phase of sinusoidal signals is influenced by several factors such as frequency drift, frequency offset, and random phase fluctuations. However, we are considering the signals with the form

$$s(t) = A(t)\sin(\Phi(t)), \tag{4.14}$$

where $A(t)$ is amplitude of the signal and $\Phi(t)$ is the total phase as

$$\Phi(t) = 2\pi f\left(t + \Delta t^{\mathrm{RJ}}(t) + \Delta t^{\mathrm{PJ}}(t)\right), \tag{4.15}$$

where $\Delta t^{\mathrm{RJ}}(t)$ and $\Delta t^{\mathrm{PJ}}(t)$ are the random jitter and periodic jitter, respectively, which were formally defined with mathematical expressions in Sect. 4.2.

## 4.4 Jitter Sources in Ranging Systems

All the jitter types described in Sect. 4.2 occur in signals carrying the data such as in a telecommunication system. The modulation signal of the ranging camera contains certain types of jitter since it carries only an unchanging datum (i.e., the phase only), therefore, the BUJ and DDJ which depend on transmission data do not occur in the modulation signal of ranging cameras. The jitter sources in ToF cameras are related to the behavior of the camera sensor, light source, and transmission media. Let us consider the jitter sources for the two major components: random and deterministic.

### *4.4.1 Sources for Random Jitter*

Random jitter is due to random noise inherent in electronic circuits and typically follows a Gaussian distribution. Also, stochastic sources such as white noise, thermal noise, shot (Poisson) noise, flicker noise, and other high-order noise processes are led to present the random jitter [12, 13].

#### 4.4.1.1 Thermal Noise

Thermal noise is associated with electron flow in conductors and increases with bandwidth, temperature, and noise resistance. When limited to a finite bandwidth, thermal noise has a nearly Gaussian distribution. This type of noise is also referred to as Johnson-Nyquist noise [14].

#### 4.4.1.2 Shot or Poisson Noise

Shot or Poisson noise is a type of electronic noise that can be modeled by a Poisson process. In electronic shot noise originates from the discrete nature of the electric charge. This also occurs in photon counting in optical devices which is associated with the particle nature of light.

#### 4.4.1.3 Flicker Noise

Flicker noise is proportional to the reciprocal of the frequency. It is only of concern in low frequency measurements. It is normally found in resistors, diodes, switches, and transistors, among other components. Typically, it must be measured empirically (by observation).

### *4.4.2 Sources for Deterministic Jitter*

Generally, the deterministic jitter occurs due to the imperfections in the behavior of the device or transmission media of the system. In addition, due to the signal modulation, crosstalk and reflections on signals can all lead to deterministic jitter on the light and shutter modulation signals [15].

#### 4.4.2.1 Crosstalk

Crosstalk arises by the accidental coupling of a magnetic and/or electric field to an adjacent conductor carrying a signal. The unwanted components are added to the original signal altering its bias determined by the amount of the interference signal.

#### 4.4.2.2 Electromagnetic Interference

Electromagnetic interference results from conducted and radiated emissions undesirably radiated from the system. Switching power supplies that radiate strong, high frequency magnetic and electric fields conducting a large amount of electrical noise are some common sources that lack adequate shielding and output filtering in the system. These sources are related to periodic jitter.

#### 4.4.2.3 Reflection

Reflection is caused by interference of the signal with itself occurring when impedance mismatches exist. A portion of the energy is reflected back to the transmitter if an impedance mismatch exists at the receiver, coming from uncontrolled stubbing and incorrect terminations. Likewise, if the mismatch is at the transmission end, it ingests part of the reflected energy, and the receiver gets reflected the remainder. Ultimately, the receiver gets the delayed signal out of phase with the original one which is algebraically added with the first arriving signal.

#### 4.4.2.4 Data Dependent Phenomena

Data Dependent phenomena is often caused by the frequency response of a cable or device. Amplitude offset errors, turn on delays, and saturation may be some of the causes of DCD. ISI occurs when the frequency component of the data is propagated at different rates by the transmission medium.

## 4.5 Analysis of Jitter in Signals

Total jitter, namely both deterministic and random, is mixed. In a signal with jitter, the zero crossing transitions of the cycles are seen to vary in time from the ideal clock timing. Jitter can be characterized, depending on the type of jitter, by parameters such as the amplitude and frequency of the jitter. Jitter amplitude is a measure of the typical time displacement in units of time. If the jitter values are repetitive, then the frequency of the repetition is of interest. The time displacement of a digital signal is equivalent to variations in the signal's phase.

Controlling jitter is a valuable task because jitter can degrade the performance of electronic systems. Measurement of jitter can be performed in the time-domain or frequency domain with the help of statistical and signal processing techniques. In addition, various kinds of jitter measuring instruments are commercially available such as the pulse/pattern generator, spectrum analyzer, real-time sampling oscilloscope, digital communications analyzer, and the bit error ratio tester [16, 17]. All of these instruments require an accurate reference clock to trigger the time measurement unit to measure jitter. The rest of this section briefly describes three common approaches to analyze the jitter in signals; (1) graphical, (2) time and frequency domains, and (3) statistical and signal processing techniques.

### *4.5.1 Graphical Approaches*

Jitter in a signal can be analyzed by graphical representation techniques such as the *eye diagram, histogram,* and *bathtub plot* with each having limitations [18–21]. The accuracy of these techniques can only be guaranteed if the jitter measurement is carried out on real-time measuring instruments with a very large memory depth to acquire sufficient data.

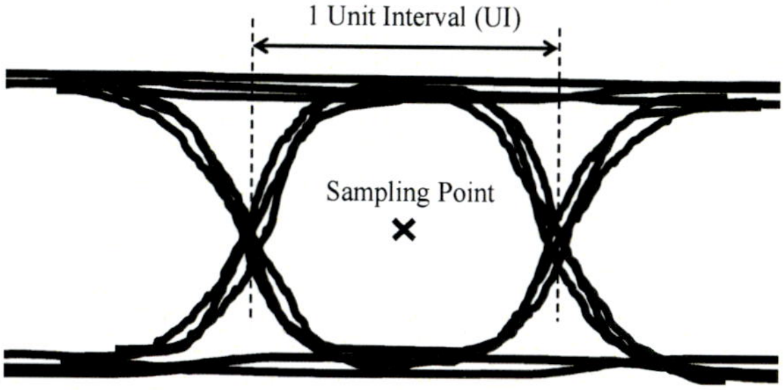


**Fig. 4.3** A picture of an eye diagram

#### 4.5.1.1 Eye Diagram

The effect of both noise and jitter can be represented by the eye diagram, in which many data voltage levels (bit periods) and edge transitions are superimposed over a one unit interval (UI) range as shown in Fig. 4.3 [9]. An eye diagram gives an overall statistical qualitative view of the quality of the signal under test. It helps for both quantitative and qualitative insight regarding the data signal path itself and what might be affecting signal integrity. The horizontal and vertical eye openings are important characteristics of the eye diagram that aid in measuring the signal quality. The horizontal eye opening is expressed as a percentage of the bit interval while vertical eye opening is expressed as a percentage of the full eye height. A large, wide-open eye in the center (marked by **x**, see Fig. 4.3) shows the ideal location for sampling each bit. At the sample point, the waveform should have settled to its low or high value and this indicates that the data stream has very little noise and jitter. As the sum of skew and noise increases, the eye closes and data is corrupted. An eye diagram with a small opening (eye closure) indicates that the data is affected greatly by jitter. Lowering the height enters the input voltage threshold while closing the width decreases the time available for accurate reception. Both opening and closure of the eye add up to 100% [22].

#### 4.5.1.2 Histogram

The histogram plots the frequency of occurrence of jitter presentable versus the range of values (bins) of a chosen parameter (time or magnitude). This is also known as a jitter histogram. It shows the shape of the statistical distribution of parameter values as illustrated in Fig. 4.4. The histogram provides a level of insight that the eye diagram cannot, such as a separating random jitter from deterministic jitter types. Histograms are essential data sets for jitter separation routines required by various digital bus standards [20]. If the histogram is normalized such that the integral is unity, it may become an approximate probability density function (PDF) of jitter. However, this PDF does not always give the actual PDF, sometimes it makes large errors (e.g., when the number of bins is not sufficient).

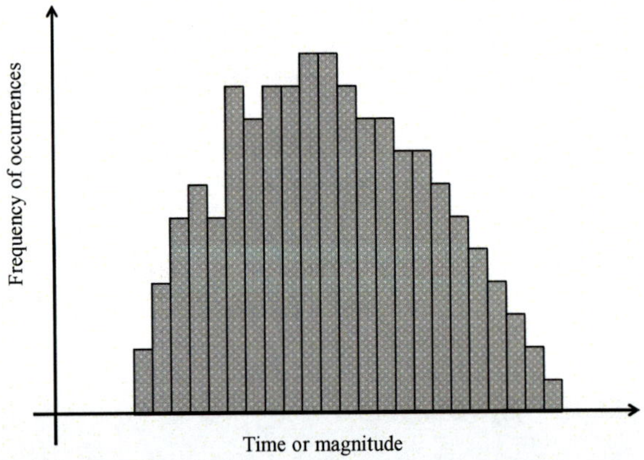


**Fig. 4.4** A picture of a histogram

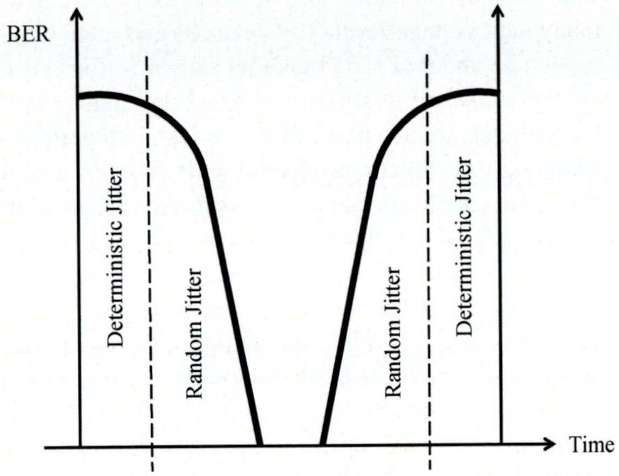


**Fig. 4.5** A picture of a bathtub curve

#### 4.5.1.3 Bathtub Curve

Bathtub curve is usually created with bit error ratio testers (BERT), therefore it is referred to as a BERT scan and presumably is only relevant to a communication channel. A BERT generates data to pass through a device under test (DUT) and then measures the transmitted data and compares for errors, thus determining the bit error ratio (BER) as depicted in Fig. 4.5. As the measurement location (i.e., sampling point) is swept across a unit interval (UI), a plot of BER as a function of the UI is constructed. This plot typically resembles a cross section of a bathtub, thus the name *bathtub curve*.

When the sampling point nears the transition points, this curve is fairly flat which is dominated by deterministic jitter. As the sampling point moves inward from both ends of the UI, the BER drops very steeply which is caused by random jitter. Thus, the bathtub curve can also be used to separate random and deterministic jitter in communication signals [20].

In addition to these graphical analyses, there are other jitter measurement methods in the time domain and frequency domain with statistical and signal processing approaches.

### 4.5.2 Time and Frequency Domains

Jitter can be calculated in the time-domain and/or frequency domain by Fourier transform (FT) and power spectrum density (PSD). Cosart et al. [23] described the measurement of phase modulation and noise using time-domain techniques. They presented a model for the relationship between zero crossings and phase on a signal and concluded measurements can be made on data signals eliminating the need for external data clock recovery. Later, a time-domain jitter separation method was introduced by Li et al. [24] as a general-purpose automated search and non-linear fitting algorithm for both deterministic and random jitter components. They fitted the tail parts of the jitter histogram with non-linear jitter models and calculated the bit error rate for each category. To reduce deterministic jitter, adding pre-emphasis (or de-emphasis) helps by improving the signal's high frequency content (i.e., improves SNR). For random jitter, there comes a point where these techniques would not provide enough compensation. The solution is re-clocking, which requires a phase-locking technique.

An analytic signal method to extract both RMS and peak-to-peak jitter values from phase-locked loops (PLLs) was presented by Yamaguchi et al. [25]. This method relied on the extension of a real signal into an analytic signal by utilizing the Hilbert transform. The authors considered the jitter as the variation of the edge position (or zero crossing) at each cycle. Later, Yamaguchi et al. [4] extended this method for measuring the cycle-to-cycle period jitter in PLL output signals and verified the relationship with the power spectral density of the total phase. The cycle-to-cycle period jitter spreads power into the sidebands of the phase noise spectrum. This method has been used for measuring the jitter of the bit clock signal in telecommunication devices operating at 10 Gbps [26].

Jitter can also be analyzed in the frequency domain by using the Fourier transform, then the energy of the resulted jitter spectrum can be used to calculate the jitter components by application of one of several techniques [26–28]. Some require a jitter-free sampling clock signal (or reference signal) to perform spectral analysis on the signal. As an example, the fast Fourier transform (FFT) and time lag correlation (TLC) functions have been used by Pang et al. [28] to measure jitter in simulation experiments. Yamaguchi et al. [27] proposed an algorithm based on FFT and probability density function *blind separator* to estimate both random jitter (RJ) and deterministic jitter (DJ). They assumed that the probability density function noise is distributed at high frequencies while both RJ and DJ are distributed around low frequencies (or the DC component or 0 Hz term).

The time lag correlation (TLC) functions and FFT with TLC functions have been used by Dou and Abraham [21] and Pang et al. [28], respectively, on simulation experiments. Both of them have used a reference clock signal for the ideal signal. To reduce the number of equations, Pang proposed using the FFT which reduces the calculation time. The relationship between the time-domain jitter measurements and the power spectrum of the phase jitter by using fundamental Fourier properties with basic random variables analysis was presented by Moon et al. [29]. They concluded that the phase jitter for a specific number of clock cycle delays in any self-referenced measurements can be extended to define other jitter measurements.

### *4.5.3 Statistical and Signal Processing Techniques*

It is well known the statistical analysis and digital signal processing are common methods to investigate the signal filtering in any field of use. In the literature, the statistical analysis for jitter measurement is used in the time domain or frequency domain. The effects of independent additive noise on timing jitter for pulse amplitude modulation data signals were evaluated by Franks and Bubrouski [30]. They have shown the minimum variance of the timing wave was increased by a Gaussian additive noise term similar in form to the constant term in the noise-free case.

For the random jitter in sampling systems, Souders et al. [31] have used the sample mean as the Markov estimator and approximated in the frequency domain by a simple filter function. They have proven the Markov estimator is asymptotically converged to the population median and it is, therefore, an unbiased estimator for monotonic waveforms sampled with jitter distributions having a median of zero.

Most of the literature on statistical analysis for the noises and signal measurements in ToF cameras assume that the signals are in Gaussian distribution [32, 33]. However, an individual statistical distribution of the modulated samples is considered to formulate a statistical model for the parameters of amplitude, phase, and intensity by Mufti and Mahony [34]. They used the effect of photon shot noise to analyze ToF measurements under various SNR conditions, which is a key parameter for the reliability of the range measurements. The authors have concluded that the proposed SNR estimator based on maximum likelihood estimation has better performance than based on Gaussian assumption. They used the Rice distribution [35], which has the uniqueness and unbiased maximum likelihood estimators for large data sets [36].

Moreover, a statistical model for simulation of direct time of flight for 3D imaging and ranging applications has been presented by Nguyen et al. [37]. Four different peak-detection algorithms have been used for quantifying the data and then have determined the effect of signal, noise, and jitter levels on the depth resolution achievable by the system. The comparison between the tail fitting algorithm and spectrum analysis on jitter estimation was investigated by Sharma et al. [38]. They concluded that using the tail fitting algorithm in presence of DDJ caused an error while in spectral analysis; the error percentage was large in the presence of high frequency PJ.

Tarczynski and Allay [39] proposed two algorithms named the weighted sample and weighted probability density functions to find digital alias-free estimation signal spectrum. They limited the usage of the algorithms on two factors: the level of the spectrum estimation error and sampling instant jitter. They concluded the spectrum estimation error resembles a wide-band noise while the jitter introduces bias to spectrum estimation.

A method for jitter estimation in ADC has been introduced by Wu et al. [40]. They used a single high frequency test without coherent sampling to significantly save the hardware and data acquisition time. Later, this method has been extended by Wu et al. [41] for jitter and noise separation in the ADC test, based on the property that jitter is modulated by the slope of the input signal. Wu et al. [42] have applied this method to separate the random and periodic jitter in ADC output by involving only one FFT and one IFFT operation.

## 4.6 Impact of Jitter on Range Imaging

Commercially available AMCW ToF cameras have low resolution and relatively high noise level due to electronic sources such as shot noise, reset noise, amplifier noise, crosstalk, analogue to digital converters (ADC) quantization, clock jitter, and due to multipath light interference. These noise sources directly affect the signal to noise ratio of the range measurements. Many noise sources in these cameras such as harmonic contamination [43, 44], non-linearity and multipath interference [45–48], and light scattering [49, 50] are well studied.

Researchers have barely investigated jitter related to the AMCW ToF ranging systems. A couple of studies that affect the jitter on range measurements were found and both of them are related to random jitter [1, 2]. The investigations on the periodic jitter in ToF range imaging cameras are not studied in the literature.

The possibility of the jitter and/or drift between the modulation signals and the camera shutter was investigated by Streeter et al. [1]. The variations in phase and frequency were modeled in a linear model by using an algorithm which is used for jitter correction. The model is solved after computing the jitter parameters of the current frame with respect to the reference frame by using linear regression. The root-mean-square error (RMSE) values with respect to the reference image are analyzed to consider the effect of those filtering methods. They concluded that the combination of the filters *Bilateral* and *Non-local means* performs well with their de-jitter algorithm, which depends only on the assumption that jitter affects equally all pixels in the scene.

Seiter et al. [2] investigated the impact of two common types of cycle-to-cycle jitter distributions, namely random Gaussian and discrete jitter distributions on the performances of ToF measurements. They proved a discrete jitter has a strong impact while the random Gaussian jitter has a negligible impact on the performances, by analyzing the standard deviation of the obtained measurements.

There is therefore a disagreement between those two papers, on the importance of different types of jitter on range measurement.

In addition, due to the aforementioned properties of the ToF cameras, the depth sensors will be further contributed in various applications where do not much use at present such as smartphones [51], smart buildings [52]. These applications may need very accurate range measurements. Thus, the investigation of the jitter on ToF measurements (range imaging data) is timely, because they can cause errors in the measurements. Once the error sources are examined, the noise and jitter performance can be improved and hence these effects on range data can be corrected. The ultimate goal of this book is to investigate the jitter measuring, extracting and influencing range measurements in full-field AMCW ToF range imaging cameras.

## 4.7 Chapter Remarks and Further Readings

The noise and jitter in a signal can be analyzed by different approaches in time and frequency domains with the help of statistical and signal processing techniques. Detailed descriptions of mathematical models for jitter types can be found in [6, 9]. Many of the noise sources such as non-linearity, multipath interference, and harmonic cancellation are well investigated in the literature and those who are interested can be referred to [15, 53]. However, the jitter related to the ToF ranging systems is barely investigated. We found only a couple of studies and there is a disagreement between those two studies regarding the importance of different types of the jitter in ToF range imaging cameras [1, 2].

Thus, there is a huge gap in the knowledge of jitter investigation in the AMCW ToF range imaging cameras such as extraction and how effect, type, and amount of jitter have most effect on range measurements. All of these are unanswered investigations in the literature. In addition, these investigations are important in the depth sensor manufacturing industries as well. Therefore, the jitter investigation of the AMCW ToF range imaging is highly significant for the exploration of reducing the aforementioned gap in ToF ranging systems. Thus, the rest of the chapters of this book contribute to minimizing this gap as much as possible.

In contrast, in Chap. 5, we will propose a common methodology (camera independent) to extract the jitter in the AMCW ToF ranging systems without using a sampling clock to trigger the time measurement unit. Instead, we use signal processing techniques in the proposing methodology.

## References

1. Lee Streeter, Michael J Cree, Adrian Dorrington, et al. A strategy for the correction of effects of jitter in AMCW lidar images. In *Image and Vision Computing New Zealand (IVCNZ), 2013 28th International Conference of*, pages 500–505. IEEE, 2013.

2. Johannes Seiter, Michael Hofbauer, Milos Davidovic, and Horst Zimmermann. Investigation of the distance error induced by cycle-to-cycle jitter in a correlating time-of-flight distance measurement system. *Optical Engineering*, 53(7):073104–073104, 2014.
3. Eduard Säckinger. *Analysis and design of transimpedance amplifiers for optical receivers*. John Wiley & Sons, 2017.
4. Takahiro J Yamaguchi, Mani Soma, David Halter, Rajesh Raina, Jim Nissen, and Masahiro Ishida. A method for measuring the cycle-to-cycle period jitter of high-frequency clock signals. In *VLSI Test Symposium, 19th IEEE Proceedings on. VTS 2001*, pages 102–110. IEEE, 2001.
5. Roland E Best. *Phase-locked loops: design, simulation, and applications*. McGraw-Hill Education, 2007.
6. Nicola Da Dalt and Ali Sheikholeslami. *Understanding Jitter and Phase Noise: A Circuits and Systems Perspective*. Cambridge University Press, 2018.
7. Dallas Semiconductor. Clock (CLK) jitter and phase noise conversion. Technical report, Application Note 3359, 2007.
8. David C Lee. Analysis of jitter in phase-locked loops. *IEEE Transactions on Circuits and Systems II: Analog and Digital Signal Processing*, 49(11): 704–711, 2002.
9. Mike Li. *Jitter, Noise, and Signal Integrity at High-speed*. Prentice Hall Press, Upper Saddle River, NJ, USA, first edition, 2007. ISBN 9780132429610.
10. Andy Kuo, Roberto Rosales, Touraj Farahmand, Sassan Tabatabaei, and André Ivanov. Crosstalk bounded uncorrelated jitter (BUJ) for high-speed interconnects. *IEEE Transactions on Instrumentation and measurement*, 54(5):1800–1810, 2005.
11. Andy Kuo, Touraj Farahmand, Nelson Ou, Sassan Tabatabaei, and André Ivanov. Jitter models and measurement methods for high-speed serial interconnects. In *Test Conference, 2004. Proceedings. ITC 2004. International*, pages 1295–1302. IEEE, 2004.
12. H. Rapp, M. Frank, F.A. Hamprecht, and B. Jähne. A theoretical and experimental investigation of the systematic errors and statistical uncertainties of time-of-flight-cameras. *Int. J. Inte. Sys. Tech. and Appl.*, 5(3/4), 2008.
13. John Patrin and Mike Li. Comparison and correlation of signal integrity measurement techniques. *DesignCon 2002*, 2002.
14. John B Johnson. Thermal agitation of electricity in conductors. *Nature*, 119(2984):50, 1927.
15. Dragos Falie and Vasile Buzuloiu. Noise characteristics of 3D time-of-flight cameras. In *2007 International Symposium on Signals, Circuits and Systems*, volume 1, pages 1–4. IEEE, 2007.
16. Johnnie Hancock and Steve Draving. Jitter-understanding it, measuring it, eliminating it; part 2: Jitter measurements. *High Frequency Electronics*, pages 20–28, 2004.
17. Ronen Cohen. Jitter and signal noise in frequency sources. Technical report, Raltron Electronics Corp, Miami, Florida 33172, U.S.A., 2005.
18. Mitsuru Shinagawa, Yukio Akazawa, and Tsutomu Wakimoto. Jitter analysis of high-speed sampling systems. *IEEE Journal of Solid-State Circuits*, 25(1):220–224, 1990.
19. Turker Kuyel. Method and system for measuring jitter, October 28 2003. US Patent 6,640,193.
20. Johnnie Hancock et al. Jitter-understanding it, measuring it, eliminating it; part 1: Jitter fundamentals. *High Frequency Electronics*, 4(4):44–50, 2004.
21. Qingqi Dou and Jacob A Abraham. Jitter decomposition by time lag correlation. In *Proceedings of the 7th International Symposium on Quality Electronic Design*, pages 525–530. IEEE Computer Society, 2006.
22. Eduard Säckinger. *Broadband circuits for optical fiber communication*. John Wiley & Sons, 2005.
23. L. D. Cosart, L. Peregrino, and A. Tambe. Time domain analysis and its practical application to the measurement of phase noise and jitter. *IEEE Transactions on Instrumentation and Measurement*, 46(4):1016–1019, 1997.
24. Mike P. Li, Jan Wilstrup, Ross Jessen, and Dennis Petrich. A new method for jitter decomposition through its distribution tail fitting. In *null*, page 788. IEEE, 1999.
25. Takahiro J Yamaguchi, Mani Soma, Masahiro Ishida, Toshifumi Watanabe, and Tadahiro Ohmi. Extraction of peak-to-peak and RMS sinusoidal jitter using an analytic signal method. In *VLSI Test Symposium, 2000. Proceedings. 18th IEEE*, pages 395–402. IEEE, 2000.

26. Takahiro J Yamaguchi, Mani Soma, Louis Malarsie, Masahiro Ishida, and Hirobumi Musha. Timing jitter measurement of 10 Gbps bit clock signals using frequency division. In *VLSI Test Symposium, 2002.(VTS 2002). Proceedings 20th IEEE*, pages 207–212. IEEE, 2002.
27. Takahiro J Yamaguchi, Masahiro Ishida, Harry X Hou, Dave Armstrong, Koji Takayama, and Mani Soma. An FFT-based jitter separation method for high-frequency jitter testing with a 10x reduction in test time. In *Test Conference, 2007. ITC 2007. IEEE International*, pages 1–8. IEEE, 2007.
28. Hong Pang, Jian Zhu, and Wei Huang. Jitter decomposition by fast Fourier transform and time lag correlation. In *Communications, Circuits and Systems, 2009. ICCCAS 2009. International Conference on*, pages 365–368. IEEE, 2009.
29. Un-Ku Moon, Karti Mayaram, and John T Stonick. Spectral analysis of time-domain phase jitter measurements. *IEEE Transactions on Circuits and Systems II: Analog and Digital Signal Processing*, 49(5):321–327, 2002.
30. L Franks and J Bubrouski. Statistical properties of timing jitter in a PAM timing recovery scheme. *IEEE Transactions on Circuits and Systems*, 21(4):489–496, 1974.
31. T Michael Souders, Donald R Flach, Charles Hagwood, and G Yang. The effects of timing jitter in sampling systems. In *Instrumentation and Measurement Technology Conference, 1989. IMTC-89. Conference Record., 6th IEEE*, pages 199–203. IEEE, 1989.
32. Tobias Moller, Holger Kraft, Jochen Frey, Martin Albrecht, and Robert Lange. Robust 3D measurement with PMD sensors. In *In: Proceedings of the 1st Range Imaging Research Day at ETH*, pages 3–906467, 2005.
33. Bernhard Büttgen, Thierry Oggier, Michael Lehmann, Rolf Kaufmann, and Felix Lustenberger. CCD/CMOS lock-in pixel for range imaging: Challenges, limitations and state-of-the-art. *1st range imaging research day*, pages 21–32, 2005.
34. Faisal Mufti and Robert Mahony. Statistical analysis of signal measurement in time-of-flight cameras. *{ISPRS} Journal of Photogrammetry and Remote Sensing*, 66(5):720 – 731, 2011. ISSN 0924-2716. https://doi.org/10.1016/j.isprsjprs.2011.06.004.
35. S.O. Rice. Mathematical analysis of random noise-conclusion. *Instruments and Measurements, IEEE Transactions on*, 24:46–156, 1945.
36. Carlo FM Carobbi and Marco Cati. The absolute maximum of the likelihood function of the Rice distribution: Existence and uniqueness. *IEEE Transactions on Instrumentation and Measurement*, 57(4):682–689, 2008.
37. KQK Nguyen, EMD Fisher, AJ Walton, and I Underwood. An experimentally verified model for estimating the distance resolution capability of direct time of flight 3D optical imaging systems. *Measurement Science and Technology*, 24(12):125001, 2013. https://doi.org/10.1088/0957-0233/24/12/125001.
38. Vijender Kumar Sharma, Jai Narayan Tripathi, Rajkumar Nagpal, Sujay Deb, and Rakesh Malik. A comparative analysis of jitter estimation techniques. In *Electronics, Communication and Computational Engineering (ICECCE), 2014 International Conference on*, pages 125–130. IEEE, 2014.
39. Andrzej Tarczynski and Najib Allay. Spectral analysis of randomly sampled signals: suppression of aliasing and sampler jitter. *IEEE Transactions on Signal Processing*, 52(12):3324–3334, 2004.
40. Minshun Wu, Guican Chen, and Degang Chen. ADC jitter estimation using a single frequency test without requiring coherent sampling. *IEICE Electronics Express*, 9(18):1485–1491, 2012.
41. Minshun Wu, Zhiqiang Liu, Li Xu, and Degang Chen. Accurate and cost-effective technique for jitter and noise separation based on single-frequency measurement. *Electronics Letters*, 52 (2):106–107, 2015a.
42. Minshun Wu, Zhiqiang Liu, and Degang Chen. Extracting random jitter and sinusoidal jitter in ADC output with a single frequency test. *IEICE Electronics Express*, 12(20):20150742–20150742, 2015b.
43. T. Kahlmann, F. Remondino, and H. Ingensand. Calibration for increased accuracy of the range imaging camera SwissRanger$^{TM}$. In *Image Engineering and Vision Metrology*, volume 36 of *3*, pages 136–141, 2006.

44. Lee Streeter and Adrian A Dorrington. Simple harmonic error cancellation in time of flight range imaging. *Optics Letters*, 40(22):5391–5394, 2015. https://doi.org/10.1364/OL.40.005391.
45. John P Godbaz, Michael J Cree, and Adrian A Dorrington. Understanding and ameliorating non-linear phase and amplitude responses in AMCW lidar. *Remote Sensing*, 4(1):21–42, 2011.
46. Daniel Freedman, Yoni Smolin, Eyal Krupka, Ido Leichter, and Mirko Schmidt. Sra: Fast removal of general multipath for tof sensors. In *Computer Vision–ECCV 2014*, pages 234–249. Springer, 2014.
47. David Jiménez, Daniel Pizarro, Manuel Mazo, and Sira Palazuelos. Modeling and correction of multipath interference in time of flight cameras. *Image and Vision Computing*, 32(1):1–13, 2014.
48. R. Whyte, L. V. Streeter, M. J. Cree, and A. A. Dorrington. Resolving multiple propagation paths in time of flight range cameras using direct and global separation methods. *Optical Engineering*, 54(11):1–9, 2015. https://doi.org/10.1117/1.OE.54.11.113109.
49. James Mure-Dubois and Heinz Hügli. Real-time scattering compensation for time-of-flight camera. In *Proceedings of the ICVS Workshop on Camera Calibration Methods for Computer Vision Systems*, 2007.
50. Henrik Schäfer, Frank Lenzen, and Christoph S Garbe. Model based scattering correction in time-of-flight cameras. *Optics express*, 22(24):29835–29846, 2014.
51. What is a time-of-flight camera and which phones have it. https://www.pocket-lint.com/phones/news/147024-what-is-a-time-of-flight-camera-and-which-phones-have-it, 2019a. [Accessed May 2021].
52. Terabee introduces latest 3d time-of-flight camera. https://lidarmag.com/2019/04/16/terabee-introduces-latest-3d-time-of-flight-camera/, 2019b. [Accessed May 2021].
53. S. Foix, G. Alenya, and C. Torras. Lock-in time-of-flight (tof) cameras: A survey. *IEEE Sensors Journal*, 11(9):1917–1926, Sept 2011. ISSN 1530-437X. https://doi.org/10.1109/JSEN.2010.2101060.

# Part II
# Jitter Extraction and Measurement

The next four chapters comprise Part II of the book on jitter extraction and measurement. Camera independent methodology to extract the periodic and random jitter present in the light source of the AMCW ToF range imaging cameras is proposed in Chap. 5. The proposed methodology is verified first by using generated simulation data and then by experimentally. The simulation results based on this proposed methodology are included in Chap. 6. For the experimental verification, the measurement of jitter present in two commercial AMCW ToF cameras is reported in Chap. 7. Chapter 8 is an investigation of the possibility of re-purposing a vastly cheaper software defined radio technology for jitter extraction in the above two cameras.

# Chapter 5
# Proposed Methodology for Jitter Measurement

**Abstract** As described in Chap. 4, jitter in ToF ranging systems has barely been investigated in the literature. This chapter proposes a camera independent methodology for extraction of the periodic and random jitter in the light source of the AMCW ToF range imaging cameras. Also, it is assumed that the noise in the signals is zero. This methodology can be used for jitter extraction in any analogue or digital signal without requiring a reference or sampling clock signal to trigger the time measurement unit. In addition, the methodology can be evaluated for the standard geometrical shapes of signals such as sinusoidal, triangular, and rectangular (square).

**Keywords** Time-zero crossings · Signal smoothing · Moving average · Savitzky–Golay · Quadratic (parabolic) interpolation · Discrete Fourier transform · Nyquist frequency · Spectral leakage · Data windowing · Coherent gain · Incoherent gain · Gaussian distribution · Random process · Regression · Power-law

## 5.1 Background

There are only two studies that the authors are aware of that investigate the effect of jitter in ranging with AMCW ToF range imaging cameras [1, 2]. These studies considered only the effect of random jitter on measurement in specific ToF cameras. The author is not aware of any investigations on the extraction of periodic jitter in the light signal of ToF cameras in the literature. It is a very challenging task to completely extract the sources of jitter in range imaging applications, because it is impossible to get to the internal signals without disassembling the camera, and even then, specific signals may not be accessible. Therefore, we focus on jitter in the emitted light signal which is sensible since the light signal of the camera is relatively straightforward to access. However, it was noted in the last chapter that an accurate reference clock is needed to enable jitter extraction in a signal. Since we have no access to the reference clock, we develop a method here to use Fourier analysis to generate the reference clock (ideal signal) for the camera from



G. Anthonys, *Timing Jitter in Time-of-Flight Range Imaging Cameras*,
https://doi.org/10.1007/978-3-030-94159-8_5

its measured data without involving any other signal or data and this is the novelty of the proposed algorithm. Fourier analysis and signal processing techniques for generating the reference clock signal are described in the following sections.

## 5.2 Proposed Methodology

The proposed methodology to extract the jitter components consists of three main processes as shown in Fig. 5.1. Three processes consist of steps ① to ②, steps ③ to ⑥, and step ⑦, respectively. This proposed method can be used for real-time jitter investigation for all AMCW ToF range imaging cameras since the light sources of these cameras are directly accessible. The corresponding steps for the three processes are shown as a flowchart in Fig. 5.2. The detailed description for step ① to step ⑥ of this flowchart will be discussed in the next subsections while step ⑦ is separately explained in Sect. 5.3.

Different techniques can be used to achieve each of the above step and some of them are introduced in relevant subsections. As usual, execution times and accuracy for each technique for the corresponding step will be varied. The reader can perform those methods and compare the results within the step.

In step ①, we capture the data signal from the light source of the camera. In general, this signal consists of noisy data at the intersections of the time axis (zero crossing points). Therefore, in order to clearly identify the zero crossings of the data signal, it is needed to smooth the signal with minimal impact on zero crossings.

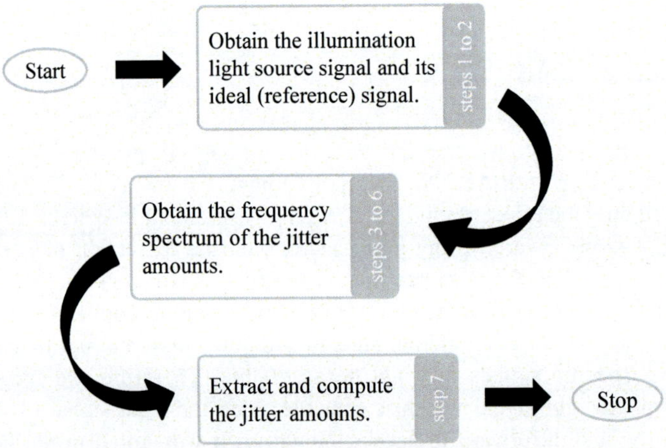


**Fig. 5.1** Proposed methodology for jitter extraction in ToF range imaging cameras

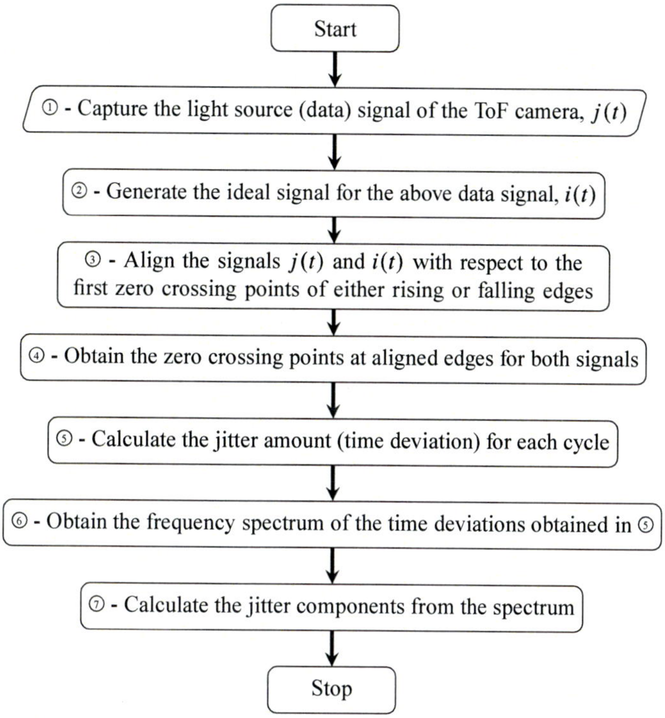


**Fig. 5.2** Main steps for the proposed methodology shown in Fig. 5.1

## 5.2.1 *Signal Smoothing*

Signal smoothing is how we observe important patterns in the data while leaving out that are unimportant such as noise. Smoothing can be done in many ways, but in very basic and general terms it means that these most smoothing algorithms are based on the *shift* and *multiply* technique. The set of adjacent points in the original data are multiplied sample-by-sample by a set of coefficients that defines the smooth shape [3]. Here we briefly discuss three types of common methods. Among them, the appropriate method can be chosen for the requirement.

### 5.2.1.1 Moving Average

Moving average is the simplest technique to attenuate additive noise and is based on the assumption that noise is not going to change the underlying structure of the signal. For each point in a signal, it calculates the average of its consecutive points. Let $y_k, k = 1, 2, \ldots$ are the sample points of a signal with noise and let $y_k^{(s)}$ are the corresponding smoothed points. Then, the moving average of an odd number of consecutive window size $2m + 1$ for $i = -m$ to $m$ can be defined as

$$y_k^{(s)} = \frac{\sum_{i=-m}^{m} y_{k+i}}{2m + 1}. \tag{5.1}$$

Usually, the odd number $2m + 1$ is called filter width. When the width is greater, the smoothing effect is more intense but is also likely to cause an excessive smoothing of the original signal. Moving average is the optimal solution for linear signals which are affected by additive uniform noise. However, these kinds of convenient constraints are unlikely to find in real world applications.

#### 5.2.1.2 Weighted Moving Average

In the moving average method, each point in the window is treated with the same importance. A more reasonable approach is to value points that are further away from $y_k$ less. Therefore, in the weighted moving average, it is introducing a weight $\mathcal{W}_i$ for each time step in the window as

$$y_k^{(s)} = \sum_{i=-m}^{m} \mathcal{W}_{m+i}\, y_{k+i} \quad \text{with} \quad \sum_{i=-m}^{m} \mathcal{W}_i = 1. \tag{5.2}$$

Choosing weights $\mathcal{W}_i$ gives in a variety of interesting effects. The most popular choice when using weighted moving averages is to use a higher weighting for recent values [3].

#### 5.2.1.3 Savitzky-Golay

The main idea of this method is to find the weighting coefficients (i.e., convolution integers) to carry out the smoothing operation by fitting the data $y_k$ to a polynomial for each value of the data with $2m + 1$ neighboring points (including the point to be smoothed), with $m$ being equal or greater than the order of the polynomial. Savitzky and Golay have shown that a moving polynomial fit can be numerically handled in exactly the same way as a weighted moving average since the coefficients of the smoothing procedure are constant for all $y_k$ [4]. Let $C_i$ for $i = -m, -m+1, \ldots, m$ be $2m + 1$ weighting coefficients for data $y_k$. Then, by using the weighted moving average method, the smoothed data $y_k^{(s)}$ can be computed by [5]

$$y_k^{(s)} = \frac{\sum_{i=-m}^{m} C_i\, y_{k+i}}{\sum_{i=-m}^{m} C_i}. \tag{5.3}$$

The Savitzky-Golay algorithm finds the weights $C_i$ that are optimal in the sense that they minimize the least-squares error in fitting a polynomial to frames of noisy data. The advantages of the SG algorithm include that it smooths the fluctuations and increases the signal-to-noise ratio without significant distortion of the analyzed data. Also, the SG algorithm uses all data and does not exclude outliers, and the smoothed values can be expressed as a linear transformation of the input values [6].

However, it is important to optimize the two parameters: filter length and order of polynomial of the SG algorithm. Some various methods have been described in the literature [7–11]. They have focused on the optimization of one parameter at a time and it is very difficult to optimize both parameters at the same time. However, all of these methods have been tested only for specific applications. The reader can test these methods and compare them with the results in this book. When a signal is smoothed, it is impossible to obtain a method that eliminates all the noise without losing any valuable information. However, we used the trial and repetition method by visual inspection because it was quick and reliable.

For step ②, we generate the corresponding sinusoidal (or triangular or rectangular) reference signal for the camera. For that, the relevant frequency and amplitude of the ideal signal have to be found. First, we obtain the Fourier spectrum of the smoothed data signal, find the relevant frequency at the maximum magnitude, and perform quadratic interpolation with its neighbors to find the frequency of the ideal signal.

### 5.2.2 *Frequency Estimation for the Reference Signal*

If the sampling rate is sufficiently high with respect to the frequency of the cross-correlation, the quadratic (parabolic) interpolation would give unbiased estimates [12]. Also, this method is fast and simple to implement. A parabolic curve between three adjacent points is placed in the quadratic interpolation method. Figure 5.3 illustrates the peak with its three closest neighbor samples $(f_0, y_0)$, $(f_1, y_1)$, and $(f_2, y_2)$, from the Fourier spectrum of a smoothed signal and let the coordinate of the peak be $(f', y')$. The term "a.u." of the $y$ axis stands for an arbitrary unit, for example, this is either current or voltage. Then, a quadratic polynomial (or parabola) for the peak can be expressed, namely [13].

$$\begin{aligned} y = y_0 &+ \frac{(y_1 - y_0)}{(f_1 - f_0)}\left(\frac{f - f_0}{h}\right) \\ &+ \frac{(y_2 - y_1)(f_1 - f_0) - (y_1 - y_0)(f_2 - f_1)}{2(f_2 - f_1)(f_1 - f_0)(f_2 - f_0)}\left(\frac{f - f_0}{h}\right)\left(\frac{f - f_1}{h}\right) \end{aligned} \tag{5.4}$$

At the peak, it should satisfy,

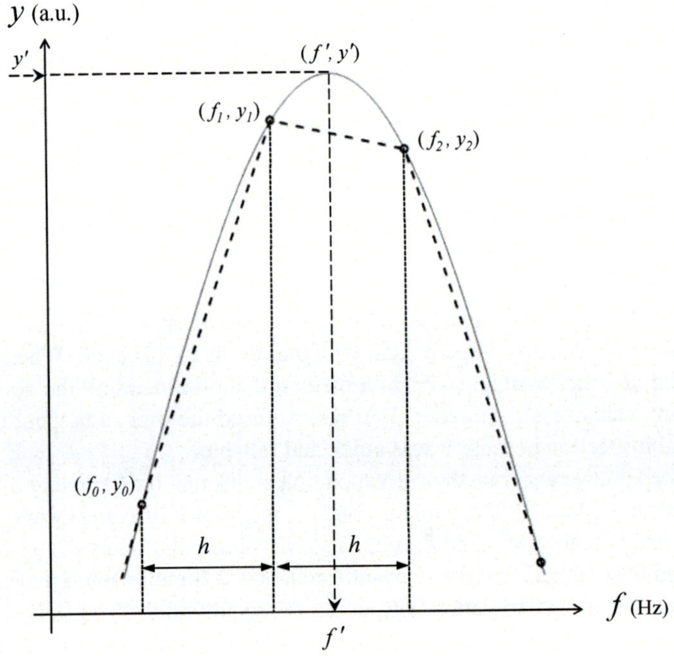


**Fig. 5.3** Estimation of the frequency by quadratic interpolation

$$\left.\frac{dy}{df}\right|_{(f',y')} = 0 \tag{5.5}$$

and gives

$$f' = f_0 + \frac{h}{2}\left(\frac{y_2 - 4y_1 + 3y_0}{y_2 - 2y_1 + y_0}\right). \tag{5.6}$$

Thus, the corresponding frequency of the reference signal can be calculated from (5.6). Note that if the sample rate does not properly match the signal frequencies, then spectral leakage will occur in the Fourier spectrum of the signal. That is, spectral leakage results in a single-tone frequency bin spread among several frequencies in the Fourier domain and causes errors in determining the ideal signal frequency. The magnitude of the error depends on the location of the actual peak of the cross-correlation function with respect to the location of the samples. Boucher and Hassab [14] developed general expressions of the mean and variance of the time delay that is estimated using parabolic interpolation as a function of the spectral characteristics of the signal.

### 5.2.3 Amplitude Estimation for the Reference Signal

It is well known that the magnitude of the fundamental of the harmonic in the FFT or the maximum magnitude of the signal can be used as an amplitude for the reference signal. But we use a special technique since this special method involves the whole signal rather than the previous methods. Each method gives approximately the similar values. As an example, let us consider a smoothed signal obtained from step ① as shown in Fig. 5.4. In order to find the amplitude for the ideal signal $A'$, we first take the averages of the peaks and dips within a threshold value, separately, namely

$$A^{+}_{avg} = \frac{1}{K_1} \sum_{k_1=1}^{K_1} A^{+}_{k_1}, \tag{5.7a}$$

$$A^{-}_{avg} = \frac{1}{K_2} \sum_{k_2=1}^{K_2} A^{-}_{k_2}, \qquad K_1, K_2 \in \mathbb{Z}^{+} \tag{5.7b}$$

where $A^{+}_{k_1}$ (; $A^{-}_{k_2}$) and $K_1$ (; $K_2$) are the amplitude of the $k_1$th (; $k_2$th) sample and the total number of samples of the peaks (; dips), respectively. Then, take mean among them as

$$A' = \frac{A^{+}_{avg} + |A^{-}_{avg}|}{2}. \tag{5.8}$$

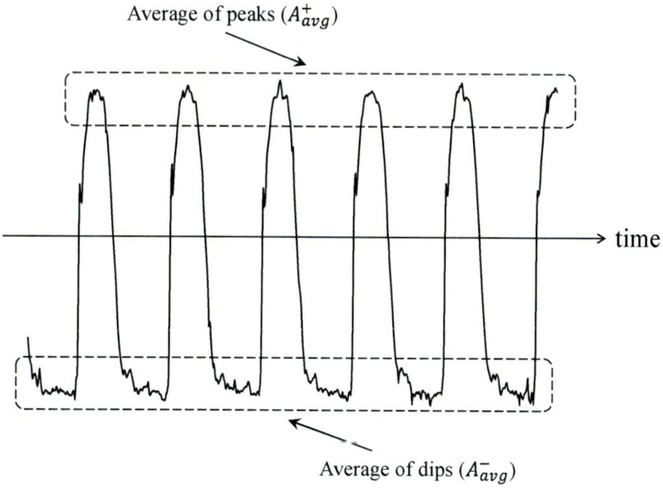


**Fig. 5.4** Estimation of the amplitude by averaging peaks and dips

Therefore, the corresponding amplitude of the reference signal can be calculated from (5.8). Note that the accuracy of the signal's amplitude is not important since the zero crossings on the time axis are not dependent on the amplitude of the signal. However, this estimation can be used to compare the ideal and data signals visually.

Next, both signals (data and its reference) are aligned with respect to the first zero crossing point of either rising or falling edges as for step ③ in our proposed algorithm. We choose rising edges. In order to find the zero crossing points, the two corresponding values that sit on either side of the zero crossing for each cycle are chosen. The shape of the signal is very close to the straight line near the zero crossings. By using linear interpolation, the zero crossing points at rising edges for both signals are estimated, this being step ④. The linear interpolation method provides faster results than other techniques [15].

## 5.2.4 *Identifying Zero Crossings*

Figure 5.5 shows an example of the linear interpolation technique. Let the points $(t_1, v_1)$ and $(t_2, v_2)$ be the coordinates of the corresponding points that sit either side of the zero crossing for a cycle. Also, let $(t', v')$ be the relevant point of the zero crossing of that cycle (note that $v' = 0$). By considering the geometrical relation of the two triangles, it is not difficult to obtain

$$(v' - v_1)(t_2 - t_1) = (v_2 - v_1)(t' - t_1) \tag{5.9}$$

and then, the corresponding zero crossing point at the rising edge can be computed by simplifying (5.9) with $v' = 0$, thus,

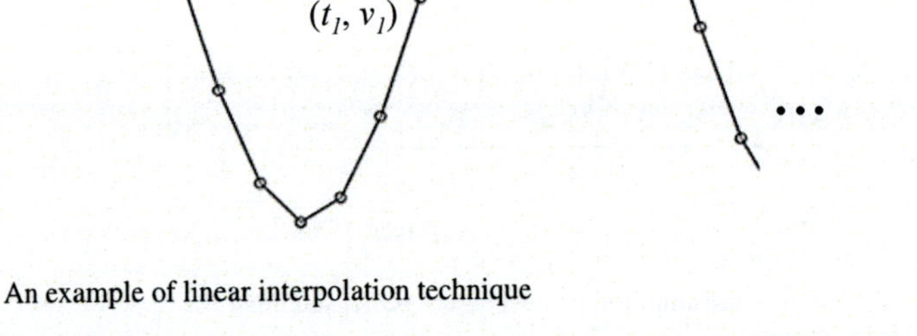


**Fig. 5.5** An example of linear interpolation technique

$$t' = \frac{t_1 \nu_2 - \nu_1 t_2}{\nu_2 - \nu_1}, \qquad \nu_2 \neq \nu_1. \tag{5.10}$$

We separately use (5.10) to obtain more approximate zero crossings at the rising edge for each signal.

In step ⑤, the jitter $\Delta t(t)$ of each rising edge of the signal is calculated at the time difference between the zero crossing points (obtained in step ④) of the data and reference signals of the camera. For step ⑥, the frequency domain spectrum is obtained by using the discrete Fourier transform (DFT) of the jitter signal. The number of cycles captured for the camera $N_{\Delta t}^{cal}$ depends on the modulation frequency $f_m$, the number of samples $N_s$ in the acquired time-domain signal, and the sample rate of the acquisition $F_s$. This is

$$N_{\Delta t}^{cal} = \frac{f_m N_s}{F_s}. \tag{5.11}$$

Note that because the length of the signal is somewhat arbitrary there is no guarantee of complete sampled cycles of any periodic jitter present in the signal. The DFT assumes periodicity of the time-domain signal, which (5.11) violates. Therefore, we take the nearest integer value $N_{\Delta t}$ by taking the floor of (5.11) as

$$N_{\Delta t} = \text{floor}\left(N_{\Delta t}^{cal}\right) = \left\lfloor N_{\Delta t}^{cal} \right\rfloor, \tag{5.12}$$

and this is the corresponding total number of jitter samples for the experiment (with respect to the acquisition setup used).

### 5.2.5 *Spectral Leakage*

The real problem is that a $N$-point DFT assumes the signal is repeated every $N$-points throughout all time. But for arbitrary $N_{\Delta t}$ from (5.12) and periodic jitter frequency, this assumption is violated. This discontinuity of the signal leads to spectral leakage in the spectrum, which is the phenomenon whereby the energy of a single sinusoidal signal (frequency) is spread out into neighboring bins of the spectrum. That is, the non-integral cycle frequency component of the signal does not correspond exactly to one of the spectra frequency lines as shown in Fig. 5.6. Therefore, the obtained spectrum is not the actual spectrum of the original signal, but a spread version over the full range of DFT coefficients.

We can see multiple frequencies in Fig. 5.6, due to the strange behavior at the period's boundary. As in the case of the DFT, this spectral leakage is caused by finite-length sampling that occurs for any practical application. Increasing the sampling frequency, thereby generating longer discrete-time sequences for equivalent sampling times, reduces spectral leakage. But, the user cannot increase the modulation frequency to whatever need (some ToF cameras allow to change

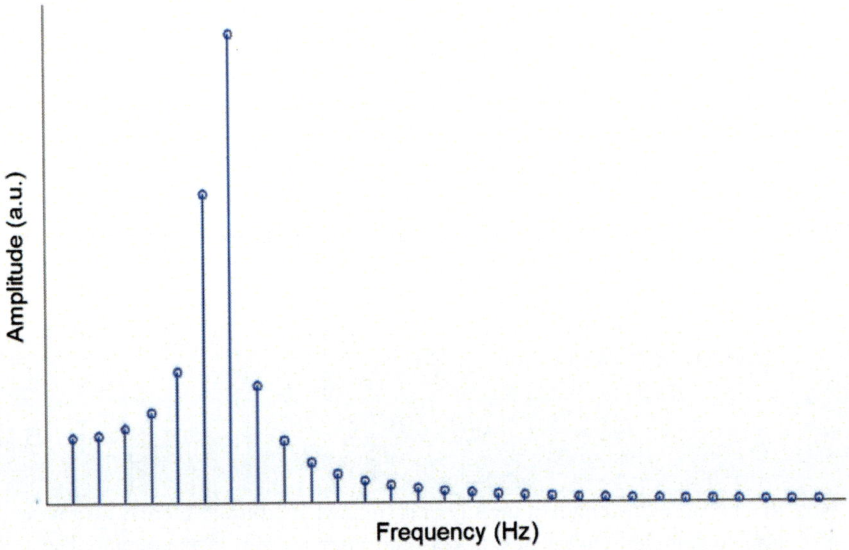


**Fig. 5.6** The effect of spectral leakage

the frequency within a small range, see Sect. 2.5.1). One common method for minimizing spectral leakage in the spectral content is called *data windowing*.

## 5.2.6 Window Functions for Minimizing Spectral Leakage

In *data windowing*, the original signal is modified by multiplying with a windowing function to smoothly attenuate both ends of the data towards zero and reaches a peak near at mid of the frequency axis. There are many different types of windowing functions used to reduce spectral leakage when performing a Fourier Transform in time-domain data and converting it into the frequency domain. Each window function is designed with a specific purpose [16]. Here some of the popular methods are briefly described.

### 5.2.6.1 Uniform

The uniform window (sometimes called "rectangular") results from the direct truncation of the signal and has a value of 1.0 across the entire measurement time axis, is defined as

$$w_U(n) = \begin{cases} 1, & |n| \leq \frac{N-1}{2} \text{ and } N \text{ is odd} \\ 0, & \text{otherwise} \end{cases} \tag{5.13}$$

This creates no frequency or amplitude distortion when the measured signal is periodic. The Fourier transform of this rectangular window function represents a sinc function. Generally, the rectangular window has the narrowest main lobe [16].

#### 5.2.6.2 Hann

The Hann window can be seen as one period of a cosine "raised" so that its negative peaks just touch zero (sometimes referred to as "raised cosine" or "hanning"). The Hann window [17] is

$$w_H(n) = \begin{cases} 0.5 + 0.5\cos\left(\frac{2\pi n}{N-1}\right), & |n| \leq \frac{N-1}{2} \\ 0, & \text{otherwise} \end{cases} \tag{5.14}$$

The Hann window is better to use in general data analysis for a good trade-off between frequency and amplitude accuracy [18]. Both leakage and ripple are reduced and selectivity is improved by this window for most applications [17, 19].

#### 5.2.6.3 Hamming

This is an optimized form of the Hann window [20]. Generalized Hamming window can have a step discontinuity at their endpoints, but no impulsive points. Like the Hann window, the Hamming window is also one period of a raised cosine and is defined by

$$w_{Hm}(n) = \begin{cases} 0.54 + 0.46\cos\left(\frac{2\pi n}{N-1}\right), & |n| \leq \frac{N-1}{2} \\ 0, & \text{otherwise} \end{cases} \tag{5.15}$$

In addition, a generalized Hamming window can have a step discontinuity at their endpoints, but no impulsive points. Hamming window is used in applications of telephone communications or transmission profiles.

#### 5.2.6.4 Blackman

This is a straightforward generalization of the Hamming window. In Blackman window/indexBlackman window, one more cosine weighted is added with respective to the Hamming window, is defined by [20]

$$w_B(n) = \begin{cases} 0.42 - 0.5\cos\left(\frac{2\pi n}{N-1}\right) + 0.08\cos\left(\frac{4\pi n}{N-1}\right), & |n| \le \frac{N-1}{2} \\ 0, & \text{otherwise} \end{cases} \tag{5.16}$$

Now, three degrees of freedom to work with instead of two. However, the main-lobe width of this window is thrice that of the uniform window. The periodic version is the preferred method when using a Blackman window in spectral analysis because the DFT assumes the periodic extension of the input signal [18].

## 5.2.7 *Window Correction Factors*

Because the signal is windowed, both the amplitude and energy of the spectrum (area under the curve) are reduced. This must be taken account of. The resultant amplitudes and energies from the spectrum must be multiplied by a correction factor. So, window correction factors are used to try and compensate for the effects of applying a window to data. There are both amplitude and energy correction factors. The two factors are computed by considering the signal's coherent and incoherent gains, respectively.

### 5.2.7.1 Coherent Gain

The coherent gain of a window is defined as the sum of the coefficients of the window function over the interval [21]. Thus, the amplitude correction factor $\mathrm{A}^{cf}$ is obtained by taking the reciprocal of the coherent gain, namely

$$\mathrm{A}^{cf} = \left(\int_0^1 w(nt)\,dt\right)^{-1}, \quad n = 0, 1, \ldots, N-1, \tag{5.17}$$

where $w(nt)$ is the window function. As an example, for the Hann window, the amplitude correction factor $\mathrm{A}_H^{cf}$ can be computed from (5.14) and (5.17) as

$$\begin{aligned} \mathrm{A}_H^{cf} &= \left(\frac{1}{2}\int_0^1 \left(1 + \cos\left(\frac{2\pi nt}{N-1}\right)\right) dt\right)^{-1} \\ &= \left(\frac{1}{2}\left[t + \left(\frac{N-1}{2\pi n}\right)\sin\left(\frac{2\pi n t}{N-1}\right)\right]_0^1\right)^{-1} = 2.0. \end{aligned} \tag{5.18}$$

#### 5.2.7.2 Incoherent Gain

The incoherent gain is defined as the root of the sum of squares of the coefficients of the window function over the interval [21]. The energy correction factor, $E^{cf}$, is obtained by the reciprocal of the incoherent gain as

$$E^{cf} = \left(\sqrt{\int_0^1 \left(w(nt)\right)^2 dt}\right)^{-1}, \quad n = 0, 1, \ldots, N-1. \tag{5.19}$$

Thus, from (5.14) and (5.19), the energy correction factor for the Hann window $E_H^{cf}$ is given as

$$\begin{aligned} E_H^{cf} &= \left(\sqrt{\frac{1}{4}\int_0^1 \left(1 + \cos\left(\frac{2\pi nt}{N-1}\right)\right)^2 dt}\right)^{-1} \\ &= \left(\sqrt{\frac{1}{4}\left[\frac{3t}{2} + \left(\frac{N-1}{\pi n}\right)\sin\left(\frac{2\pi n t}{N-1}\right) + \left(\frac{N-1}{8\pi n}\right)\sin\left(\frac{4\pi n t}{N-1}\right)\right]_0^1}\right)^{-1} \\ &= \sqrt{8/3} \approx 1.63. \end{aligned} \tag{5.20}$$

Similarly, the amplitude and energy correction factors for other types of window can be computed as shown in Table 5.1

Now, the amplitude and energy can be corrected by multiplying with the appropriate correction factor in the frequency domain. Thus, each spectral line of the windowed frequency spectrum is multiplied by the above corresponding fixed factor. If the energy of the signal lands in a single bin (which is very unlikely to occur), then the periodic jitter can be computed by multiplying the corresponding peak with amplitude correction factor $A^{cf}$.

But, if general, the energy of the signal is spread over multiple bins due to leakage and to correctly calculate the amplitude of the signal one must include the full energy of the signal. Therefore, in this case, (5.19) must be used to correctly calculate the amplitude, hence the amount of jitter. The relevant portion of the peak

**Table 5.1** Window correction factors

| Window function | Correction factor for | |
|---|---|---|
| $w(n)$ | Amplitude, $A^{cf}$ by (5.17) | Energy, $E^{cf}$ by (5.19) |
| Uniform | 1.0 | 1.0 |
| Hann | 2.0 | 1.63 |
| Hamming | 1.85 | 1.59 |
| Blackman | 2.80 | 1.97 |

(with neighboring bins) can be obtained by choosing the interpolation level (see Fig. 5.8) between the periodic and random jitter in the windowed spectrum. The interpolation level (IL) is computed by

$$\mathrm{IL} = \mu \pm \alpha\, \sigma, \qquad \alpha \in \mathbb{N} \tag{5.21}$$

where $\mu$ and $\sigma$ are the mean and standard deviation of the windowed spectrum, respectively, and $\alpha$ is chosen by visual inspection.

Finally, in the last step ⑦, the corresponding jitter parameters for the camera are extracted as described in Sect. 5.3.

## 5.3 Jitter Extraction Using the Spectrum

Since the ultimate goal of this session is to extract the jitter in the light source signal of the AMCW ToF cameras, the random jitter and periodic jitter are focused. Other types of jitter are not meaningful to the light source since they are data dependent. The first part of this section explains the extraction and calculation of the corresponding parameters of the random and periodic jitter from the jitter spectrum. The second part describes further analysis of the jitter spectrum by using the technique of curve fitting with the logarithmic scale.

### *5.3.1 Calculation of the Jitter Components*

Figure 5.7 shows a sinusoidal signal (ideal) and the signal with jitter. As previously described, the jitter amount for the $n$th cycle is given by $\Delta t_n = t_{n+1} - t_n + \Delta t_{n-1} - T$.

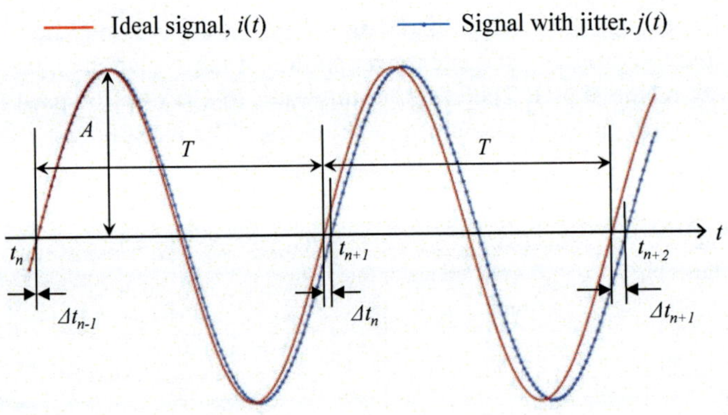


**Fig. 5.7** Ideal and jittery signals

Let the ideal signal of frequency $f$ and amplitude $A$ be (see (4.14))

$$i(t) = A\sin(2\pi f t). \tag{5.22}$$

It is assumed both the phase shift and background offset of the signal are zero (see Sect. 2.2.3). The signal with jitter is

$$j(t) = A\sin\left(2\pi f(t + \Delta t^{RJ}(t) + \Delta t^{PJ}(t))\right), \tag{5.23}$$

where $\Delta t^{RJ}(t)$ and $\Delta t^{PJ}(t)$ are the random jitter (RJ) and periodic jitter (PJ), respectively. Once again, the phase shift and background offset of the signal are assumed zero. The RJ can be characterized with its RMS $\sigma_{RJ}$, thus,

$$\Delta t^{RJ}(t) = \varepsilon(t), \tag{5.24}$$

where $\varepsilon(t)$ is a random process which is a function of time and has a Gaussian distribution $\mathcal{N}$ with mean $\mu_{\mathrm{RJ}}$ and variance ${\sigma_{\mathrm{RJ}}}^2$, that is $\varepsilon \sim \mathcal{N}\left(\mu_{\mathrm{RJ}}, {\sigma_{\mathrm{RJ}}}^2\right)$.

Periodic jitter is defined as the repetition of the jitter at a certain period in a signal. In a signal, there can be multiple PJs of various frequencies and these follow the superposition rule, thus, from (4.8) [22]

$$\Delta t^{PJ}(t) = \sum_{p=1}^{P} {A_{\mathrm{PJ}}}^{(p)} \sin\left(2\pi {f_{\mathrm{PJ}}}^{(p)} t\right), \qquad P \in \mathbb{Z}^+, \tag{5.25}$$

where ${A_{\mathrm{PJ}}}^{(p)}$ and ${f_{\mathrm{PJ}}}^{(p)}$ are the amplitude and frequency of the $p$th component of the PJ, respectively. Note that it was assumed that no phase difference in the periodic jitter (i.e., from (4.8) with $\phi_{\mathrm{PJ}}^{(p)} = 0$). From a signal perspective, periodic jitter is the same as any periodic signal in terms of frequency and phase, but its amplitude is jitter in units of timing. Then, from (5.24) and (5.25), the total jitter $\Delta t(t)$ can be expressed as

$$\Delta t(t) = \varepsilon(t) + \sum_{p=1}^{P} {A_{\mathrm{PJ}}}^{(p)} \sin\left(2\pi {f_{\mathrm{PJ}}}^{(p)} t\right). \tag{5.26}$$

In the presence of both RJ and PJ, one first has to find the PJ by discrete Fourier transform (DFT). The DFT spectrum measures the distribution of energy values as a function of frequency and each point represents the square root of energy of that corresponding frequency component. The DFT of the total jitter signal is

$$\mathscr{F}_k = \sum_{n,k=0}^{N-1} \Delta t_n(t)\, e^{-i\frac{2\pi nk}{N}}, \tag{5.27}$$

where $\mathscr{F}_k$ is the DFT of the $k$th sample of the jitter signal. From the spectrum, we can easily identify the frequencies ${f_{PJ}}^{(p)}$ of each component of the PJ by peaks in the spectrum. Other jitter parameters ${A_{PJ}}^{(p)}$ of the PJ, and ${\sigma_{RJ}}^2$ for RJ can be computed by measuring the energy of the corresponding portions in the spectrum.

As an example, Fig. 5.8 shows the spectrum of the total jitter $\mathscr{F}_k$, against its frequency $f$, with the RJ and PJ at a single frequency. The PJ appears as a clear high peak at its frequency $f_{PJ}$. The RJ is spread evenly throughout the spectrum. The line $\widehat{\mathscr{F}}_k^{(1)}$ is the interpolated level of the RJ across that part of the spectrum that the PJ contributes to. Each point of the spectrum is the square root of energy of that frequency component of the signal. In the frequency domain, the energy of a spectral line is calculated by taking the square of the DFT magnitude [23]. Thus, the total energy in the spectrum is obtained by

$$\mathrm{E}\,(\Delta t) = \mathrm{E}^{cf} \sum_{k=0}^{N-1} |\mathscr{F}_k|^2, \tag{5.28}$$

then, the amplitude of the total jitter of the signal is given by

$$\Delta t = \sqrt{2\,\mathrm{E}\,(\Delta t)}, \tag{5.29}$$

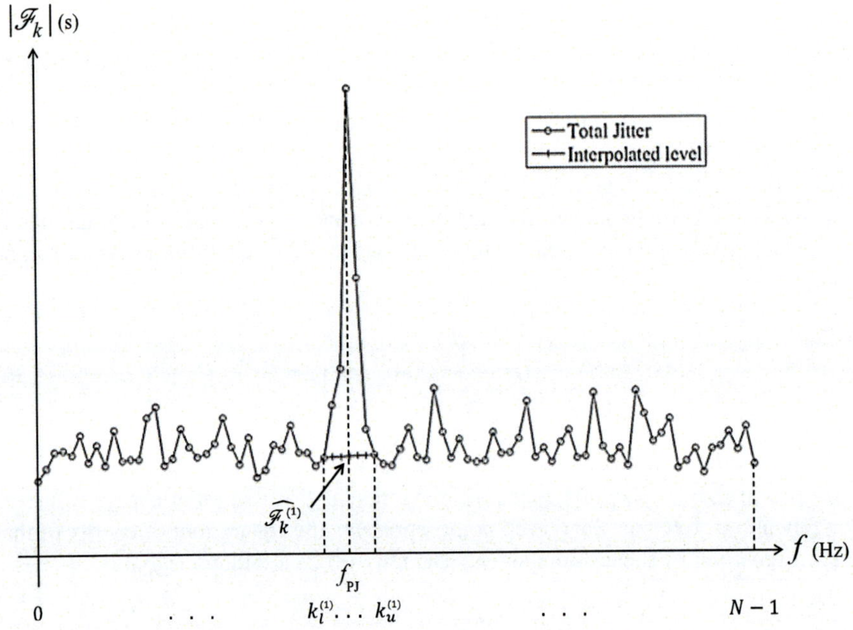


**Fig. 5.8** An example of a spectrum of the total jitter

where the factor $\sqrt{2}$ is the conversion of the RMS of a sinusoidal to its amplitude. The corresponding energy of the PJ is in its peak only, thus

$$\mathrm{E}\left(\Delta t^{PJ}\right) = \mathrm{E}^{cf} \sum_{p=1}^{P} \sum_{k=k_l^{(p)}}^{k_u^{(p)}} \left( |\mathscr{F}_k^{(p)}|^2 - |\widehat{\mathscr{F}}_k^{(p)}|^2 \right), \tag{5.30}$$

where $\mathscr{F}_k^{(p)}$ is the DFT of the $k$th sample of the $p$th component of the PJ and $\widehat{\mathscr{F}}_k^{(p)}$ is the interpolated level of the RJ across that part of the spectrum that $p$th component of the PJ contributes to. This part needs to be subtracted from the corresponding component of the PJ peak to be able to estimate the energy of the PJ. Terms $k_l^{(p)}$ and $k_u^{(p)}$ are the corresponding left (lower) and right (upper) indices of the interpolated level of the spectrum, respectively (see Fig. 5.8 when $p = 1$). The amplitude of the PJ is

$$A_{\mathrm{PJ}} = \sqrt{2\,\mathrm{E}\left(\Delta t^{PJ}\right)}. \tag{5.31}$$

Hence, by using (5.28) and (5.30), the energy corresponding to the RJ is given by

$$\mathrm{E}\left(\Delta t^{RJ}\right) = \mathrm{E}\left(\Delta t\right) - \mathrm{E}\left(\Delta t^{PJ}\right) - \mathrm{E}\left(\widetilde{\Delta t^{PJ}}\right), \tag{5.32}$$

where $\mathrm{E}\left(\widetilde{\Delta t^{PJ}}\right)$ is the corresponding energy of the other spikes of the spectrum. Thus, the RMS of the RJ is

$$\sigma_{RJ} = \sqrt{\mathrm{E}\left(\Delta t^{RJ}\right)}. \tag{5.33}$$

### 5.3.2 Curve Fitting Analysis for the Spectrum

The resulting jitter spectrum is not linearly distributed due to the presence of various jitter components in the signal. Therefore, the behavior of the jitter spectrum needs to be analyzed in order to get a better understanding of the jitter in the light source of the camera. We can analyze the jitter spectrum by using a standard curve fitting technique called *regression*.

Regression is a powerful tool for fitting non-linear data. Non-linear data can perform by transforming it into a straight line using the logarithmic scale and then applying the linear regression. Linear regression of the transformed data yields a slope and intercept that can be used to determine the parameters of interest [13].

As an example, Fig. 5.9 shows the log-log scale of the jitter amplitude $\mathscr{F}$ against jitter frequency $f$. We define two linear models with different slopes and use the maximum among them as

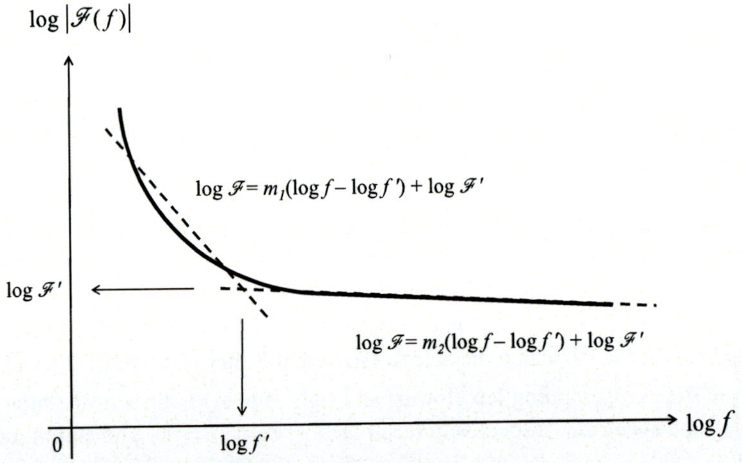


**Fig. 5.9** An example of the curve fitting for a jitter spectrum

$$\log \mathscr{F} = \max \{m_1(\log f - \log f') + \log \mathscr{F}', \quad m_2(\log f - \log f') + \log \mathscr{F}'\}, \tag{5.34}$$

where $m_1, m_2$ are *slopes* of the two linear models. $f'$ and $\mathscr{F}'$ are constants corresponding to the frequency and amplitude at the intersection point, respectively. By rearranging and taking the anti-logging of (5.34) for either $i = 1$ or 2, gives

$$\mathscr{F} = \mathscr{F}' \left(\frac{f}{f'}\right)^{m_i}, \quad i = 1 \oplus 2, \tag{5.35}$$

and this is a power-law function of the form $\mathscr{F} \propto f^m$ (with $m = m_1 \oplus m_2$) that represents the straight line on a log-log graph. The power-law function for the jitter amount $\mathscr{F}$ against the frequency $f$ is given by

$$\mathscr{F} = \eta f^m \quad \text{with} \quad \eta = \frac{\mathscr{F}'}{(f')^m}, \tag{5.36}$$

where $\eta$ and $m$ are the exponent and power of the power-law function, which represent the *intercept* and *slope* of the log-log graph, respectively.

Now, all the steps in the proposed methodology for our requirement are explained. Before applying these steps for the experiments on ToF range imaging cameras, the proposed methodology is first evaluated by using the generated simulation data which is described in the next chapter.

## 5.4 Chapter Remarks and Further Readings

An efficient method to extract the relevant types of jitter in the light source signal of the AMCW ToF range imaging cameras is proposed in this chapter. The most important advantage of this method is that it does not require a reference or sampling clock signal to measure jitter. Another dominant advantage is, this methodology can be used for jitter extraction in analogue or digital signals with any kind of standard geometrical shapes (e.g., sinusoidal, triangular, and rectangular (square)) and where there is no access to the clock signal. The Fourier analysis and signal processing techniques are used in the methodology.

In order to smooth the noisy signal, the Savitzky-Golay algorithm was used. More details on smoothing techniques can be found in [3, 6]. The zero crossings of the time axis are computed by the linear interpolation method. However, more details on other interpolation techniques can be found in [15]. Several interpolation methods and numerical comparisons with examples are included in this reference.

More details on the spectral leakage can be found by Lyons [21]. The reader can understand the minimization of the spectral leakage with signal processing applications better with the book by Prabhu [16].

The random and periodic jitter is only considered on the extraction of the jitter in illumination signal since other types of jitter are data dependent which is not meaningful to the light source. Before applying the proposed methodology with real applications, first, it is evaluated on the generated simulation data in Chap. 6.

## References


1. Lee Streeter, Michael J Cree, Adrian Dorrington, et al. A strategy for the correction of effects of jitter in AMCW lidar images. In *Image and Vision Computing New Zealand (IVCNZ), 2013 28th International Conference of*, pages 500–505. IEEE, 2013.
2. Johannes Seiter, Michael Hofbauer, Milos Davidovic, and Horst Zimmermann. Investigation of the distance error induced by cycle-to-cycle jitter in a correlating time-of-flight distance measurement system. *Optical Engineering*, 53(7):073104–073104, 2014.
3. Jeffrey S Simonoff. *Smoothing methods in statistics*. Springer Science & Business Media, 2012.
4. A. Savitzky and M. J. E. Golay. Smoothing and differentiation of data by simplified least squares procedures. *Analytical Chemistry*, 36(8):1627–1639, 1964. https://doi.org/10.1021/ac60214a047.
5. Sophocles Orfanidis. *Introduction to Signal Processing*. Prentice Hall, 1996.
6. Aleksandra Kawala-Sterniuk, Michal Podpora, Mariusz Pelc, Monika Blaszczyszyn, Edward Jacek Gorzelanczyk, Radek Martinek, and Stepan Ozana. Comparison of smoothing filters in analysis of EEG data for the medical diagnostics purposes. *Sensors*, 20(3):807, 2020.
7. Per-Olof Persson and Gilbert Strang. Smoothing by Savitzky-Golay and Legendre filters. In *Mathematical systems theory in biology, communications, computation, and finance*, pages 301–315. Springer, 2003.
8. Gabriel Vivó-Truyols and Peter J Schoenmakers. Automatic selection of optimal Savitzky-Golay smoothing. *Analytical chemistry*, 78(13):4598–4608, 2006.

9. Sunder Ram Krishnan and Chandra Sekhar Seelamantula. On the selection of optimum Savitzky-Golay filters. *IEEE transactions on signal processing*, 61(2):380–391, 2012.
10. An-Xin Zhao, Xiao-Jun Tang, Zhong-Hua Zhang, and Jun-Hua Liu. The parameters optimization selection of Savitzky-Golay filter and its application in smoothing pretreatment for FTIR spectra. In *2014 9th IEEE Conference on Industrial Electronics and Applications*, pages 516–521. IEEE, 2014.
11. Mohammad Sadeghi and Fereidoon Behnia. Optimum window length of Savitzky-Golay filters with arbitrary order. *arXiv preprint arXiv:1808.10489*, 2018.
12. I Céspedes, Y Huang, J Ophir, and S Spratt. Methods for estimation of subsample time delays of digitized echo signals. *Ultrasonic imaging*, 17(2):142–171, 1995.
13. Steven C Chapra and Raymond P Canale. *Numerical Methods for Engineers*, volume 7. McGraw-Hill New York, 2015.
14. R Boucher and J Hassab. Analysis of discrete implementation of generalized cross correlator. *IEEE Transactions on Acoustics, Speech, and Signal Processing*, 29(3):609–611, 1981.
15. Einar Maeland. On the comparison of interpolation methods. *IEEE transactions on medical imaging*, 7(3):213–217, 1988.
16. KM Muraleedhara Prabhu. *Window functions and their applications in signal processing*. CRC press, 2013.
17. Fredric J Harris. On the use of windows for harmonic analysis with the discrete Fourier transform. *Proceedings of the IEEE*, 66(1):51–83, 1978.
18. Alan V. Oppenheim, Ronald W. Schafer, and John R. Buck. *Discrete-time Signal Processing (2nd Ed.)*. Prentice-Hall, Inc., Upper Saddle River, NJ, USA, 1999. ISBN 0-13-754920-2.
19. S Gade, N Thrane, H Konstantin, and J Wismer. Application note: Time windows. *Bruel and Kjær, Denmark*, 1987.
20. R. B. Blackman and J. W. Tukey. The measurement of power spectra from the point of view of communications engineering Part I. *The Bell System Technical Journal*, 37(1):185–282, Jan 1958. ISSN 0005-8580. https://doi.org/10.1002/j.1538-7305.1958.tb03874.x.
21. Richard G Lyons. *Understanding Digital Signal Processing*. Pearson Education, Inc., Boston, MA, USA, third edition, 2011.
22. Mike Li. *Jitter, Noise, and Signal Integrity at High-speed*. Prentice Hall Press, Upper Saddle River, NJ, USA, first edition, 2007. ISBN 9780132429610.
23. Randolph B Randall. *Frequency Analysis*. Brül & Kjor, third edition, 1987.

# Chapter 6
# Evaluation of the Proposed Methodology

**Abstract** As described in Chap. 5, we proposed camera independent methodology for periodic and random jitter measurement in the light source of AMCW ToF range imaging cameras. Before experimentally testing the proposed methodology, this chapter evaluates the methodology by using the generated simulation data. The standard geometrical shapes of signals (sinusoidal, triangular, and rectangular) with various numerous values for the characteristic parameters of the signal and jitter are considered in this evaluation. The corresponding results with discussion are included at the end of the chapter.

**Keywords** Sinusoidal signal · Triangular signal · Rectangular (square) signal · Periodic jitter · Random jitter · Periodic jitter frequency · Periodic jitter amplitude · Root-mean-square · Simulation data · Fitted curve

## 6.1 Introduction

A methodology for jitter extraction in ToF range imaging systems that is camera independent is proposed in Chap. 5. Before experimentally testing the proposed methodology shown in Fig. 5.2, we can evaluate it with the simulation data first. We generated the reference signal and jittery signal with the random and/or periodic jitter as described in Sect. 6.2. For each case, the same frequency is used for the two signals as in the homodyne operation of the AMCW ToF ranging cameras. For convenience, the noise is not considered in the generated signals. Therefore, the smoothing technique does not require to perform in the simulation. In addition, we use the Hann window [1] with the corresponding energy factor (i.e., $E_H^{cf} = 1.63$ from Table 5.1) in order to correct the values of jitter parameters.



G. Anthonys, *Timing Jitter in Time-of-Flight Range Imaging Cameras*,
https://doi.org/10.1007/978-3-030-94159-8_6

## 6.2 Simulation Setup

Five simulated signals with various frequencies (i.e., $f_i = i \times 10\,\text{MHz}, i = 1, \ldots, 5$) and unit amplitude (i.e., $A = 1$) are considered throughout the simulation setup. Three types of setup including the numerate amount of random jitter, periodic jitter at a single frequency, and both types that are separately injected into each simulated signal corresponding to the frequency $f_i$ are considered. Figure 6.1 illustrates the simulation setup with the three types of geometrical shapes of signals. The jitter amounts are chosen as two, four, six, eight, and ten times the lowest value (5 ps) for each of $\sigma_{\text{RJ}}$ and $A_{\text{PJ}}$, for the setup denoted as SET- *RJ* and SET-*PJ*, respectively, as shown in Table 6.1. All parameters are repeated for each type of the signal's shape, i.e., for sinusoidal, triangular, and rectangular.

In addition, an arbitrarily chosen specific value for the frequency of the periodic jitter as $f_{\text{PJ}} = 4.37\,\text{MHz}$ is used throughout the simulations. The combination of these two types of jitter values are chosen for the setup denoted in SET-*Both*. Furthermore, by using (5.11) and (5.12) with the parameters $f_m = f_i, N_s = 2^{16}$ and $F_s = 8$ GSa/s, the number of jitter samples for each signal frequency is also obtained in columns four and five, respectively. In order to control the uncertainties of the results, 20 sets for each simulation setup were used.

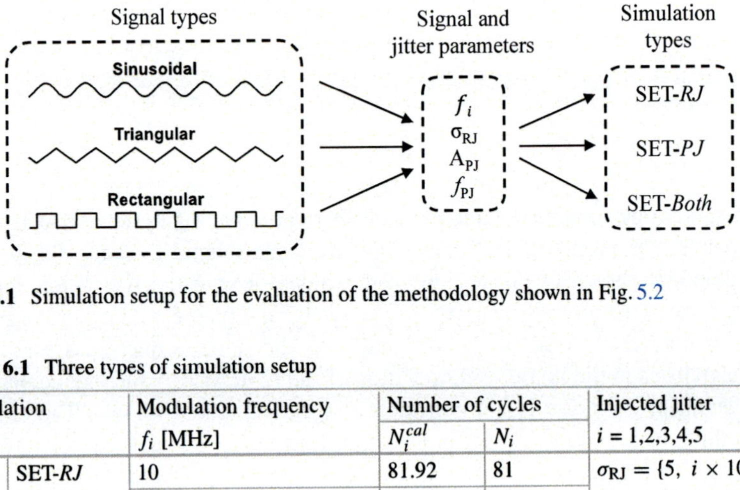


**Fig. 6.1** Simulation setup for the evaluation of the methodology shown in Fig. 5.2

**Table 6.1** Three types of simulation setup

| Simulation types | | Modulation frequency $f_i$ [MHz] | Number of cycles | | Injected jitter |
|---|---|---|---|---|---|
| | | | $N_i^{cal}$ | $N_i$ | $i = 1,2,3,4,5$ |
| 20 | SET-*RJ* | 10 | 81.92 | 81 | $\sigma_{\text{RJ}} = \{5,\ i \times 10\}$ ps |
| sets | SET-*PJ* | 20 | 163.84 | 163 | $A_{\text{PJ}} = \{5,\ i \times 10\}$ ps |
| each | SET-*Both* | 30 | 245.76 | 245 | $f_{\text{PJ}} = 4.37\,\text{MHz}$ |
| | | 40 | 327.68 | 327 | |
| | | 50 | 409.60 | 409 | |

## 6.3 Results and Discussion

First, each simulation signal with jitter was constructed by injecting a known quantity of jitter amounts for the corresponding time-domain signal (see Sect. C.1 in Appendix C for Matlab script). By using steps ① to ⑥ in Sect. 5.2, the jitter spectrum for each set of the relevant simulation setup is obtained, as for step ⑥. Finally, the mean spectrum of the total jitter along the 20 sets was obtained after taking the magnitude of spectra for the relevant simulation setup, and this mean spectrum is used to calculate the jitter components, as for step ⑦.

In order to validate the proposed methodology for the general case, three standard shapes of signals are performed separately. Let us first consider the signals with a sinusoidal shape.

### 6.3.1 *Simulation with Sinusoidal Signals*

The spectrum of the total jitter when injecting the RJ (denoted as SET-*RJ*) for the values of RMS $\sigma_{\mathrm{RJ}} = 5$ and 50 ps for the five simulated signals corresponding to the frequency $f_i$ is shown in Fig. 6.2. When the modulation frequency is relatively low (10 MHz and 20 MHz), some peaks are present in the spectra. These are due to the random fluctuation in simulation and the limited number of simulations (20 sets) used in the setup (see Table 6.1). In the limit of a very large number of simulations one would expect the spectra to tend to be at line in Fig. 6.2.

In order to illustrate the results in the same plot we padded the amplitude by zeros at the end of first four simulated signals (i.e., for the signal of $f_i = 10, \ldots, 40$ MHz), to the length of the number of bins of the spectrum of the 50 MHz signal, i.e., 409 (see Table 6.1). Similarly, when injecting the PJ (denoted as SET-*PJ*) and both RJ with PJ together (denoted as SET-*Both*) for the five simulated signals corresponding to the frequency $f_i$ are shown in Figs. 6.3 and 6.4, respectively.

By comparison, we noted that the amplitude level of the spectrum in Figs. 6.2b, 6.3b, and 6.4b is approximately 10 times that of Figs. 6.2a, 6.3a, and 6.4a, respectively. It clearly seems the high peak at the frequency of the PJ (at 4.37 MHz) in both Figs. 6.3 and 6.4 has the value of $(4.39 \pm 0.02)$ MHz. Note that the interpolation level between the PJ and RJ for Fig. 6.4 is obtained by visual inspection and using (5.21). Table 6.2 shows the corresponding results of the extracted jitter parameters for Figs. 6.2, 6.3, and 6.4.

We compare the actual and measured jitter amounts for each of three simulation setups, separately. Figures 6.5 and 6.6 show the comparison of the measured and actual jitter values with different amounts (see Table 6.1) for 10 and 50 MHz signals for simulations SET-*RJ* and SET-*PJ*, respectively. In addition, one-to-one curve and the fitted curve through the data points with the relevant linear regression for the measured jitter amounts are also included in each simulation result in the figure. The slope and intercept of this expression are good indicators for the accuracy and

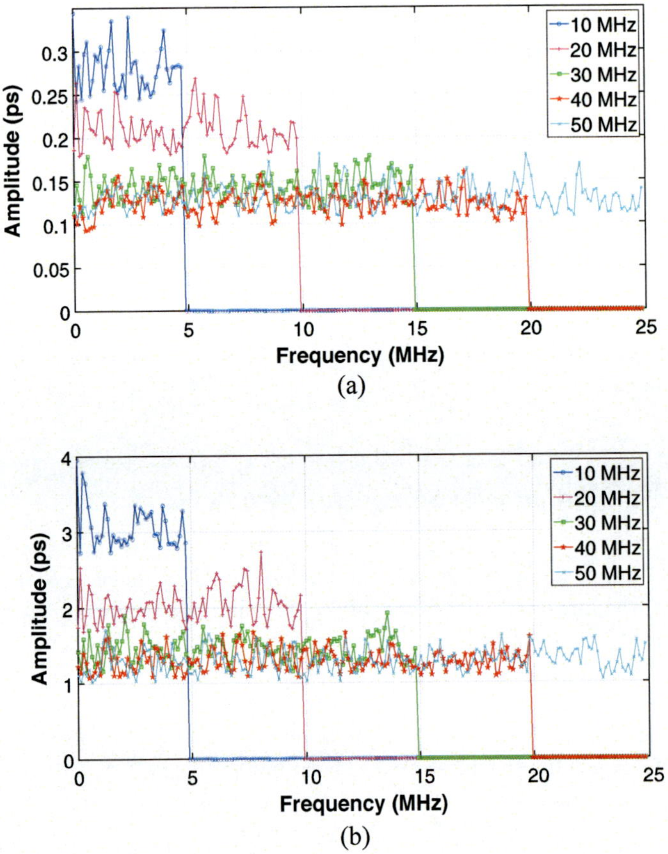


**Fig. 6.2** Spectrum of the total jitter when injecting the random jitter as $\sigma_{RJ}$ (ps): **(a)** 5 and **(b)** 50, for setup denoted as SET-*RJ* in Table 6.1

offset error, respectively, of the measured jitter parameters. When the values of slope and intercept are closer to one and zero, respectively, that indicates the matching of the actual and measured amounts in each simulation. For SET-*RJ*, the deviation between the measured and actual values increases (see Fig. 6.5) when the random jitter amount increases, and note that the offset errors are −0.3853 and 0.1850 for 10 MHz and 50 MHz signals, respectively. We believe this is due to the presence of more zero crossings with a 50 MHz signal. Statistical uncertainty is reduced due to a larger sample size. For the case of PJ (SET-*PJ*), the jitter extraction in the absence of any other noise is accurate with very small measurement uncertainty for both $f_{PJ}$ and $A_{PJ}$. The offset errors are 0.0004 and 0.0008 for 10 MHz and 50 MHz, respectively (see Fig. 6.6).

Similarly SET-*Both* on a 50 MHz signal is shown in Fig. 6.7. Figure 6.7a compares the varies of the PJ with fixed RJ of $\sigma_{RJ} = 5$ ps while Fig. 6.7b compares the varies of the RJ with fixed PJ of $A_{PJ} = 5$ ps at 4.37 MHz. Among these two plots,

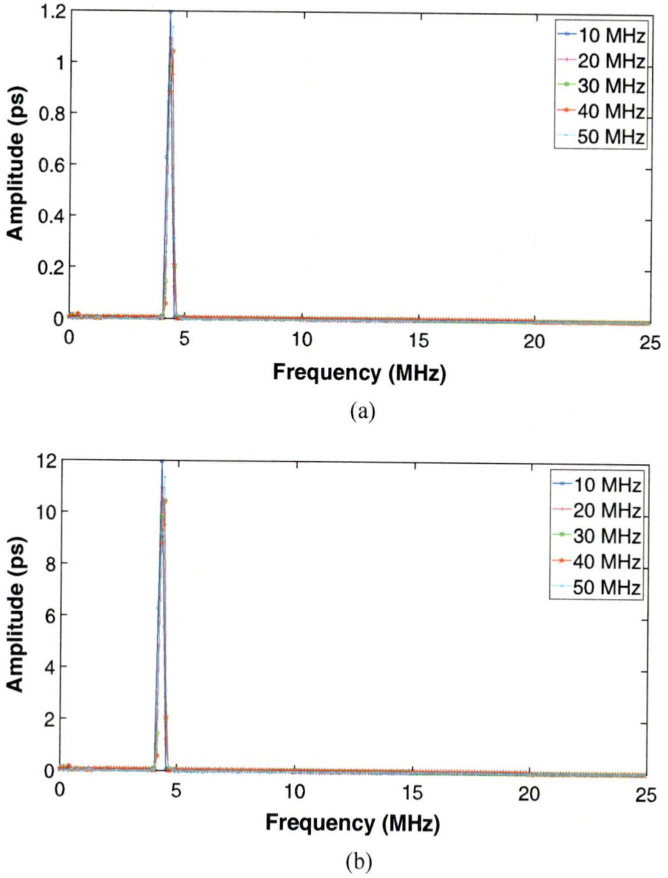


**Fig. 6.3** Spectrum of the total jitter when injecting the periodic jitter as $A_{PJ}$: (**a**) 5 ps and (**b**) 50 ps, at $f_{PJ} = 4.37$ MHz for setup denoted as SET-*PJ*

the offset error in Fig. 6.7a is relatively larger [0.1252] than the same in Fig. 6.7b, [0.0737], that is, the amounts deviate from zero. We believe it is because the PJ affects a particular frequency, while the RJ affects all frequencies of the signal with spreading of the energy through the whole jitter signal. Therefore, the effect of the RJ on the calculation of PJ is less than the effect of the PJ on the calculation of RJ, when we inject the same amount of each. The measured and actual PJ with fixed RJ is closer (slope is 0.9849) than the measured and actual RJ with fixed PJ (slope is 0.8621), see Fig. 6.7.

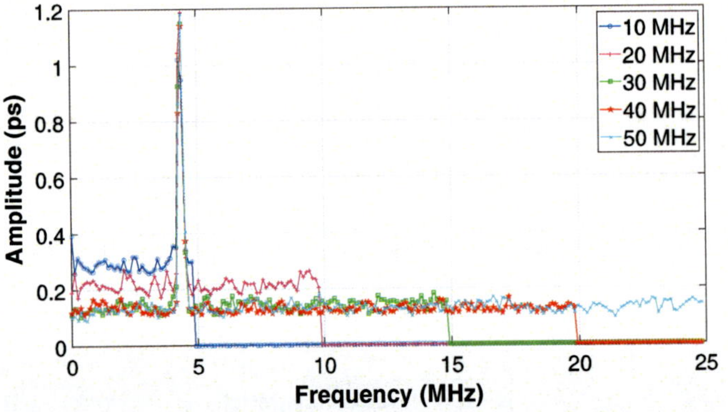


(a)

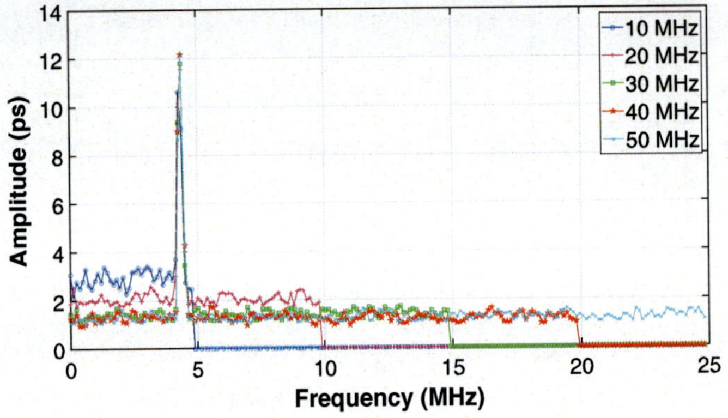


(b)

**Fig. 6.4** Spectrum of the total jitter when injecting the RJ and PJ as **(a)** $\sigma_{RJ} = 5\,\text{ps}$, $A_{PJ} = 5\,\text{ps}$, $f_{PJ} = 4.37\,\text{MHz}$ and **(b)** $\sigma_{RJ} = 50\,\text{ps}$, $A_{PJ} = 50\,\text{ps}$, $f_{PJ} = 4.37\,\text{MHz}$, for setup denoted as SET-*Both* in Table 6.1

**Table 6.2** Extracted jitter amounts from Figs. 6.2, 6.3, and 6.4 by considering the five modulation frequencies

| Simulation setup and injected jitter parameters | | Computed jitter parameters $\sigma_{RJ}$ (ps) | $f_{PJ} = 4.37\,\text{MHz}$ | $A_{PJ}$ (ps) |
|---|---|---|---|---|
| SET-*RJ* | $\sigma_{RJ} = 5\,\text{ps}$ | $4.9 \pm 0.2$ | – | – |
| | $\sigma_{RJ} = 50\,\text{ps}$ | $45.7 \pm 1.6$ | – | – |
| SET-*PJ* | $A_{PJ} = 5\,\text{ps}$ | – | $4.39 \pm 0.02$ | $4.8 \pm 0.1$ |
| | $A_{PJ} = 50\,\text{ps}$ | – | | $47.6 \pm 0.7$ |
| SET-*Both* | $\sigma_{RJ} = 5\,\text{ps}$, $A_{PJ} = 5\,\text{ps}$ | $3.2 \pm 0.1$ | $4.39 \pm 0.02$ | $4.8 \pm 0.1$ |
| | $\sigma_{RJ} = 50\,\text{ps}$, $A_{PJ} = 50\,\text{ps}$ | $31.6 \pm 1.1$ | | $49.3 \pm 1.2$ |

**Table 6.3** Computed jitter amounts for the triangular signals by considering the five modulation frequencies

| Simulation setup and parameter | | Injected jitter amounts in picoseconds | | | | | |
|---|---|---|---|---|---|---|---|
| | | 5 | 10 | 20 | 30 | 40 | 50 |
| SET-*RJ* | $\sigma_{RJ}$ (ps) | $4.8 \pm 0.1$ | $9.7 \pm 0.4$ | $18.95 \pm 0.6$ | $28.7 \pm 0.8$ | $38.3 \pm 1.2$ | $47.5 \pm 1.5$ |
| SET-*PJ* | $A_{PJ}$ (ps) | $4.8 \pm 0.1$ | $9.5 \pm 0.1$ | $19.0 \pm 0.3$ | $28.5 \pm 0.4$ | $38.1 \pm 0.5$ | $47.6 \pm 0.7$ |
| | $f_{PJ}$ (MHz) | $4.40 \pm 0.02$ | | | | | |
| SET-*Both* | | Fixed amount of the RJ with various amplitudes of the PJ at 4.37 MHz | | | | | |
| | $\sigma_{RJ} = 5$ ps | $4.7 \pm 0.2$ | $4.7 \pm 0.2$ | $4.9 \pm 0.2$ | $5.1 \pm 0.2$ | $5.3 \pm 0.2$ | $5.5 \pm 0.3$ |
| | $A_{PJ}$ (ps) | $5.0 \pm 0.1$ | $9.9 \pm 0.1$ | $19.5 \pm 0.2$ | $29.1 \pm 0.4$ | $38.6 \pm 0.6$ | $48.2 \pm 0.7$ |
| | $f_{PJ}$ (MHz) | $4.40 \pm 0.02$ | | | | | |
| | | Fixed amount of the PJ at 4.37 MHz with various amounts of the RJ | | | | | |
| | $A_{PJ} = 5$ ps | $5.0 \pm 0.1$ | $4.8 \pm 0.1$ | $4.9 \pm 0.3$ | $5.4 \pm 0.8$ | $5.9 \pm 0.7$ | $6.56 \pm 0.97$ |
| | $f_{PJ}$ (MHz) | $4.40 \pm 0.02$ | $4.40 \pm 0.02$ | $4.40 \pm 0.02$ | $4.42 \pm 0.01$ | $4.32 \pm 0.07$ | $4.42 \pm 0.07$ |
| | $\sigma_{RJ}$ (ps) | $4.7 \pm 0.2$ | $9.5 \pm 0.3$ | $18.8 \pm 0.7$ | $28.0 \pm 1.1$ | $38.06 \pm 1.03$ | $47.0 \pm 1.6$ |
| | Large amounts of RJ and PJ | Computed $\sigma_{RJ} = (47.6 \pm 1.3)$ ps, $A_{PJ} = (49.4 \pm 0.4)$ ps, and $f_{PJ} = (4.40 \pm 0.02)$ MHz when injecting $\sigma_{RJ} = 50$ ps, $A_{PJ} = 50$ ps, and $f_{PJ} = 4.37$ MHz, respectively. | | | | | |

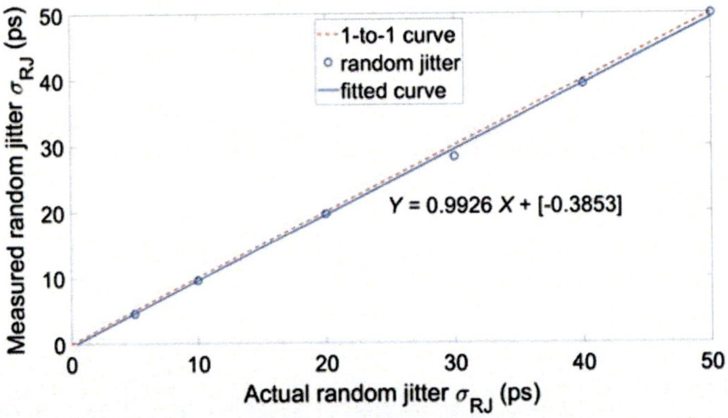


(a)

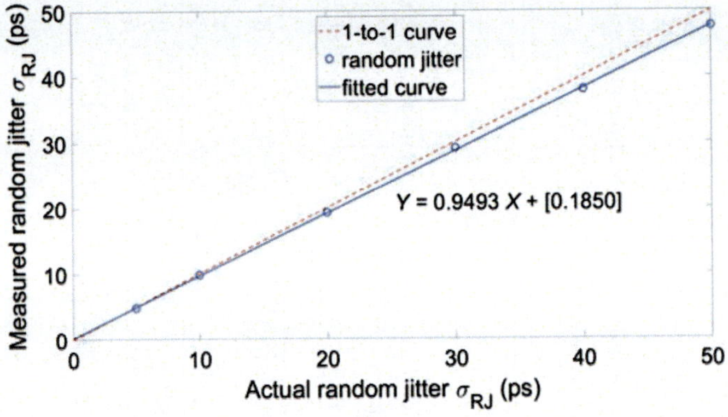


(b)

**Fig. 6.5** Comparison of the actual and measured random jitter in the signal of frequencies (**a**) 10 MHz and (**b**) 50 MHz, on the simulation setup denoted as SET-*RJ* in Table 6.1

## *6.3.2 Simulation with Triangular Signals*

All the above simulations were performed using sinusoidal signals; now we need to clarify the proposed methodology for other standard shapes of signals, such as triangular and rectangular. We repeat the analysis for the triangular shape of signals with the same parameters used in Table 6.1. The spectra of the total jitter for the three simulation setups when injecting the jitter amounts of 5 ps and 50 ps for each corresponding modulation frequency $f_i$ are shown in Fig. 6.8, separately. The maximum jitter levels of each right sub-figure are about ten times their corresponding left sub-figures. When the modulation frequency is relatively low (10 MHz and 20 MHz), some peaks are apparent in the spectra for SET-*RJ* because the number of cycles in those signals is relatively small.

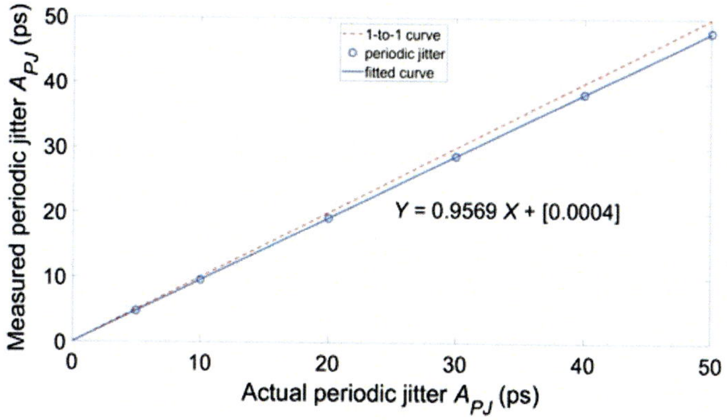


(a)

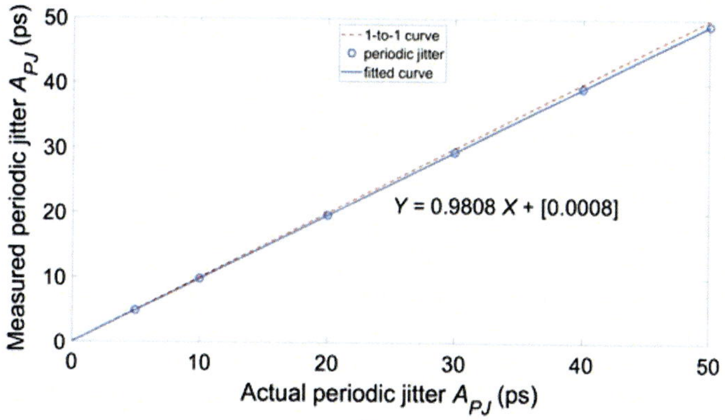


(b)

**Fig. 6.6** Comparison of the actual and measured periodic jitter at 4.37 MHz of the signal with modulation frequency $f_i$ (**a**) 10 MHz and (**b**) 50 MHz, on the simulation setup denoted as SET-*PJ* in Table 6.1

Now, let us consider the extracted jitter parameters corresponding to each simulation setup shown in Fig. 6.8. When the injected RJ is $\sigma_{\mathrm{RJ}} = 5$ ps and 50 ps, we have $\sigma_{\mathrm{RJ}} = (4.76 \pm 0.13)$ ps and $(47.5 \pm 1.5)$ ps, respectively, while for the injected PJ $A_{\mathrm{PJ}} = 5$ ps and 50 ps, we have $A_{\mathrm{PJ}} = (4.76 \pm 0.07)$ ps and $(47.6 \pm 0.7)$ ps, respectively. In addition, when both jitter types are injected, we found $\sigma_{\mathrm{RJ}} = (4.72 \pm 0.15)$ ps with $A_{\mathrm{PJ}} = (4.98 \pm 0.08)$ ps and $\sigma_{\mathrm{RJ}} = (47.61 \pm 1.31)$ ps with $A_{\mathrm{PJ}} = (49.44 \pm 0.40)$ ps when each jitter type is 5 ps and 50 ps, respectively. For each corresponding result, we found the PJ at $f_{\mathrm{PJ}} = (4.40 \pm 0.02)$ MHz when we injected it at $f_{\mathrm{PJ}} - 4.37$ MHz.

One-to-one and fitted curves through the data points with the relevant linear regression for selected modulation frequencies corresponding to the simulation setup denoted by SET-*RJ* and SET-*PJ* are shown in Figs. 6.9 and 6.10, respectively. As before, we include only the results corresponding to the chosen two modulation

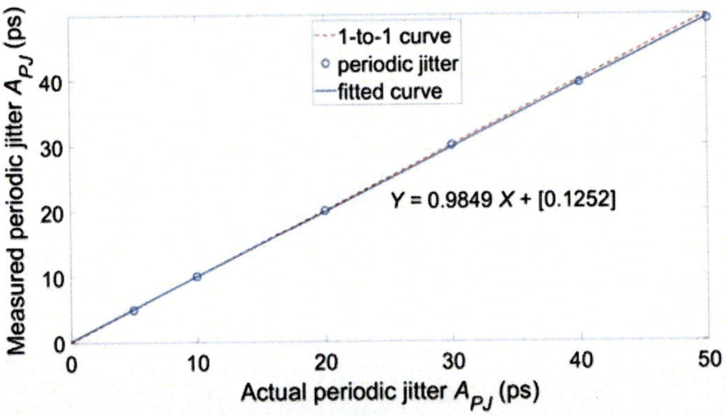


(a)

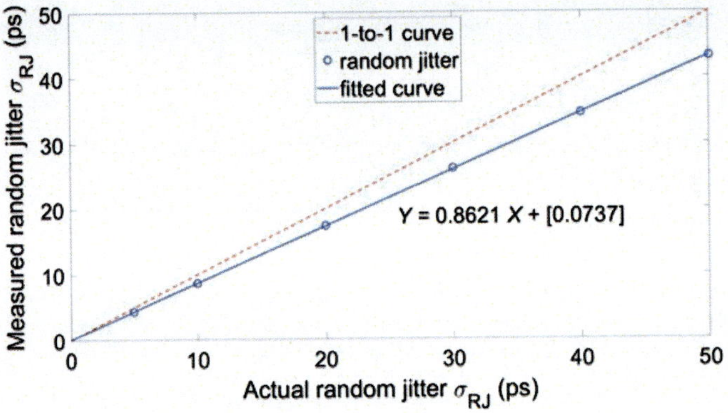


(b)

**Fig. 6.7** Comparison of the actual and measured (**a**) periodic jitter with fixed $\sigma_{RJ} = 5$ ps and (**b**) random jitter with fixed $A_{PJ} = 5$ ps at 4.37 MHz, for the signal $f_i = 50$ MHz, separately, for setup denoted as SET-*Both*

frequencies, 10 MHz and 50 MHz. By comparing the values of slopes and intercepts for each result of 10 MHz and 50 MHz, separately, we can verify that the presence of the periodic jitter gives accurate results (each slope of 0.97 and 0.96 are approximately 1 with each intercept of [0.0001] and [0.0003] are essentially zero, see Fig. 6.10, while the random jitter only is less accurately determined slopes of {0.88 and 0.88} are noticeably different from 1, with intercepts are not nearly to zero {[−0.08] and [0.04]}, see Fig. 6.9).

Figure 6.11 shows the corresponding results for the simulation setup denoted as SET-*Both*. The sub-figures (a) and (b) correspond to the modulation frequencies 10 MHz and 50 MHz, respectively, while left and right sub-figures correspond to the periodic jitter with fixed random jitter of RMS $\sigma_{RJ} = 5$ ps and the random jitter with fixed periodic jitter of $A_{PJ} = 5$ ps, respectively. We found the PJ at $(4.40 \pm 0.02)$ MHz each for the injected PJ (left sub-figures of Fig. 6.11a and b). However,

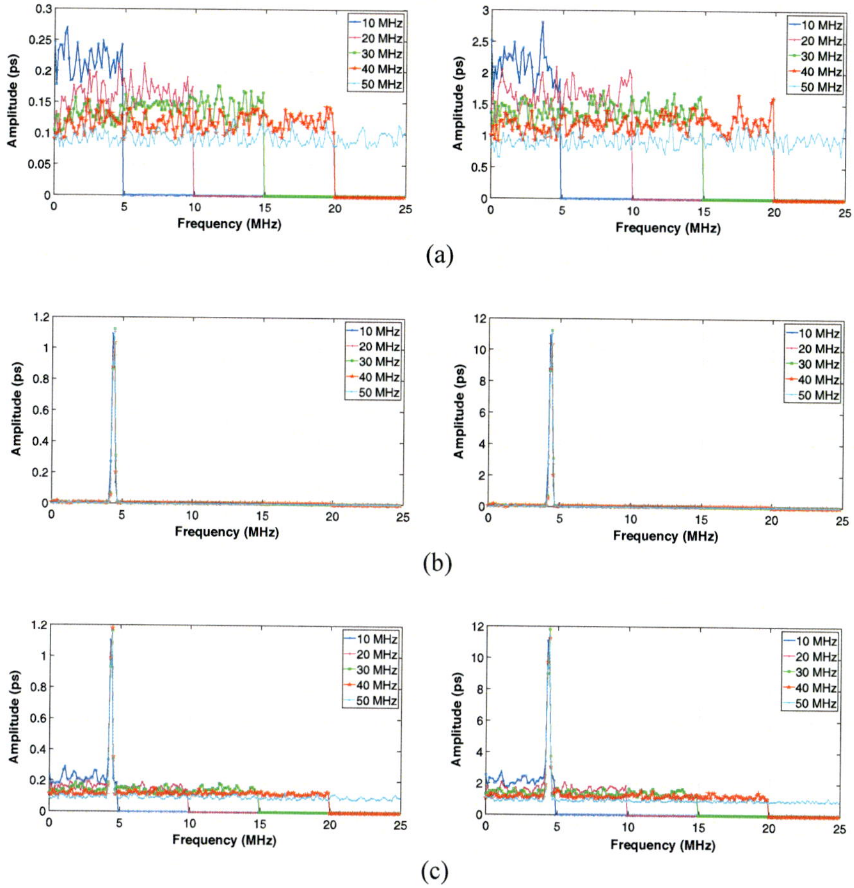


**Fig. 6.8** Spectrum of the total jitter with the triangular signal for three simulation setups denoted as (**a**) SET-*RJ*, (**b**) SET-*PJ*, and (**c**) SET-*Both*, for the jitter amounts (left) 5 ps and (right) 50 ps, separately shown in Table 6.1. Note that the PJ is injected at 4.37 MHz

when the periodic jitter is relatively less than the random jitter, the measurements are computed with less accuracy because of the difficulty to identify the interpolation level between the RJ and the PJ. The slopes are {0.88 and 0.89} with intercepts of {[0.06] and [−0.05]} when the presence of RJ with fixed PJ for the signal with 10 MHz and 50 MHz, respectively (right sub-figures of Fig. 6.11).

The frequencies of the corresponding ideal signals are found by 10.002 MHz and 50.005 MHz for the generated signals of 10 MHz and 50 MHz, respectively. The extracted jitter amounts for the triangular signals computed by considering the chosen five modulation frequencies are tabulated in Table 6.3. The uncertainties are calculated by using the expressions in Sect. E.2. It is clear that the extracted jitter amounts for the setup SET-*RJ* are less accurate than the results from setup SET-*PJ*. The accuracy of the results from the setup SET-*Both* is even less accurate and

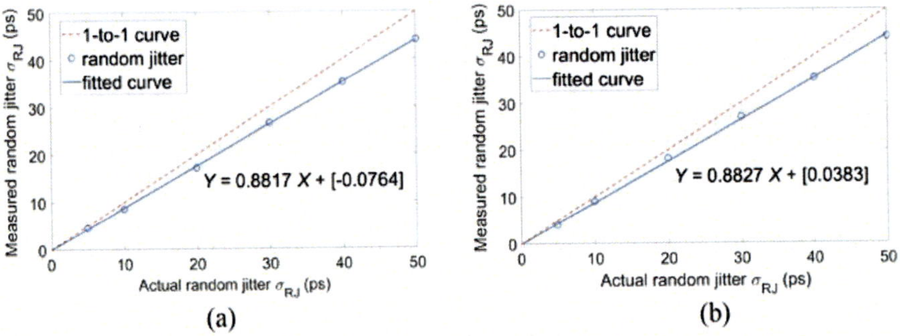


**Fig. 6.9** Comparison of the actual and measured random jitter in the triangular signal of frequencies (**a**) 10 MHz and (**b**) 50 MHz, on the simulation setup denoted as SET-*RJ* in Table 6.1

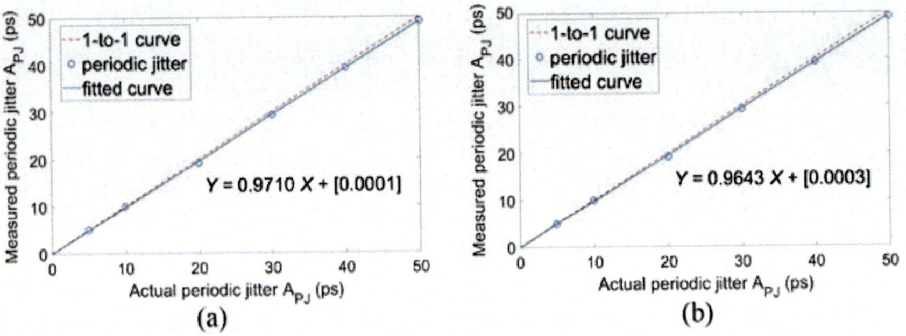


**Fig. 6.10** Comparison of the actual and measured periodic jitter at 4.37 MHz of the triangular signal with modulation frequency $f_i$ (**a**) 10 MHz and (**b**) 50 MHz, on the setup denoted as SET-*PJ* in Table 6.1

worsens with relatively less fixed PJ ($A_{\mathrm{PJ}} = 5\,\mathrm{ps}$) than the results with relatively less fixed RJ ($\sigma_{\mathrm{RJ}} = 5\,\mathrm{ps}$). The frequencies of the periodic jitter for the setups SET-*PJ* and SET-*Both* are given almost similar amounts. However, either for fixed RJ or fixed PJ (5 ps each), the calculated amounts of the RJ and PJ are larger than 5 ps each, respectively (i.e., impossible), when the amount of the other type of jitter increases (from 30 ps each). When both PJ and RJ are relatively large, then the results are relatively close to the injected jitter amounts (the last row of Table 6.3).

### *6.3.3 Simulation with Rectangular Signals*

Next, we investigate the proposed methodology for rectangular shape signals for each simulation setup. Here, the amounts for periodic and random jitter are chosen to be ten times larger than the amounts in Table 6.1, that is, $\sigma_{\mathrm{RJ}} = A_{\mathrm{PJ}} = \{50, 100, \ldots, 500\}$ ps. Figure 6.12 shows the total spectrum when injecting the minimum and maximum amounts (among the chosen) for each modulation

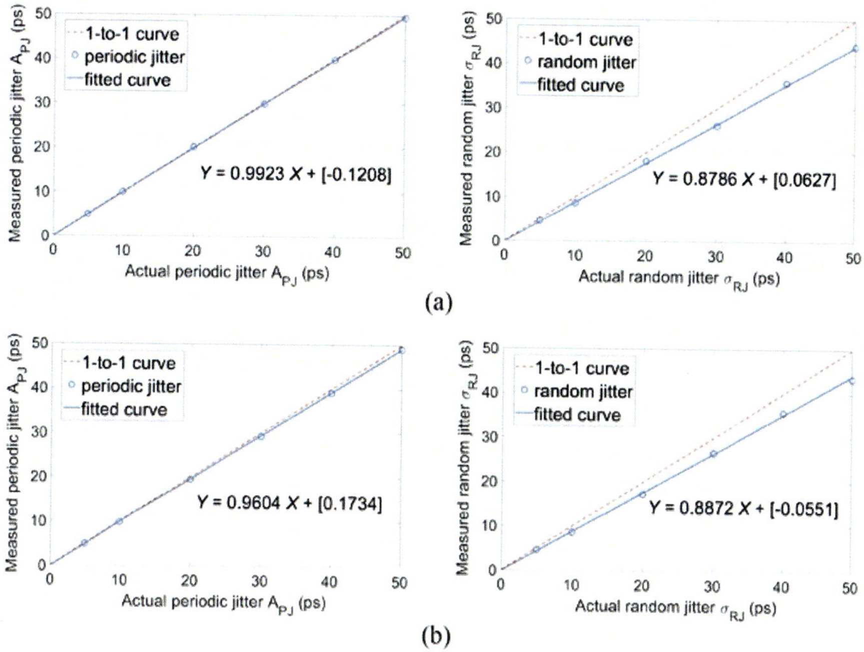


**Fig. 6.11** Comparison of the actual and measured (left) PJ with fixed RJ of $\sigma_{\mathrm{RJ}} = 5$ ps and (right) the RJ with fixed PJ of $A_{\mathrm{PJ}} = 5$ ps at 4.37 MHz, for the triangular signal with the modulation frequencies: (**a**) 10 MHz and (**b**) 50 MHz, separately, for setup denoted as SET-*Both*

frequency for the three simulation setups shown in Table 6.1. As before, for the results corresponding to SET-*RJ*, see Fig. 6.12a, there are few peaks on the low modulation frequencies. In addition, for SET-*PJ*, there are medium peaks when the injected PJ is $A_{\mathrm{PJ}} = 50$ ps (see left sub-figure of Fig. 6.12b), but the peaks are not present when the injected PJ is large $A_{\mathrm{PJ}} = 500$ ps (see right sub-figure of Fig. 6.12b).

When the injected RJ is $\sigma_{\mathrm{RJ}} = 50$ ps and 500 ps, we have $\sigma_{\mathrm{RJ}} = (77 \pm 3)$ ps and $(505 \pm 7)$ ps, respectively; while the injected PJ is $A_{\mathrm{PJ}} = 50$ ps and 500 ps, we have $A_{\mathrm{PJ}} = (71 \pm 5)$ ps and $(463 \pm 7)$ ps, respectively. When both jitter types are injected, we found $\sigma_{\mathrm{RJ}} = (72.5 \pm 1.5)$ ps with $A_{\mathrm{PJ}} = (510 \pm 1)$ ps and $\sigma_{\mathrm{RJ}} = (645 \pm 16)$ ps with $A_{\mathrm{PJ}} = (507 \pm 6)$ ps when each jitter type is 50 ps and 500 ps, respectively. For the setup SET-*Both*, the accuracies of the jitter amounts are amounts that are relatively low because the computed jitter amounts are larger than the injected amounts and the corresponding uncertainty values are also large (i.e., $\pm$ 16 ps and $\pm$ 6 ps when the injected amounts of $\sigma_{\mathrm{RJ}} = 500$ ps and $A_{\mathrm{PJ}} = 500$, respectively). For each corresponding result, we found the PJ at $f_{\mathrm{PJ}} = (4.39 \pm 0.02)$ MHz when we injected it at $f_{\mathrm{PJ}} = 4.37$ MHz.

One-to-one and fitted curves through the data points with the relevant linear regression for selected modulation frequencies corresponding to the simulation

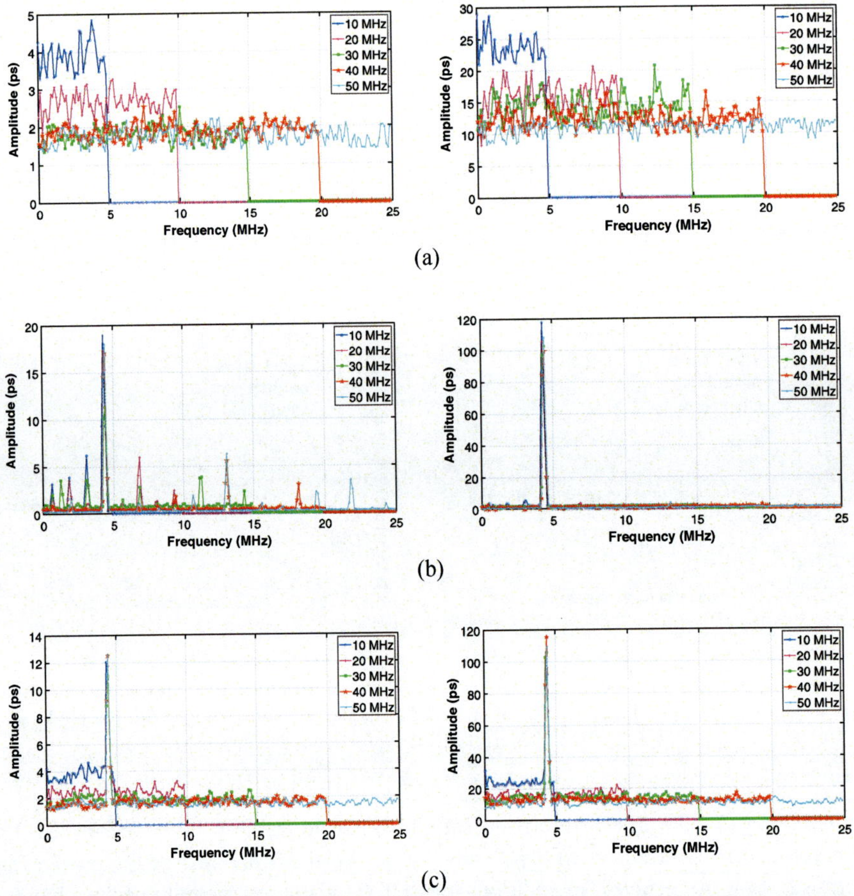


**Fig. 6.12** Spectrum of the total jitter with the rectangular signal for three simulation setups denoted as (**a**) SET-*RJ*, (**b**) SET-*PJ*, and (**c**) SET-*Both*, for the jitter amounts (left) 50 ps and (right) 500 ps, separately. Note that the injected PJ located at 4.37 MHz

setup denoted by SET-*RJ* and SET-*PJ* are shown in Figs. 6.13 and 6.14, respectively. As before, we include only the results corresponding to the chosen two modulation frequencies, 10 MHz and 50 MHz. We can see, each slope is close to 1 (approximately 0.96 and 0.99 in Fig. 6.13 with 0.97 and 0.95 in Fig. 6.14, for 10 MHz and 50 MHz, respectively) while each intercept is far from zero and this deviation is very large when the presence of random jitter only (approximately [22.11] and [19.51] in Fig. 6.13 with [5.56] and [8.97] in Fig. 6.14, for 10 MHz and 50 MHz, respectively).

Figure 6.15 shows the corresponding results for the simulation setup denoted as SET-*Both*. Left and right sub-figures correspond to the PJ with fixed random jitter of RMS $\sigma_{\mathrm{RJ}} = 50\,\mathrm{ps}$ and the RJ with fixed periodic jitter of $A_{\mathrm{PJ}} = 50\,\mathrm{ps}$, respectively, while the sub-figures (a) and (b) correspond to the modulation frequencies 10 MHz

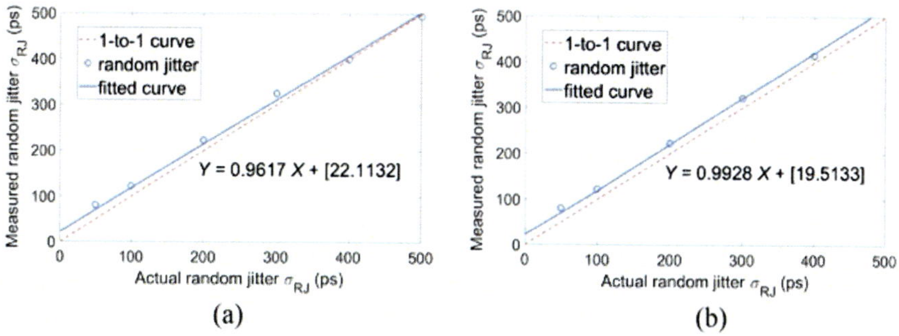


**Fig. 6.13** Comparison of the actual and measured random jitter in the rectangular signal of frequencies (**a**) 10 MHz and (**b**) 50 MHz, on the simulation setup denoted as SET-*RJ* in Table 6.1

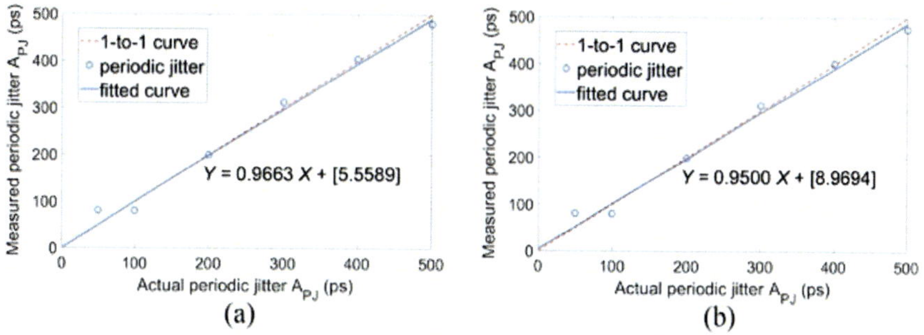


**Fig. 6.14** Comparison of the actual and measured periodic jitter at 4.37 MHz of the triangular signal with modulation frequency $f_i$ (**a**) 10 MHz and (**b**) 50 MHz, on the setup denoted as SET-*PJ* in Table 6.1

and 50 MHz, respectively. We found the PJ at (4.39 ± 0.02) MHz each for the injected PJ (left sub-figures of Fig. 6.15a and b). However, when the injected periodic jitter is fixed and relatively less than the random jitter, the frequency of the periodic jitter is difficult to locate in the spectrum. When the PJ is relatively less than the RJ, this causes the measurements to be computed with less accuracy because of the difficulty to identify the interpolation level between the RJ and the PJ. By considering each right sub-figure of Fig. 6.15, the slopes are {0.96 and 0.95} and intercepts are {[20] and [22]} when the presence of RJ with fixed PJ for the signal with 10 MHz and 50 MHz, respectively. These large deviations (intercepts are far away from zero) are due to the amounts of first two measured random jitter, i.e., when the injected RJ of $\sigma_{\mathrm{RJ}} = 50$ ps and 100 ps. As before, the frequencies of the corresponding ideal signals are found by 10.002 MHz and 50.005 MHz.

The jitter amounts for the rectangular signals computed by considering the chosen five modulation frequencies are tabulated in Table 6.4. The results for the setup SET-*RJ* have less accuracy than the results from setup SET-*PJ*. The accuracy of the results from the setup SET-*Both* is much less than the other two types of simulation and is worse with relatively less fixed PJ ($A_{\mathrm{PJ}} = 50$ ps) than the results

**Table 6.4** Computed jitter amounts for the rectangular signals by considering the five modulation frequencies

| Simulation setup and parameter | | Injected jitter amounts in picoseconds | | | | | |
|---|---|---|---|---|---|---|---|
| | | 50 | 100 | 200 | 300 | 400 | 500 |
| SET-*RJ* | $\sigma_{RJ}$ (ps) | $76.9 \pm 2.7$ | $119.2 \pm 2.56$ | $219.2 \pm 4.4$ | $315.3 \pm 3.2$ | $407.4 \pm 2.7$ | $505.4 \pm 7.0$ |
| SET-*PJ* | $A_{PJ}$ (ps) | $71.0 \pm 4.8$ | $80.77 \pm 4.02$ | $193.1 \pm 3.1$ | $303.4 \pm 4.5$ | $393.1 \pm 4.4$ | $462.5 \pm 6.7$ |
| | $f_{PJ}$ (MHz) | $4.39 \pm 0.02$ | | | | | |
| SET-*Both* | | Fixed amount of the RJ with various amplitudes of the PJ at 4.37 MHz | | | | | |
| | $\sigma_{RJ} = 50$ ps | $72.5 \pm 1.5$ | $71.7 \pm 1.2$ | $72.5 \pm 0.7$ | $72.7 \pm 0.7$ | $75.2 \pm 1.2$ | $77.0 \pm 0.8$ |
| | $A_{PJ}$ (ps) | $50.9 \pm 1.0$ | $94.9 \pm 0.5$ | $195.0 \pm 1.8$ | $292.9 \pm 2.4$ | $388.9 \pm 4.4$ | $485.5 \pm 6.0$ |
| | $f_{PJ}$ (MHz) | $4.39 \pm 0.02$ | | | | | |
| | | Fixed amount of the PJ at 4.37 MHz with various amplitudes of the RJ | | | | | |
| | $A_{PJ} = 50$ ps | $50.9 \pm 1.0$ | $48.2 \pm 0.8$ | $46.4 \pm 1.9$ | $64.2 \pm 8.3$ | $63.9 \pm 9.2$ | $69.8 \pm 10.3$ |
| | $f_{PJ}$ (MHz) | $4.39 \pm 0.02$ | $4.39 \pm 0.02$ | $4.39 \pm 0.02$ | $4.41 \pm 0.01$ | $4.0 \pm 0.4$ | $5.6 \pm 1.1$ |
| | $\sigma_{RJ}$ (ps) | $72.5 \pm 1.5$ | $119.7 \pm 2.7$ | $216.6 \pm 4.0$ | $311.1 \pm 2.5$ | $408.2 \pm 3.6$ | $493.9 \pm 6.2$ |
| | Large amounts of RJ and PJ | Computed $\sigma_{RJ} = (645 \pm 16.1)$ ps, $\sigma_{RJ} = (507.5 \pm 6.1)$ ps, and $f_{PJ} = (4.39 \pm 0.02)$ MHz when injecting $\sigma_{RJ} = 500$ ps, $A_{PJ} = 500$ ps, and $f_{PJ} = 4.37$ MHz, respectively. | | | | | |

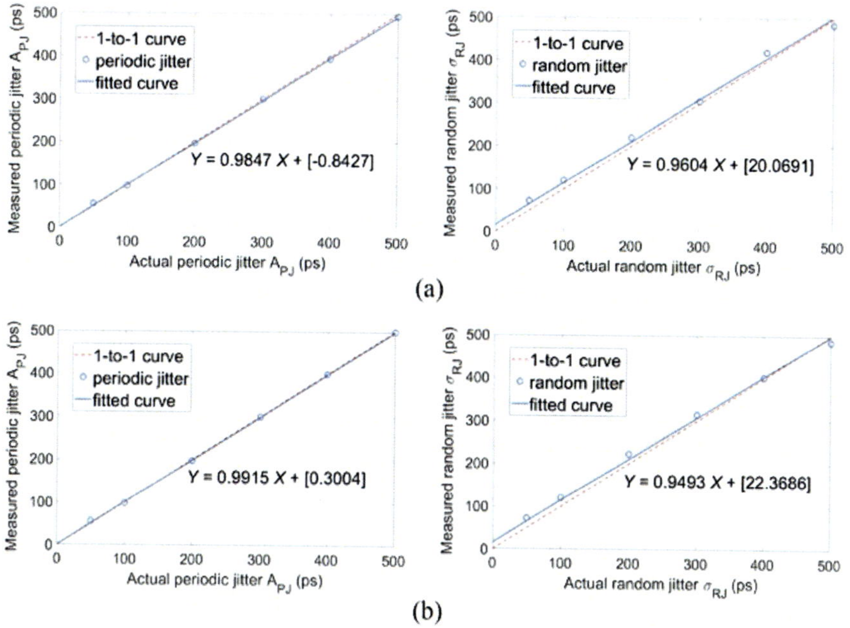


**Fig. 6.15** Comparison of the actual and measured (left) PJ with fixed RJ of $\sigma_{RJ} = 50$ ps and (right) the RJ with fixed PJ of $A_{PJ} = 50$ ps at 4.37 MHz, for the rectangular signal with the modulation frequencies: (**a**) 10 MHz and (**b**) 50 MHz, separately, for setup denoted as SET-*Both*

with relatively less fixed RJ ($\sigma_{RJ} = 50$ ps). The frequencies of the periodic jitter for the setups SET-*PJ* and SET-*Both* are given almost similar amounts except when the presence of the maximum amount of RJ with a fixed amount of PJ is given $f_{PJ} = 5.6 \pm 1.1$ (refer the last column and third row from the bottom of Table 6.4). However, either for fixed RJ or fixed PJ (50 ps each), the calculated amounts of the RJ and PJ are larger than 50 ps each, respectively (i.e., impossible), when the amount of the other types of jitter increases (from 300 ps each). If both PJ and RJ are relatively large, then the measured RJ is given a larger value than the injected amount, while the measured PJ is relatively close to the actual amount.

## 6.4 Conclusion

The simulation results show reasonable accuracy in the jitter measurements. As examples, when we inject only a RJ of RMS values, 5 and 50 ps, our approach was given the values of (4.5 ± 0.3) ps and (45.2 ± 2.4) ps, respectively. Furthermore, if we inject only the PJ of 5 and 50 ps, separately, at the frequency 4.37 MHz, the calculated parameters were given as (4.93 ± 0.04) ps and (49.3 ± 0.5) ps, respectively, at the frequency of (4.40 ± 0.04) MHz. If RJ is a large fraction of the total jitter, then it can be difficult to separate the PJ accurately, on the simulation data. Now, we believe that the proposed methodology can apply for jitter extraction in real-time applications where we have no access to a reference clock signal.

## 6.5 Chapter Remarks

Before applying the proposed algorithm with the experiments on ToF ranging cameras, we first tested it on the generated simulation data. Only the random and periodic jitter in the simulation data are considered since these are the two available types of jitter in the illumination source of ToF range imaging cameras that will be used in the next chapter. The random jitter (of known amount), periodic jitter (with known frequency and amount), and both together are separately injected to various modulation frequencies of sinusoidal signals. The approach is validated by several experiments with numerous values for the parameters of the signal and jitter. The significant accurate results are found for the proposed algorithm on generated simulation data. So, we have assured our proposed methodology for jitter measurement, and now it is needed to experiment with the proposed methodology using AMCW ToF range imaging cameras and this will be described in the next chapter.

## Reference

1. Fredric J Harris. On the use of windows for harmonic analysis with the discrete Fourier transform. *Proceedings of the IEEE*, 66 (1): 51–83, 1978.

# Chapter 7
# Jitter Extraction in ToF Cameras

**Abstract** This chapter verifies the proposed methodology in Chap. 5 by experimenting with commercial applications. First, the sub-steps of the methodology proposed in Sect. 5.2 are summarized in point form as a flowchart. Then, the experimental setup using two AMCW depth cameras is separately investigated. For each depth camera, the relevant results with discussion are also included.

**Keywords** SwissRanger 4000 · DepthSense 325 · Modulation frequency · Photodetector · Real-time digital oscilloscope · Bandwidth · Sampling rate · Signal generator · Light source signal · Peak-to-peak analysis · Connections setup · Components setup · Least-squared linear regression · Linear trend · Flicker noise

## 7.1 Introduction

Figure 7.1 is a summary of the details of the sub-steps in addition to the main steps of the methodology proposed in Sect. 5.2. We measure the jitter in modulated light sources of two AMCW ToF range imaging cameras, namely, the MESA Imaging SwissRanger 4000 [1] and SoftKinetic DepthSense 325 [2], separately. Since the emitted light source does not carry a data stream, the light signal is decomposed into random jitter (RJ) and periodic jitter (PJ). Other types of jitter (data dependent and bounded uncorrelated jitter) are only relevant to communication channels that transit data.

The light signal of each depth camera is captured by using a silicon biased photodetector, DET025A [3], and is measured by a medium speed real-time digital oscilloscope, namely, the Keysight Infiniium S-Series DSOS604A High Definition [4] with 6 GHz bandwidth and 20 GSa/s sampling rate. The number of cycles in the signal will be large when the oscilloscope has a large waveform length, refer (5.11). Now, let us consider the experimentation on the two cameras, separately.


G. Anthonys, *Timing Jitter in Time-of-Flight Range Imaging Cameras*,
https://doi.org/10.1007/978-3-030-94159-8_7

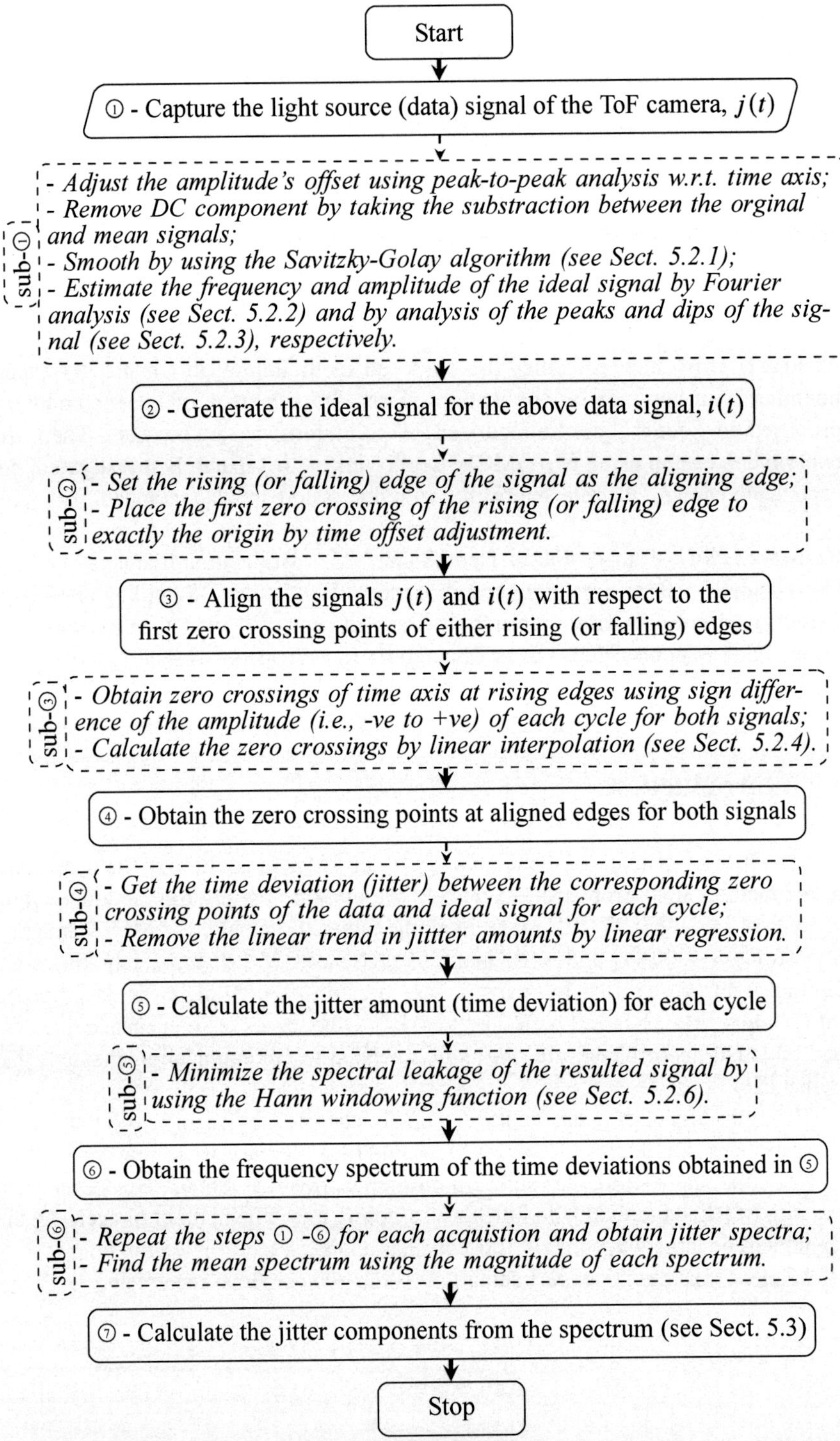


**Fig. 7.1** Summary of the sub-steps with main steps of the methodology proposed in Sect. 5.2 for jitter extraction in ToF range imaging cameras

## 7.2 MESA Imaging SwissRanger 4000

The MESA Imaging SwissRanger 4000 (SR4000) uses infrared LEDs for its illumination source (see Sect. 2.5.1). The camera's default modulation frequency (i.e., 30 MHz) with the frame rates of 50 fps is chosen for experimentation.

### *7.2.1 Experimental Setup for SR4000*

The light source signal of the SR4000 depth camera is captured by using a silicon biased photodetector, DET025A. Two coaxial cables labeled as *Medium* (1.070 ± 0.005 m) and *Long* (1.790 ± 0.005 m) that are in different lengths measured from tip to tip of the Bayonet Neill-Concelman (BNC) connectors are used for each experiment. The components are connected via relevant connectors: SubMiniature version A (SMA) and version B (SMB) with coaxial cables. For the control a 30 MHz pure sine wave (corresponding to the modulation frequency of SR4000) was generated with the HP Agilent 8648B [5] signal generator and measured with the oscilloscope. The specification of the components used is shown in Table 7.1.

The experimental setup is illustrated in Fig. 7.2 and the equipment parameters are tabulated in Table 7.2. The components and connections used for the experiment are included in Fig. 7.2. For the representation, relevant experiment of SR4000 and signal generator are denoted as EXP. SR-{...} and EXP. SG30-{...}, respectively, with {...} representing the corresponding cable as either *Medium* or *Long*.

**Table 7.1** Specification of the components used with SwissRanger 4000

| Component | Specification used | | |
|---|---|---|---|
| ToF camera | Modulation frequency ($f_m$) | Frame rates | Illumination source |
| SwissRanger 4000 | 30 MHz | 50 fps | Infrared LEDs, 850 nm |
| Light detector | Detection type | Bandwidth | Spectral range |
| Thorlabs DET025A | Silicon biased | 2 GHz | 400—1100 nm |
| Oscilloscope | Bandwidth | Sampling frequency ($F_s$) | Number of samples |
| S-Series DSOS604A | 6 GHz | 20 GSa/s | $N_s \approx 10^7$ |
| Signal generator | Sinusoidal signal of the frequency | | |
| HP Agilent 8648 | 30 MHz | | |
| Connectors | Bayonet Neill-Concelman (BNC); SubMiniature version A (SMA) and version B (SMB) | | |
| Cable | | Length in meters | |
| Coaxial (length from tip to tip of the BNC connectors) | | *Medium (M)* | *Long (L)* |
| | | 1.070 ± 0.005 | 1.790 ± 0.005 |

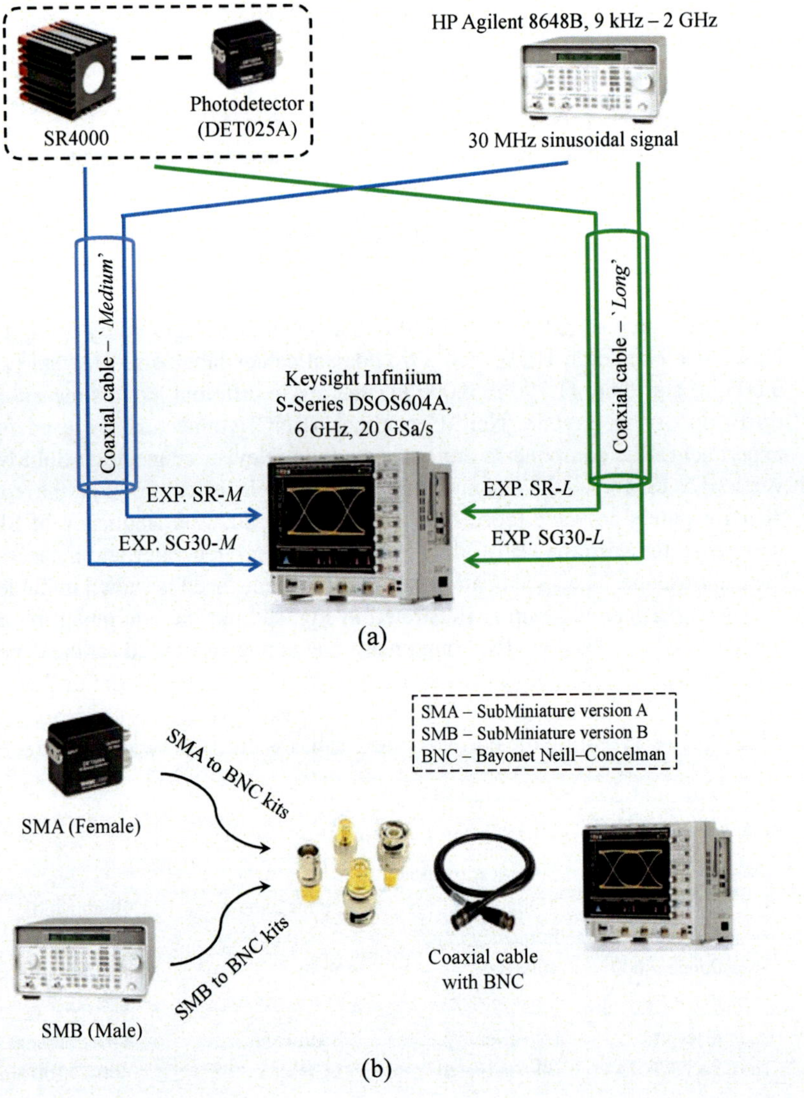


**Fig. 7.2** (**a**) Components setup and (**b**) connections setup, for the experiment with the MESA Imaging SwissRanger 4000

## *7.2.2 Results and Discussion*

For step ① of the proposed algorithm, the data signal from the light source of the camera and the 30 MHz sinusoidal signal from the signal generator were measured for each experimental setup separately. As an example, Fig. 7.3 shows the corresponding results for the SR4000 and signal generator with the *Long* coaxial

**Table 7.2** Parameters of the experimentation with SwissRanger 4000

| Device | Coaxial cable | No. of cycles $N^{cal}_{\Delta t}$ | No. of cycles $N_{\Delta t}$ | | EXP. # |
|---|---|---|---|---|---|
| MESA SR4000 with photodetector | *Medium* | 15000 | 14998 | SR-SG30 | EXP. SR-*M* |
| Signal Generator (30 MHz *sine* signal) | *Medium* | 15000 | 14998 | SR-SG30 | EXP. SG30-*M* |
| MESA SR4000 with photodetector | *Long* | 15000 | 14998 | SR-SG30 | EXP. SR-*L* |
| Signal Generator (30 MHz *sine* signal) | *Long* | 15000 | 14998 | SR-SG30 | EXP. SG30-*L* |
| Each experiment repeat 100 acquisitions | | | | | |

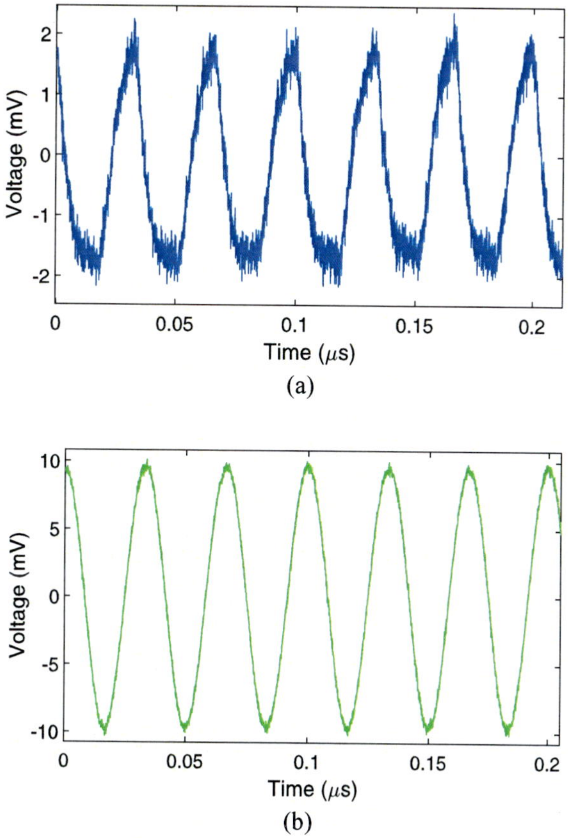


**Fig. 7.3** Data signals as for step ① in the proposed methodology in Fig. 7.1 for a single acquisition from (**a**) the light source signal of SR4000 and (**b**) signal generator (30 MHz signal), for the experiments denoted as EXP. SR-*L* and EXP. SG30-*L* in Table 7.2, respectively

cable. By comparison, it is clear that the noise in the light source signal of SR4000 is relatively higher than that from the signal generator.

Before step ②, the DC component of each signal is removed by simply taking the subtraction of the original signal from its mean. Second, each signal is smoothed by the Savitzky–Golay filtering method (with third order polynomial and frame length $m = 55$ samples in (5.3)) with minimal impact on the zero crossings at the time axis

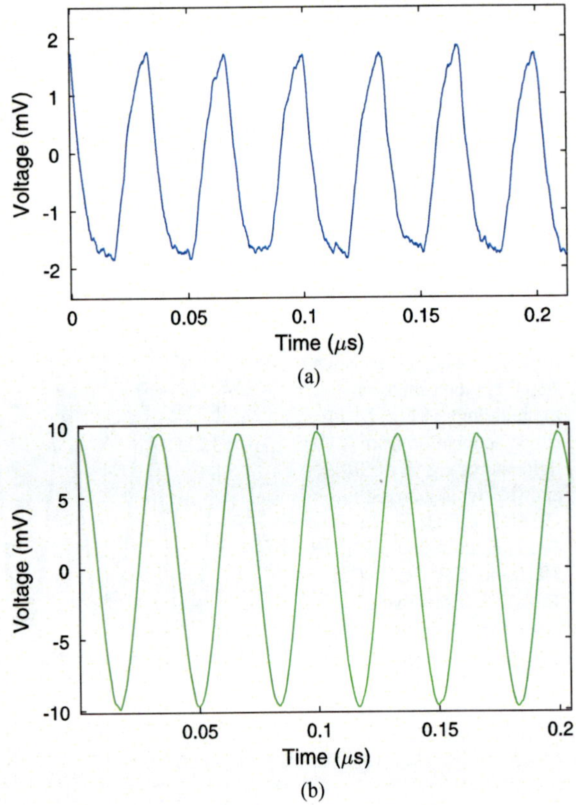


**Fig. 7.4** Smoothed data signals (using Savitzky–Golay algorithm with third order polynomial and the length of 55 samples) for the same acquisition from (**a**) the light source of SR4000 and (**b**) signal generator, for the experiments denoted as EXP. SR-*L* and EXP. SG30-*L* in Table 7.2, respectively

(see sub-①). These parameters are chosen by trial and repetition method with visual inspection. The resultant signals are shown in Fig. 7.4. The corresponding reference (ideal) signal for each setup is generated by using the Fourier analysis as step ②. For that, the relevant frequency and the amplitude for the ideal signal are found as described in Sects. 5.2.2 and 5.2.3, respectively.

Next, the starting point of the signal at either rising or falling edge is set by removing the first full or partial cycles from the smoothed signal. We chose the rising edge as the aligning edge for all experiments in this chapter. Then, in order to align the smoothed signal with respect to the first zero crossing, the rising edge of the signal is placed exactly at the origin. For that, we use the corresponding time offset and linear interpolation technique. These sub-steps are illustrated in Fig. 7.5 for the SR4000 (EXP. SR-*L*) and the signal generator (EXP. SG30-*L*).

For step ③, the data and ideal signals are aligned with respect to the first zero crossing point of rising edges as shown in Fig. 7.6. In addition, Fig. 7.7 shows the corresponding zoomed-in portion around the origin of the aligned signals shown in Fig. 7.6.

Next, as sub-③, the zero crossings of the time axis at rising edges are obtained by analyzing the sign difference of the amplitude (i.e., -*ve* to +*ve*) of each cycle for

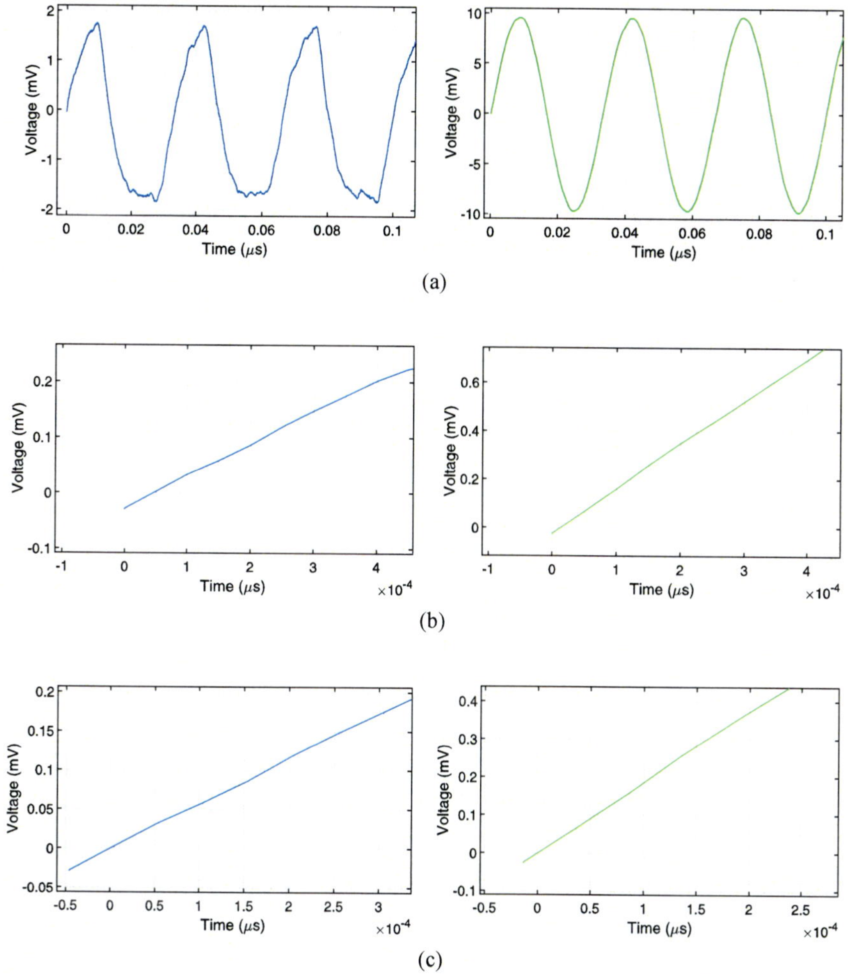


**Fig. 7.5** Sub-steps (as sub-②) for the same acquisition shown in Fig. 7.4 from (left) SR4000 and (right) signal generator, for the experiments denoted as EXP. SR-*L* and EXP. SG30-*L* in Table 7.2, respectively: (**a**) set the rising edge of smoothed signal as aligning edge, (**b**) zoom-in the portion around the origin of the aligned edge, and (**c**) set the rising edge exactly at the origin

both signals. By using linear interpolation (see Sect. 5.2.4), the zero crossing points at rising edges are estimated as for step ④. Then, the time deviation between the corresponding zero crossings of the data and ideal signals for each case is obtained as for step ⑤, as the jitter amount for each cycle of the signal. However, some of the resulted signals present with a linear trend because the ideal signal frequency is not exactly estimated. A *least-squared* linear regression [6] is performed and subtracted off to detrend the signal as shown in Fig. 7.8.

**Fig. 7.6** Aligned data and ideal signals (w.r.t. first zero crossing at rising edge) as step ③ shown in Fig. 7.1 for the same acquisition from (**a**) SR4000 and (**b**) signal generator, for the experiments denoted as EXP. SR-*L* and EXP. SG30-*L* in Table 7.2, respectively

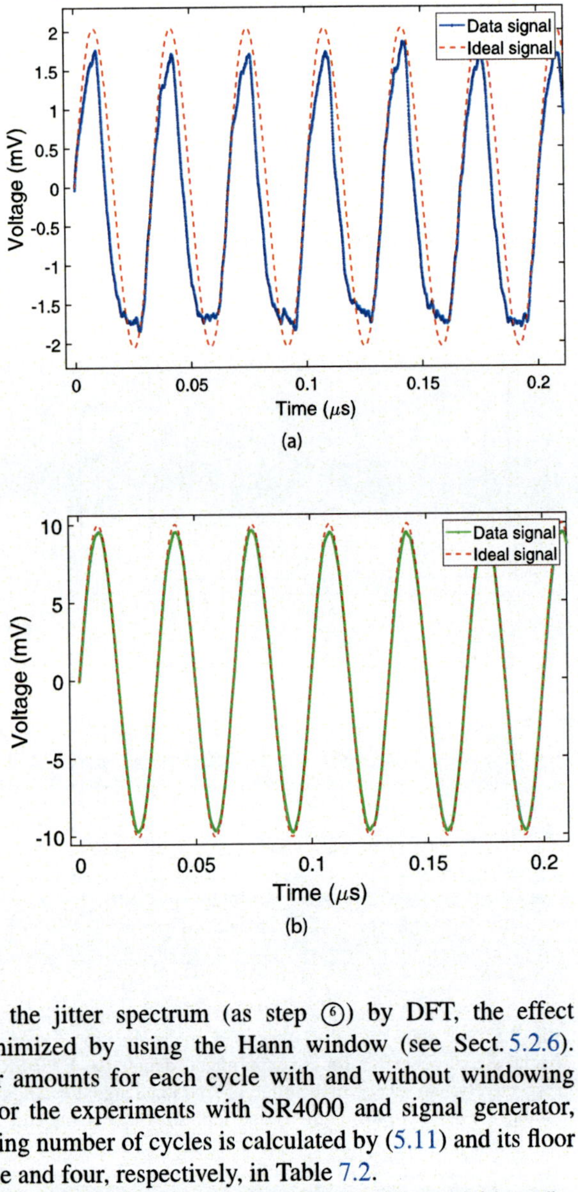


Next, before obtaining the jitter spectrum (as step ⑥) by DFT, the effect of spectral leakage is minimized by using the Hann window (see Sect. 5.2.6). Figure 7.9 shows the jitter amounts for each cycle with and without windowing for the same acquisition for the experiments with SR4000 and signal generator, separately. The corresponding number of cycles is calculated by (5.11) and its floor is included in columns three and four, respectively, in Table 7.2.

In order to reduce the uncertainty of the result, we captured 100 acquisitions (by automating, see Sect. C.2 in Appendix C) for each experimental setup and obtained the jitter spectrum for each acquisition by using the steps ① to ⑥ in the proposed algorithm as sub-⑥. Then, the total jitter spectrum is formed by taking the mean spectrum among the multiple jitter spectra for each experiment. The jitter spectra for SR4000 and signal generator with the experiments using *Medium* and *Long* coaxial cables are illustrated in Figs. 7.10 and 7.11, respectively. In addition, jitter

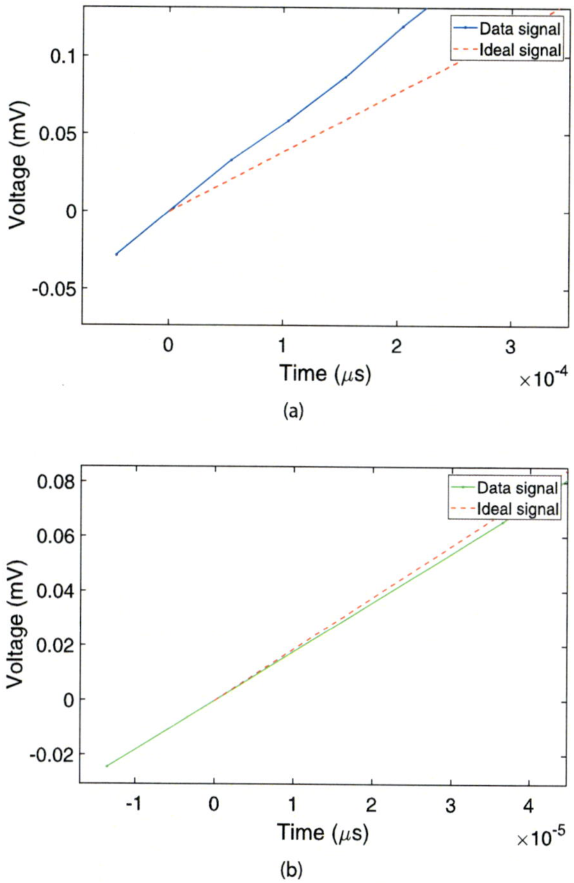


**Fig. 7.7** Zoomed-in portion around the origin of the aligned signals shown in Fig. 7.6 for **(a)** SR4000 and **(b)** signal generator, for the experiments denoted as EXP. SR-*L* and EXP. SG30-*L* in Table 7.2, respectively

components and the interpolated level between periodic and random jitter for each signal for both cases are marked in the figures.

By comparing two jitter signals (see Figs. 7.10 and 7.11) in each corresponding experiment, it is clear that there are two new high peaks at around 0 and 5 MHz on the jitter signal of SR4000 (i.e., both EXP. SR-*M* and EXP. SR-*L*) that are not present in the jitter signal of the signal generator (i.e., both EXP. SG30-*M* and EXP. SG30-*L*). There is also a broad and flat spectrum being the random jitter at the levels of approximately 1 and 0.3 ps for SR4000 and signal generator, respectively, in each case, while a harmonic of the first PJ at around 0.12 MHz (see zoomed portions in Figs. 7.10 and 7.11).

By using spectra of the total jitter (see Figs. 7.10 and 7.11), it is straightforward to obtain the corresponding frequencies of the periodic jitter $f_{\mathrm{PJ}}^{(p)}$ in SR4000. The same value is given 0.064 MHz for the first frequency of PJ in both cases, while they are bit different as 5.001 and 5.173 MHz for the second frequency of the PJ. Other jitter parameters are calculated by considering the energy of the spectrum as

**Fig. 7.8** Jitter amounts (**a**) with linear trend and (**b**) without trend by using linear regression, for a particular acquisition of SR4000

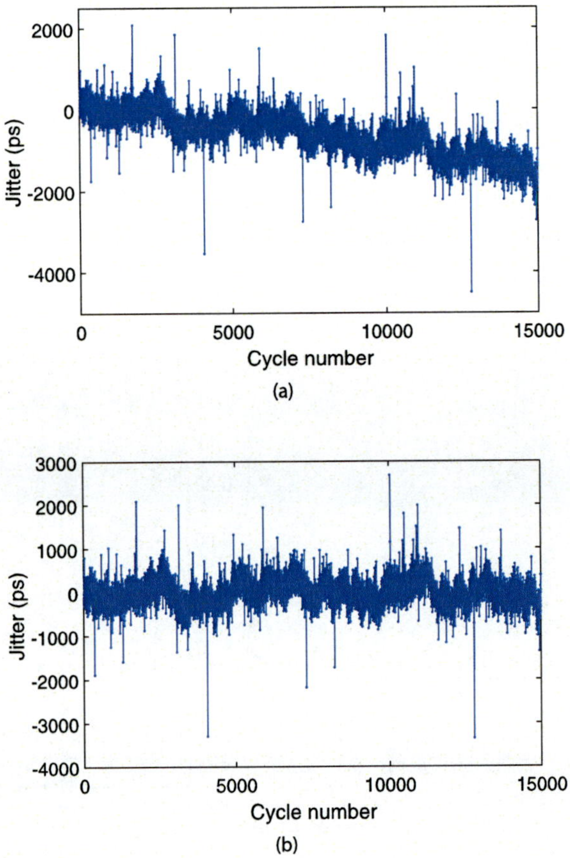


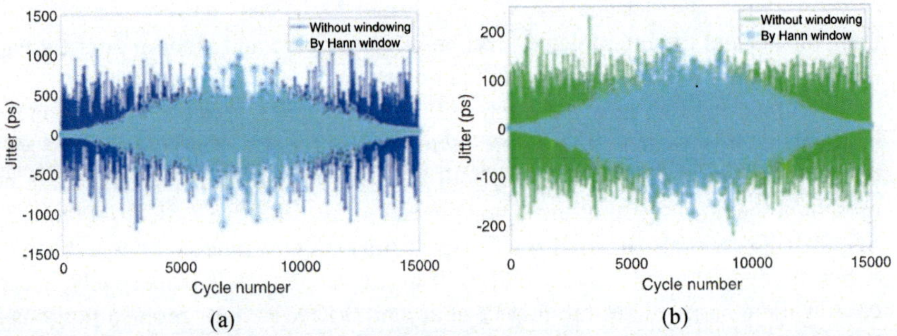


**Fig. 7.9** As step ⑤, jitter amount for each cycle (time deviation) with and without applying the Hann window for the same acquisition of (**a**) SR4000 and (**b**) signal generator, for the experiment denoted as EXP. SR-*L* and EXP. SG30-*L*, respectively

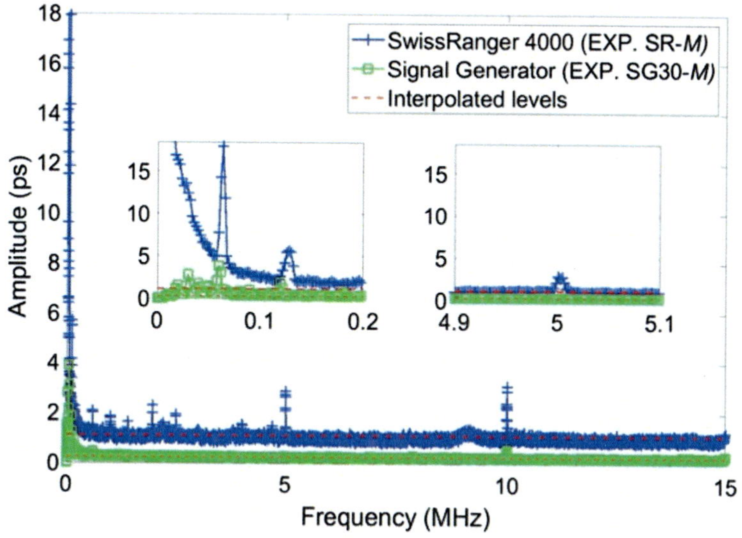


**Fig. 7.10** Full spectrum with its expansion portions at the two frequencies of PJ at the range 0–0.2 MHz and 4.9–5.1 MHz, for the SR4000 and signal generator for the experiment with *Medium* cable, denoted as EXP. SR-*M* and EXP. SG30-*M*

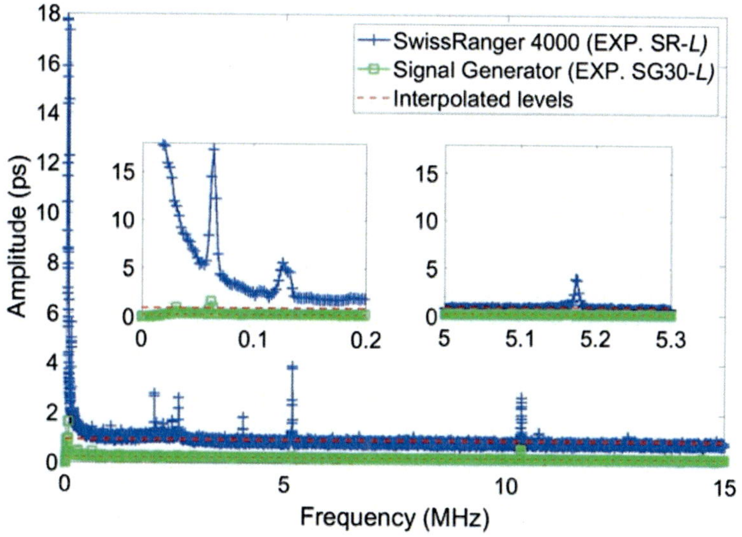


**Fig. 7.11** Full spectrum with its expansion portions at the two frequencies of PJ at the range 0–0.2 MHz and 5.0–5.3 MHz, for the SR4000 and signal generator for the experiment with *Long* cable, denoted as EXP. SR-*L* and EXP. SG30-*L*

**Table 7.3** Jitter parameters from the experiments with SwissRanger 4000

| EXP. # | Reference signal's frequency (MHz) | Periodic jitter $[f_{\text{PJ}}^{(p)}$ (MHz), $A_{\text{PJ}}^{(p)}$ (ps)] | Random jitter $\sigma_{\text{RJ}}$ (ps) |
|---|---|---|---|
| SR-*M* | 29.999645 ± 0.000003 | [0.064, 71] & [5.001, 12] | 173.6 ± 0.1 |
| SG30-*M* | 30.00001 ± 0.00006 | nil | 36.9 ± 0.1 |
| SR-*L* | 29.999668 ± 0.000003 | [0.064, 72] & [5.173, 16] | 145.7 ± 0.1 |
| SG30-*L* | 30.00001 ± 0.00006 | nil | 34.3 ± 0.1 |

**Table 7.4** Jitter parameters for the same experiments shown in Table 7.2 with a low speed oscilloscope HP Infiniium 54846B

| EXP. # | Reference signal's frequency (MHz) | Periodic jitter $[f_{\text{PJ}}^{(p)}$ (MHz), $A_{\text{PJ}}^{(p)}$ (ps)] | Random jitter $\sigma_{\text{RJ}}$ (ps) |
|---|---|---|---|
| SR-*M* | 30.00564 ± 0.00004 | [0.12, 61] & [5.04, 34] | 159.9 ± 0.3 |
| SG30-*M* | 30.00554 ± 0.00003 | nil | 35.4 ± 0.1 |
| SR-*L* | 30.00523 ± 0.00004 | [0.12, 103] & [5.04, 30] | 167.7 ± 0.5 |
| SG30-*L* | 30.00565 ± 0.00003 | nil | 33.7 ± 0.2 |

described in Sect. 5.3.1. The parameters, two amplitudes of periodic jitter $A_{\text{PJ}}^{(p)}$ and the RMS of random jitter $\sigma_{\text{RJ}}$, are calculated by using (5.31) and (5.33), respectively. The calculated jitter parameters are tabulated in Table 7.3. The uncertainties are computed with the formulas shown in Sect. E.2 in Appendix E.

The tabulated results are obtained by considering the full spectrum of the jitter signal for each relevant experiment. The corresponding frequency of the ideal signal for each experiment (a pure sine wave without jitter, extracted from the data) is included in column two. The standard error of the frequencies is also provided as a measure of precision. The SR4000 camera and signal generator frequencies are within one standard deviation of each other. The standard deviations are small compared to the ideal signal, with precision to six and seven significant digits for the SR4000 light source and signal generator, respectively. Therefore, we have measured the frequency to sufficient precision for our purposes. As usual, there is no detectable periodic jitter in the signal generator 30 MHz signal in both cases.

The lower frequency of the PJ is 0.064 MHz with an amplitude of (71.5 ± 0.3) ps, while the higher frequency is (5.09 ± 0.06) MHz with an amplitude of (14.0 ± 1.4) ps. The RMS of the RJ in SR4000 are (173.6 ± 0.1) and (145.7 ± 0.1) ps, while for the signal generator, they are (36.9 ± 0.1) and (34.3 ± 0.1) ps for the experiments with *Medium* and *Long* coaxial cables, respectively.

In addition, Table 7.4 shows the experimental results for the same setup that is measured by using a low speed digital oscilloscope (HP Infiniium 54846B [7] with 2.25 GHz bandwidth, 8 GSa/s sampling rate, and the maximum number of samples of $2^{16}$, i.e., waveform length) for the same setup shown (with 1000 acquisitions) in the last column of Table 7.2. We found the random jitter of (35.4 ± 0.1) and (33.7 ± 0.2) ps for the signal generator experimented with *Medium* and *Long* coaxial cables, respectively. This is strong evidence that the calculated random jitter corresponds to the signal generator itself only, not with including the oscilloscope.

On the other hand, it is found that for the SR4000 (see third column of Table 7.4), the frequencies of periodic jitter were at 0.12 and 5.04 MHz with amplitudes of (82.0 ± 14.8) and (32.0 ± 1.4) ps, respectively. It was unable to locate the frequency of 0.064 MHz from this low speed oscilloscope. The bin width of this low speed oscilloscope (8 GSa/s $\div$ $2^{16}$ = 122.11 kHz) is relatively larger than the medium speed oscilloscope (20 GSa/s $\div$ $10^{7}$ = 2 kHz); thus the low speed oscilloscope cannot resolve a low frequency periodic jitter. The intrinsic RMS jitter of the medium speed and low speed oscilloscopes are given as 100 fs [4] and 8 ps ± 0.005% (delay setting) [7], respectively, which are relatively very small compared to the measured jitter in the SR4000 camera.

Since the jitter amplitudes at lower frequencies are very large, it is better to plot these figures in logarithmic scales. We analyze the data with curve fitting (see Sect. 5.3.2) and obtain the corresponding log–log graphs for the total jitter. Fig. 7.12 shows the log–log plots for the jitter spectra for the SR4000 and signal generator for both experimental setups (corresponding to Figs. 7.10 and 7.11). Note that the

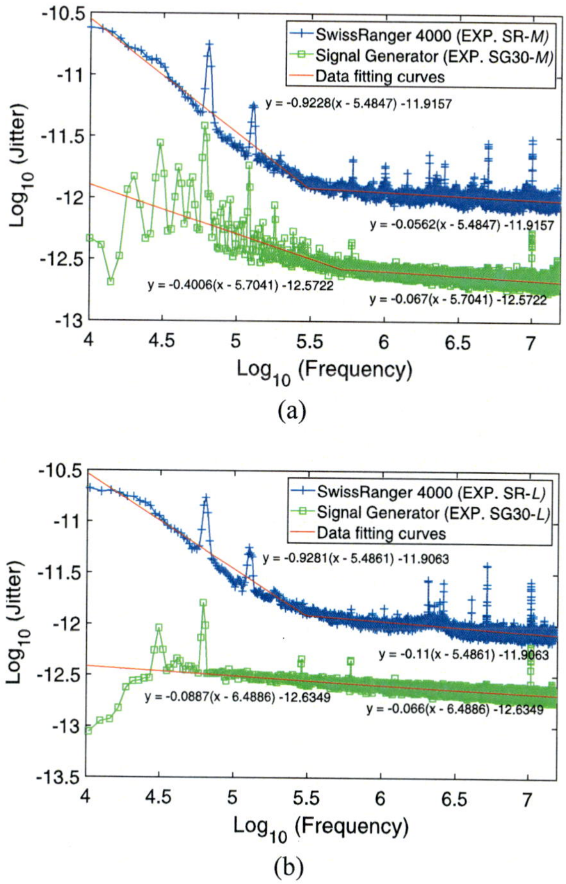


**Fig. 7.12** Log–log scale (base 10) with fitted curves for spectra shown in Figs. 7.10 and 7.11 for the SR4000 and signal generator for the experiments with **(a)** *Medium* and **(b)** *Long* cables, denoted as SET. SR -*L* and SET. SG30-*L* in Table 7.2

first few points from the data are not plotted because they are extremely large. In addition, the fitted curves (see Sect. 5.3.2) with their characteristic equations for each case are stated in the figures.

The slopes at low frequencies of the total jitter for the SR4000 are higher than at the rest of the frequencies (see Figs. 7.10 and 7.11) in each case. At the low frequencies of SR4000, the slope is $-0.94 \pm 0.02$, while it is $-0.3 \pm 0.3$ for the signal generator. This may be due to the flicker noise ($f^{-0.94\pm0.02} \approx 1/f$) in the camera. This trend is approximately similar in both cases for the SR4000 camera. In addition, there is a relatively larger slope at low frequencies for the *Medium* cable than *Long* cable for the signal generator (compare (a) and (b) of Fig. 7.12). It is believed that these are due to cables and connectors.

## 7.3 SoftKinetic DepthSense 325

The illumination signal of the SoftKinetic DepthSense 325 (DS325) is captured by using the same photodetector, DET025A. The DS325 uses eye-safe near infrared laser light for its illumination source. Here we use its fixed modulation frequency (i.e., 50 MHz) with the frame rates of 25 fps throughout the experiments. The illumination source of this camera produces approximately square waves and all signals are measured with the same oscilloscope.

### *7.3.1 Experimental Setup for DS325*

The same experimental setup used in Sect. 7.2.1 is used here, but additionally, a third coaxial cable that is shorter than the previous two cables was also used. Therefore, the three coaxial cables labeled as *Short, Medium*, and *Long* where length measured from tip to tip of the BNC connectors are listed in Table 7.5. For the control, the 50 MHz pure sine waves with the HP Agilent 8648B are generated and measured with the DSOS604A oscilloscope. The specification of the components used is presented in Table 7.5.

The experimental setup and its relevant parameters are shown in Fig. 7.13 and Table 7.6, respectively. In order to verify our proposed algorithm for general cases, both square and sine waves are generated as for the ideal signals of the camera and both cases are analyzed. These are denoted as EXP. $DS_{sq}$-$\{\dots\}$ and EXP. $DS_{sn}$-$\{\dots\}$, in the figure and table.

**Table 7.5** Specification of the components used with DepthSense 325

| Component | Specification used | | |
|---|---|---|---|
| ToF camera | Modulation frequency ($f_m$) | Frame rates | Illumination source |
| DepthSense 325 | 50 MHz | 25 fps | Infrared laser |
| Light detector | Detection type | Bandwidth | Spectral range |
| Thorlabs DET025A | Silicon biased | 2 GHz | 400 – 1100 nm |
| Oscilloscope | Bandwidth | Sampling frequency ($F_s$) | Number of samples |
| S-Series DSOS604A | 6 GHz | 20 GSa/s | $N_s \approx 5 \times 10^6$ |
| Signal generator | Sinusoidal signal of the frequency | | |
| HP Agilent 8648 | 50 MHz | | |
| Cable | Length (m), from tip to tip of the BNC connectors | | |
| | *Short (S)* | *Medium (M)* | *Long (L)* |
| Coaxial | 0.490 ± 0.005 | 1.070 ± 0.005 | 1.790 ± 0.005 |

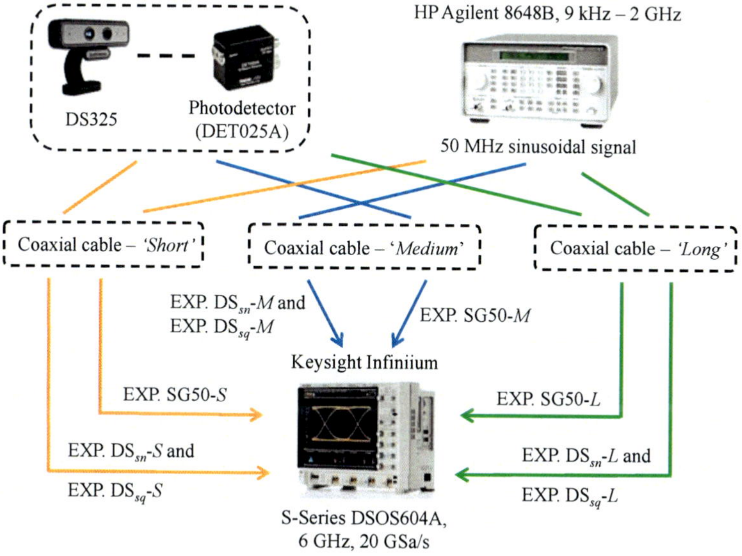


**Fig. 7.13** Experimental setup with SoftKinetic DepthSense 325

## *7.3.2 Results and Discussion*

As before, we capture the data signal from the light source of the camera and directly from the signal generator, for each experimental setup. Figure 7.14 shows the square wave from the DS325 and a sine wave from the signal generator for a single acquisition of the experiments with *Long* coaxial cable, separately. These sub-figures are adjusted by the offset amplitude. The noise in the DS325 is also relatively higher than the signal generator.

**Table 7.6** Parameters of the experimentation with DepthSense 325

| Device | Type of the ideal signal | Coaxial cable | Number of cycles ($N_{\Delta t}$) | EXP. # | |
|---|---|---|---|---|---|
| DS325 with photodetector | *sine* | *Short* | 12,325 | DS325 | EXP. DS$_{sn}$-*S* |
| | *square* | | | | EXP. DS$_{sq}$-*S* |
| | *sine* | *Medium* | | | EXP. DS$_{sn}$-*M* |
| | *square* | | | | EXP. DS$_{sq}$-*M* |
| | *sine* | *Long* | | | EXP. DS$_{sn}$-*L* |
| | *square* | | | | EXP. DS$_{sq}$-*L* |
| Signal generator HP Agilent 8648 (50 MHz *sine* signal) | | *Short* | | SG50 | EXP. SG50-*S* |
| | | *Medium* | | | EXP. SG50-*M* |
| | | *Long* | | | EXP. SG50-*L* |

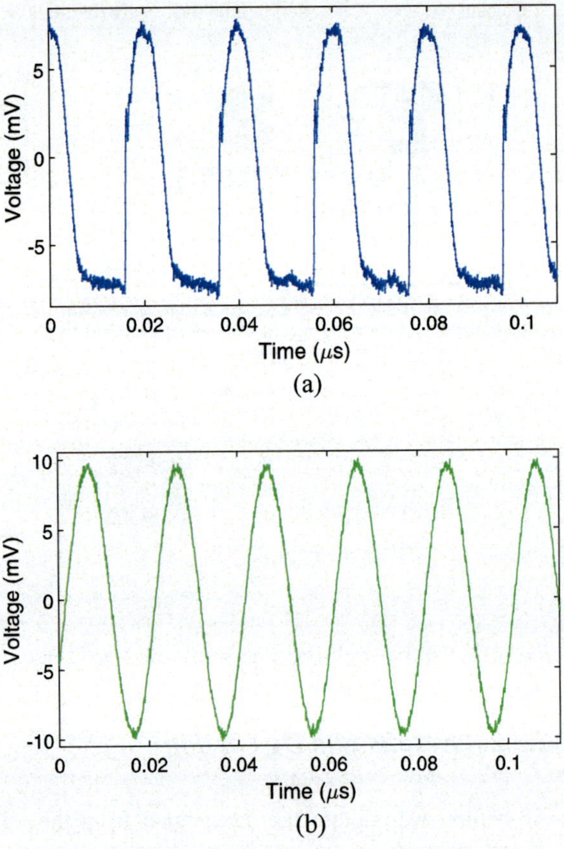


**Fig. 7.14** Data signals as for step ① in the proposed algorithm in Fig. 7.1 for a single acquisition from **(a)** the light source signal of DS325 and **(b)** 50 MHz signal from signal generator, for the experiments denoted as EXP. DS$_{sq}$-*L* and EXP. SG50-*L* in Table 7.6, respectively

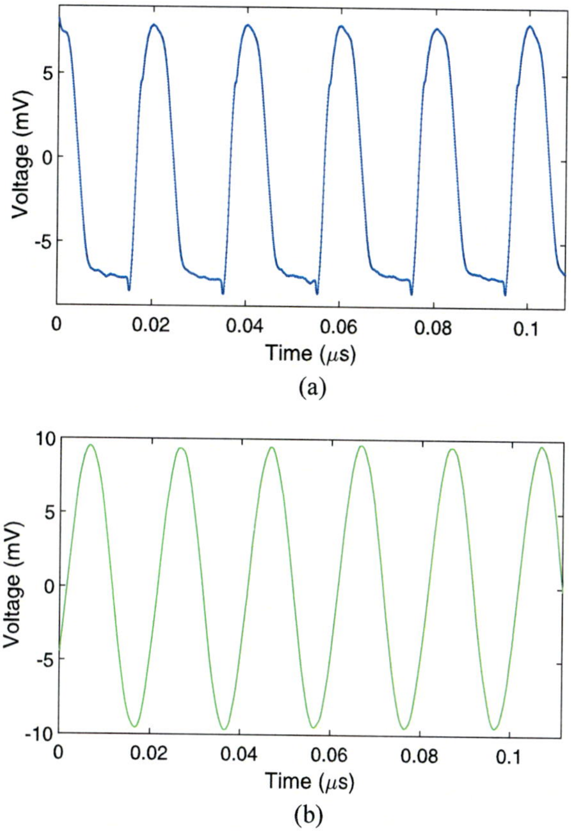


**Fig. 7.15** Smoothed data signals (using Savitzky–Golay algorithm with third order polynomial and the length of 55 samples) for the same acquisition from (**a**) the light source of DS325 and (**b**) signal generator, for the experiments denoted as EXP. $DS_{sq}$-$L$ and EXP. SG50-$L$ in Table 7.6, respectively

To generate the ideal signal, the procedure sub-① is used as shown in Fig. 7.1 including the smoothing of the signal by Savitzky–Golay filtering method (with third order polynomial and frame length $m = 55$ samples in (5.3)). The resultant signal is shown in Fig. 7.15 for both the camera and the signal generator. The reference signal (ideal) for each setup is generated by using the Fourier analysis as step ②.

Next, the first couple of full or partial cycles from the smoothed signal are removed in order to start the signal at rising edge. Then, the first zero crossing at rising edge of the signal is placed exactly at the origin by using the sub-steps denoted as sub-② and the results are illustrated in Fig. 7.16 for DS325 (SET. $DS_{sq}$-$L$) and signal generator (EXP. SG50-$L$), separately.

For step ③, the data and ideal signals are aligned with respect to the first zero crossing point of rising edges as shown in Fig. 7.17, for each case. In addition, when we consider a sine wave ideal signal for the camera (denoted as SET. $DS_{sn}$-{...}), we aligned the data and ideal signals as shown in Fig. 7.18. Next, as sub-③, the zero crossings of the time axis at rising edges are obtained by analysis of the sign difference of the magnitude (i.e., $-ve$ to $+ve$) of each cycle for both signals. By

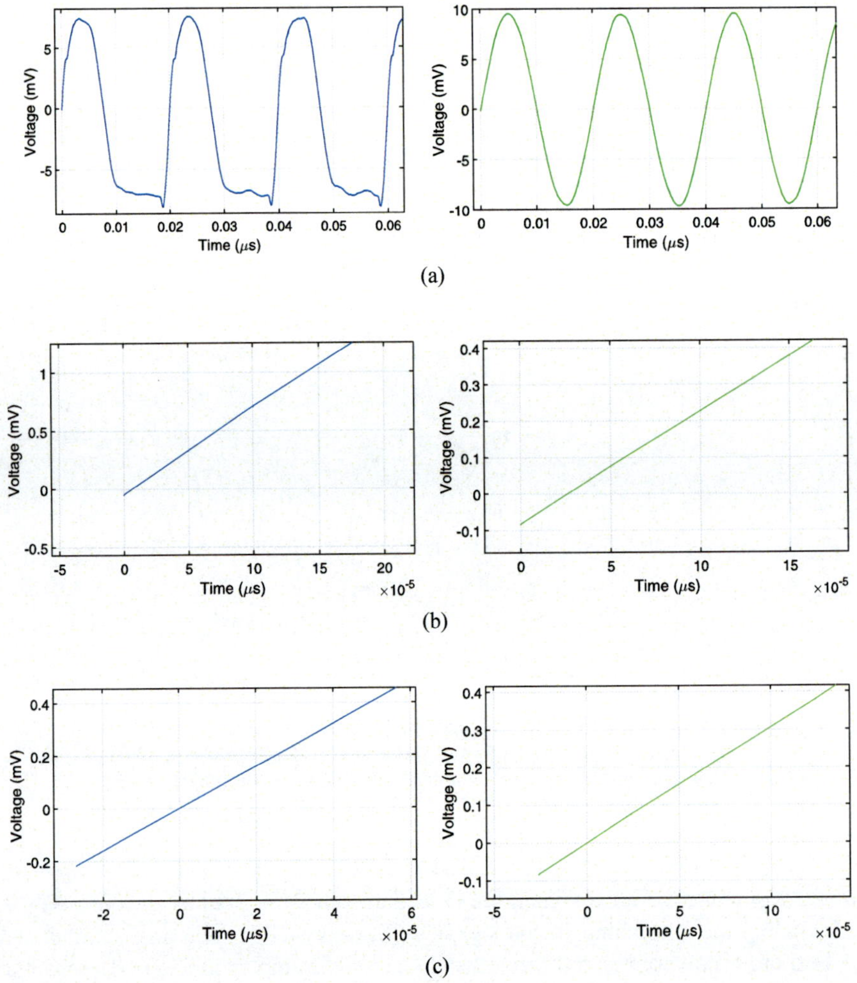


**Fig. 7.16** Sub-steps for the same acquisition shown in Fig. 7.15 from (left) DS325 and (right) signal generator, for the experiments denoted as EXP. $DS_{sq}$-$L$ and EXP. SG50-$L$ in Table 7.6, respectively: (**a**) set the rising edge of smoothed signal as aligning edge, (**b**) zoom-in the portion around the origin of the aligned edge, and (**c**) set the rising edge exactly at the origin

using linear interpolation (see Sect. 5.2.4), the zero crossing points at rising edges are estimated as for step ④. Next, the time deviation between the corresponding zero crossings of the data and ideal signals for each case is obtained as for step ⑤. This gives the jitter signal.

The effect of spectral leakage is reduced by using the Hann window. Figure 7.19 shows the jitter signal for each cycle with and without windowing for the same acquisition for the experiments with DS325 when the shape of the ideal signal is in square wave and sine wave, separately. The calculated corresponding number of

**Fig. 7.17** Aligned data and ideal signals (w.r.t. first zero crossing at rising edge) as step ③ in Fig. 7.1 for the same acquisition from (**a**) DS325 and (**b**) signal generator, for the experiments denoted as EXP. $DS_{sq}$ -*L* and EXP. SG50-*L* in Table 7.6, respectively

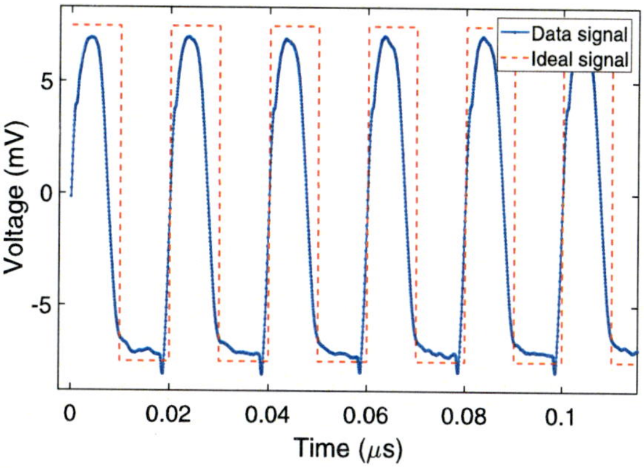


(a)

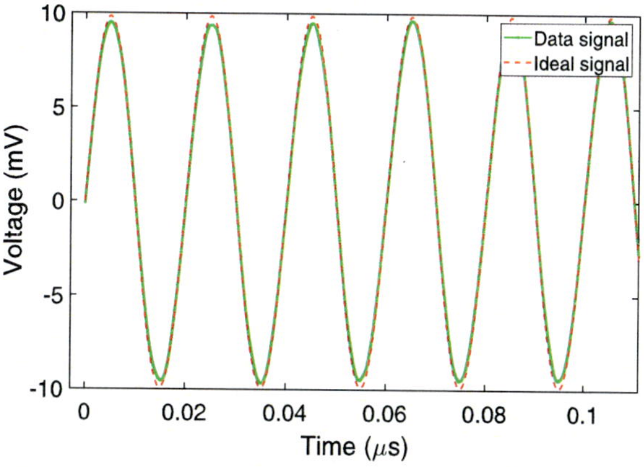


(b)

cycles for each case (by using (5.11)) and its floor are included in columns three and four, respectively, in Table 7.6.

As before, in order to control the uncertainty of the results, 100 acquisitions were captured for each experimental setup and the jitter spectrum was formed for each acquisition by using steps ① to ⑥ of the proposed algorithm. Then, the total jitter spectrum is formed by taking the mean spectrum among the multiple jitter spectra for each experiment. The jitter spectra for DS325 and signal generator with the experiments using *Short* coaxial cable are illustrated in Fig. 7.20. Each sub-figure corresponds to the shape of the generated ideal signal of the DS325 (i.e., sine wave as SET. $DS_{sn}$-{...} and square wave as EXP. $DS_{sq}$-{...}), separately.

In addition, jitter components for the DS325 and the interpolated level between periodic and random jitter for all cases are marked in the figures. Here also we can see there are two high peaks on the jitter signal of DS325 (both EXP. $DS_{sn}$-*S* and EXP. $DS_{sn}$-*S*) that are not present in the jitter signal of the signal generator.

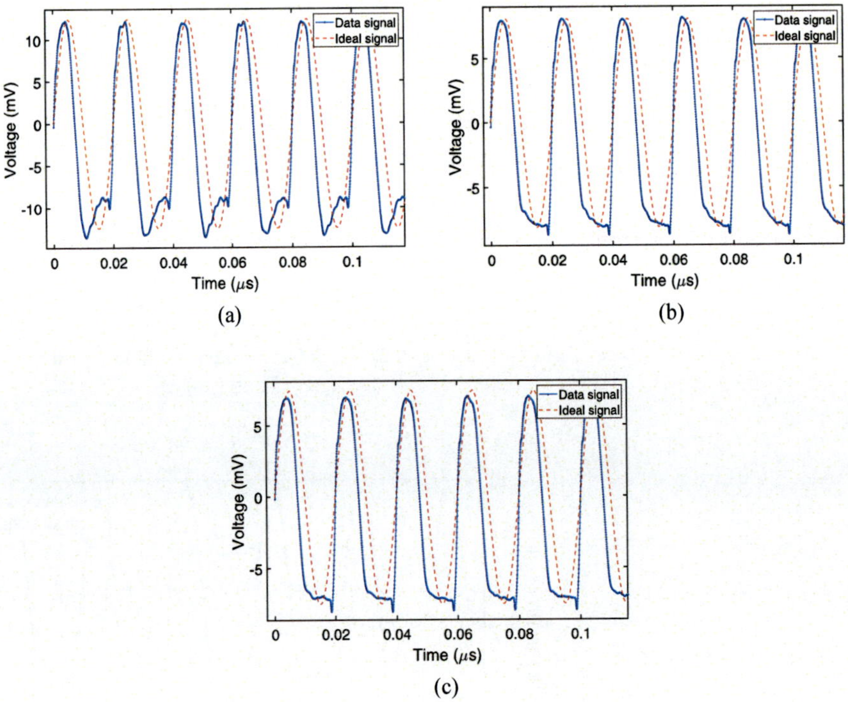


**Fig. 7.18** Aligned data and sine shape ideal signals (w.r.t. first zero crossing at rising edge) as step ③ in Fig. 7.1 for the acquisitions from DS325 for the experiments with (**a**) *Short* cable as EXP. DS$_{sn}$-*S*, (**b**) *Medium* cable as EXP. DS$_{sn}$-*M*, and (**c**) *Long* cable as EXP. DS$_{sn}$-*L* in Table 7.6

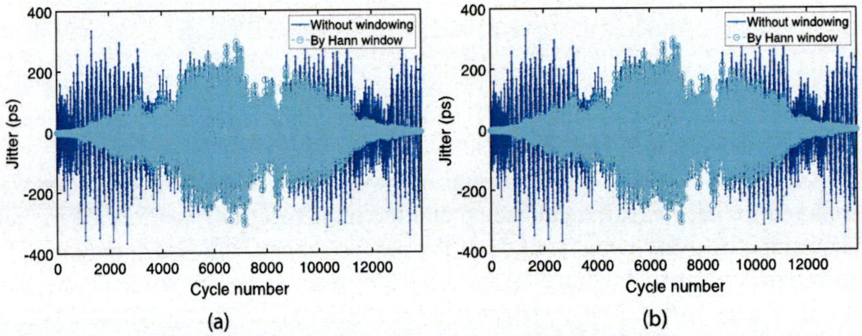


**Fig. 7.19** As for step ⑤, jitter amount for each cycle (time deviation) with and without applying Hann window for the same acquisition of DS325 when the shape of the generated ideal signal in (**a**) square wave and (**b**) sine wave, for the experiment denoted as EXP. DS$_{sq}$-*L* and EXP. DS$_{sn}$-*L* in Table 7.6, respectively

The expanded versions of these two peaks are illustrated in Fig. 7.20. Similarly the results of the jitter spectra and their expanded versions from the experiments with *Medium* and *Long* coaxial cables are illustrated in Figs. 7.21 and 7.22, respectively.

By comparing the DS325 and signal generator's two jitter signals in each corresponding experiment (see Figs. 7.20, 7.21, and 7.22), it is clear that there are two new high peaks at around 0.4 and 2.9 MHz on the jitter signal of DS325 (i.e., EXP. $DS_{sn}$ and EXP. $DS_{sq}$) that are not present in the jitter signal of the signal generator (i.e., EXP. SG50). The corresponding sub-figures show that the jitter spectra are approximately similar regardless of the shape of the ideal signal generated for DS325. This is evidence that our proposed method can be used for any standard shape of the signal (i.e., sine, square, triangular etc.). The random jitter levels are approximately 0.5 and 0.13 ps for DS325 and signal generator, respectively, in each case.

By using the spectra of the total jitter, it is straightforward to obtain the corresponding frequencies of the periodic jitter ${f_{\mathrm{PJ}}}^{(p)}$ in DS325. Other jitter parameters are calculated by considering the energy of the spectrum as described in Sect. 5.3.1. The calculated jitter parameters are tabulated in Table 7.7.

The tabulated results are obtained by considering the full spectrum of the jitter signal for each experiment. For each experiment, the corresponding frequency of the generated ideal signal indicates (third column of Table 7.7) the good accuracy of both signals for the DS325 and signal generator. The standard error of the frequencies is relatively small values that provided a measure of precision to five and six significant digits for the DS325 light source and signal generator, respectively. Therefore, we have measured the frequency to sufficient precision for our purposes. The random jitter of the signal generator is (20.0 $\pm$ 1.7) ps. The DS325 camera has (40.37 $\pm$ 0.04) ps of random jitter and the PJ at two frequencies.

The PJ occurs at (0.399 $\pm$ 0.003) MHz with the amplitude of (61 $\pm$ 4) ps and (2.905 $\pm$ 0.009) MHz with the amplitude of (36.3 $\pm$ 2.0) ps. Furthermore, the curvature at low frequencies of the total jitter for the DS325 (Figs. 7.21 and 7.22) is lower than the same for the SR4000 (Figs. 7.10 and 7.11), for *Medium* and *Long* cables, respectively.

After removing the first few points, the corresponding log–log scales of the total jitter spectra for the experiments from the three cables are shown in Fig. 7.23, separately. In addition, the fitted curves with their characteristic equations for each case are stated in the figures. For the same cable these relevant equations are almost similar for two types of the reference signal (i.e., EXP. $DS_{sn}$ and EXP. $DS_{sq}$).

As before, the slopes at low frequencies of the total jitter for the DS325 are higher than those at the rest of the frequencies in each case. This trend is approximately similar in all cases for the DS325 camera, while it is reasonably different for the signal generator with the experiments with *Long* cable than other two cables (compare the slope at low frequencies for the signal generator in Fig. 7.23). At the low frequencies for the DS325, the slope is $-0.504 \pm 0.005$, while it is $-0.5 \pm 0.3$ for the signal generator.

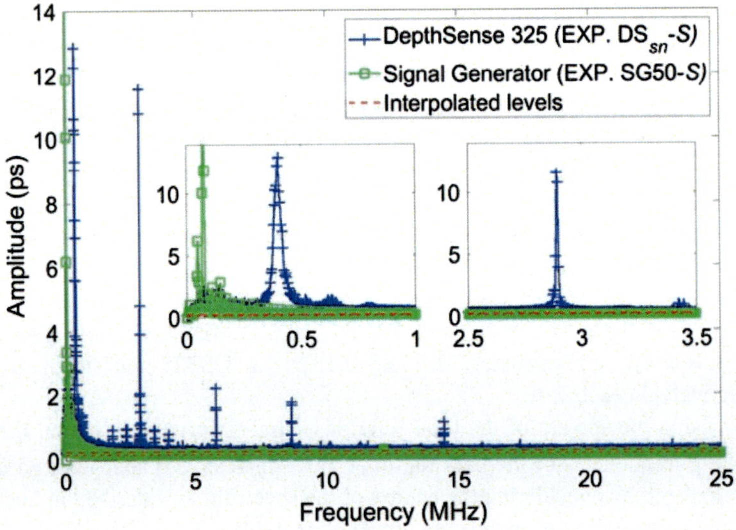


(a)

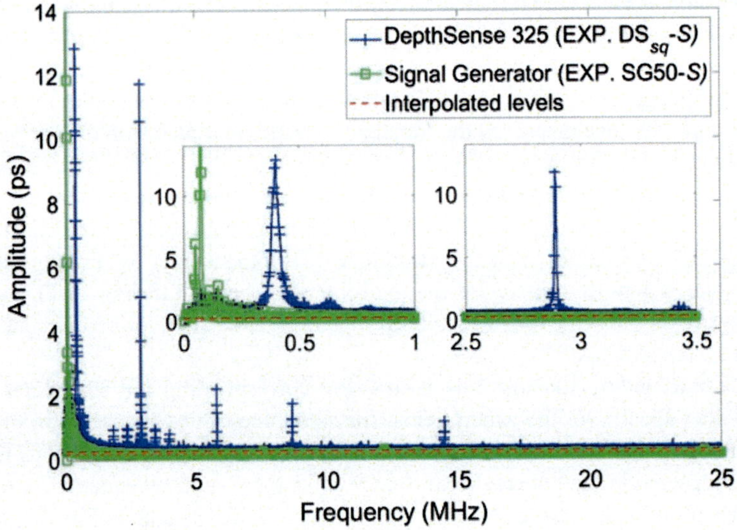


(b)

**Fig. 7.20** Full spectrum with its expansion portions at the two frequencies of PJ at the range 0–1 MHz and 2.5 – 3.5 MHz, for the DS325 and signal generator for the experiment with *Short* cable when the generated ideal signal for the DS325 (**a**) sine wave and (**b**) square wave, denoted as EXP. $DS_{sn}$-*S* and EXP. $DS_{sq}$-*S* in Table 7.6, respectively

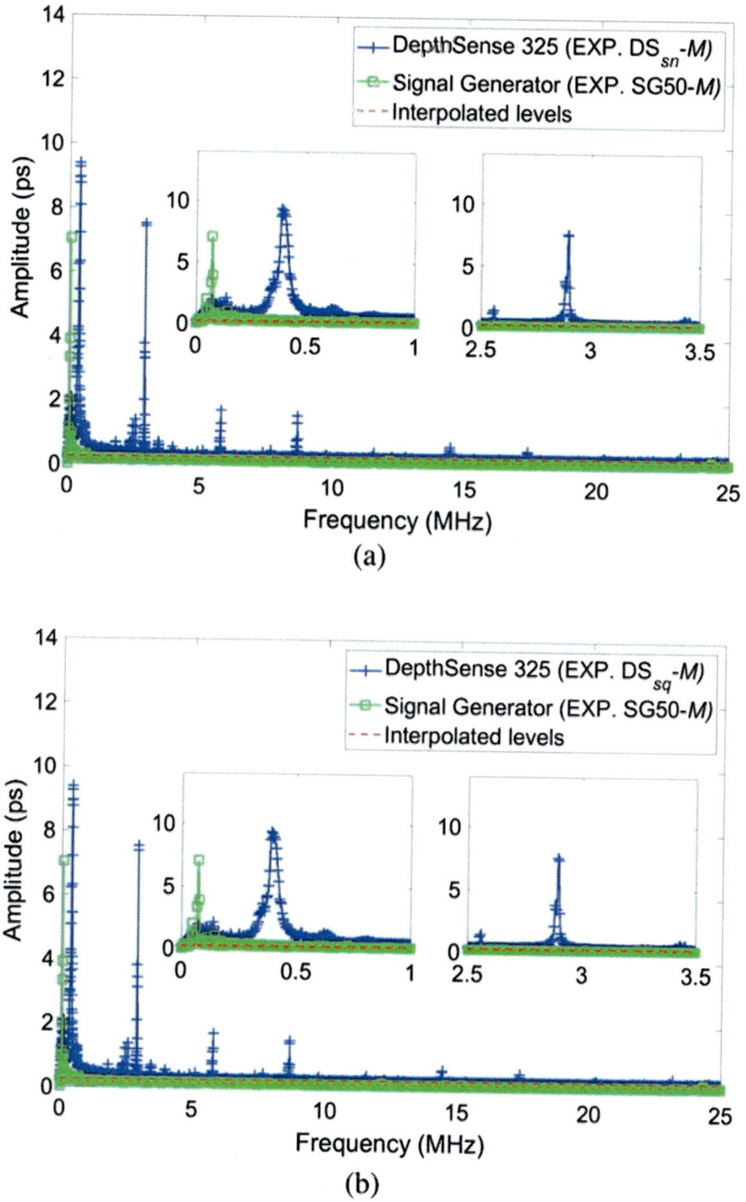


**Fig. 7.21** Full spectrum with its expansion portions at the two frequencies of PJ at the range 0–1 MHz and 2.5–3.5 MHz, for the DS325 and signal generator for the experiment with *Medium* cable when the generated ideal signal for the DS325 (**a**) sine wave and (**b**) square wave, denoted as EXP. $DS_{sn}$-*M* and EXP. $DS_{sq}$-*M* in Table 7.6, respectively

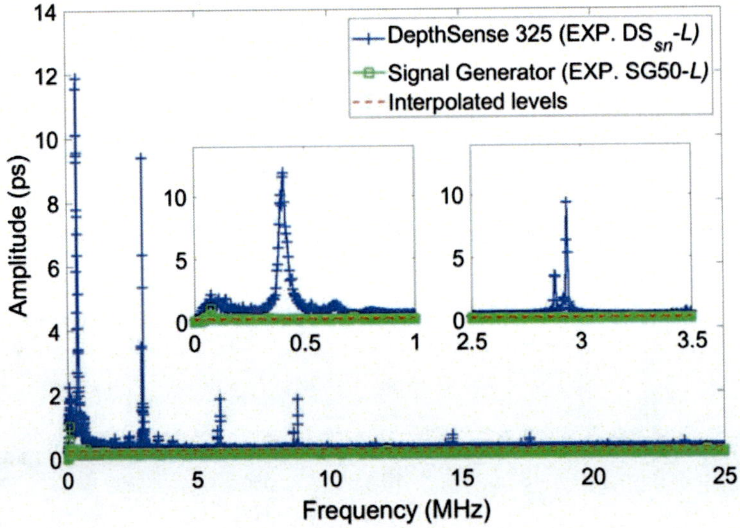


(a)

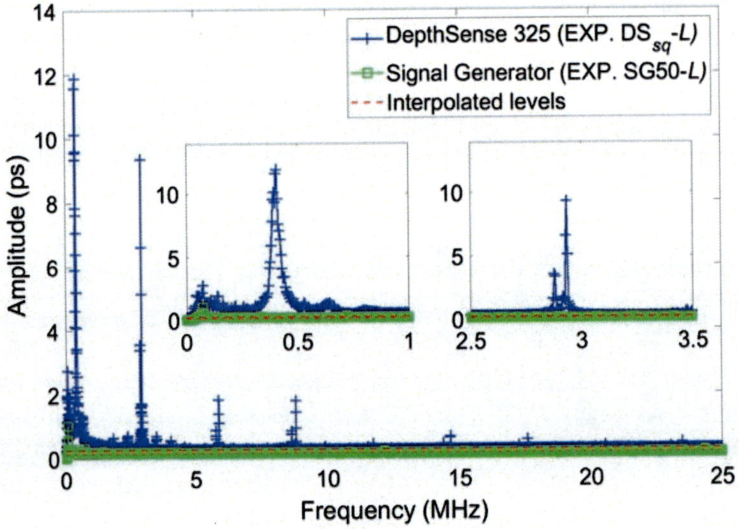


(b)

**Fig. 7.22** Full spectrum with its expansion portions at the two frequencies of PJ at the range 0–1 MHz and 2.5–3.5 MHz, for the DS325 and signal generator for the experiment with *Long* cable when the generated ideal signal for the DS325 (**a**) sine wave and (**b**) square wave, denoted as EXP. DS$_{sn}$-$L$ and EXP. DS$_{sq}$-$L$ in Table 7.6, respectively

**Table 7.7** Jitter parameters from the experiments with DepthSense 325

| EXP. # | | Reference signal's frequency (MHz) | Periodic jitter $[f_{PJ}^{(p)}$ (MHz), $A_{PJ}^{(p)}$ (ps)] | RJ $\sigma_{RJ}$ (ps) |
|---|---|---|---|---|
| EXP. DS325 | EXP. $DS_{sn}$-$S$ | $50.000461 \pm 3 \times 10^{-6}$ | [0.402, 69], [2.888, 44] | $40.48 \pm 0.04$ |
| | EXP. $DS_{sq}$-$S$ | $50.000461 \pm 3 \times 10^{-6}$ | [0.402, 69], [2.888, 44] | $40.48 \pm 0.04$ |
| | EXP. $DS_{sn}$-$M$ | $50.000399 \pm 3 \times 10^{-6}$ | [0.389, 51], [2.892, 32] | $40.35 \pm 0.04$ |
| | EXP. $DS_{sq}$-$M$ | $50.000399 \pm 3 \times 10^{-6}$ | [0.389, 51], [2.896, 32] | $40.35 \pm 0.04$ |
| | EXP. $DS_{sn}$-$L$ | $50.000422 \pm 5 \times 10^{-6}$ | [0.406, 63], [2.933, 33] | $40.31 \pm 0.04$ |
| | EXP. $DS_{sq}$-$L$ | $50.000422 \pm 5 \times 10^{-6}$ | [0.406, 63], [2.933, 33] | $40.26 \pm 0.04$ |
| EXP. SG50 | EXP. SG50-$S$ | $49.99997 \pm 6 \times 10^{-5}$ | nil | $22.21 \pm 0.08$ |
| | EXP. SG50-$M$ | $49.99998 \pm 6 \times 10^{-5}$ | nil | $19.31 \pm 0.07$ |
| | EXP. SG50-$L$ | $50.00004 \pm 6 \times 10^{-5}$ | nil | $18.75 \pm 0.07$ |

## 7.4 Chapter Remarks

In this chapter, the proposed method in Chap. 5 for jitter extraction in the light source of the ToF range imaging cameras is verified experimentally with two depth cameras, separately. In order to control the uncertainties, we captured 100 acquisitions for each experimental setup with a medium speed real-time oscilloscope, while 1000 acquisitions were collected with a low speed oscilloscope. From the methodology we used, the frequencies of the corresponding generated reference signals for each camera and the signal generator are found very accurately. They are $(29.999656 \pm 0.000003)$ and $(50.000427 \pm 0.000004)$ MHz for the MESA SR4000 and SoftKinetic DS325, respectively. For the signal generator they are $(30.00001 \pm 0.00006)$ and $(49.999997 \pm 0.000006)$ MHz for the corresponding signals of 30 and 50 MHz, respectively.

The periodic jitter at two frequencies are found in each camera. For SR4000, they are at 0.064 and $(5.09 \pm 0.06)$ MHz with the amplitudes of $(71.5 \pm 0.3)$ and $(14.0 \pm 1.4)$ ps, respectively. For DS325, they occurred at $(0.399 \pm 0.003)$ and $(2.905 \pm 0.009)$ MHz with the amplitudes of $(61 \pm 4)$ and $(36.3 \pm 2)$ ps, respectively. Among the two cameras, relatively large random jitter is available in SR4000, which is the RMS of $(159.6 \pm 0.1)$ ps and $(40.37 \pm 0.04)$ ps in DS325 camera. In contrast, evidence for the availability of the flicker noise in the MESA SR4000 camera was found. On the other hand, for the signal generator, the RMS of RJ were found as $(35.6 \pm 0.1)$ and $(20.09 \pm 0.07)$ ps for the 30 and 50 MHz signals, respectively. As usual, there is no periodic jitter in the signal generator and we believe the oscilloscope measurements are well accurate.

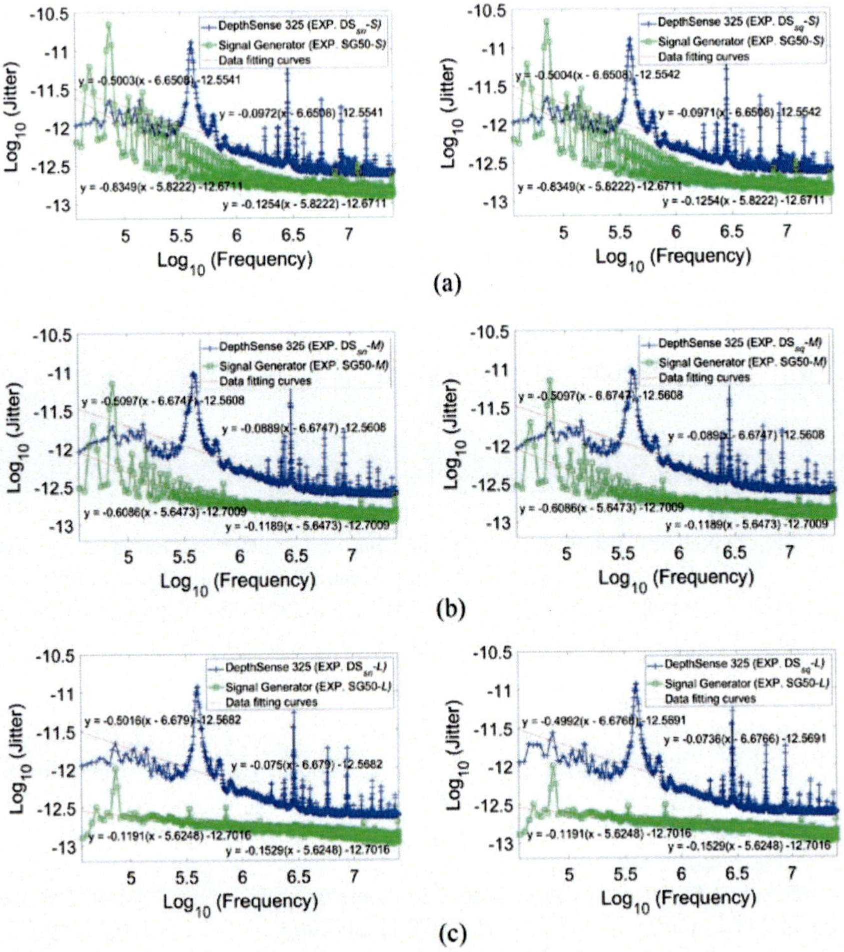


**Fig. 7.23** Log–log scale with fitted curves for spectra in Figs. 7.20, 7.21, and 7.22 for the experiment with coaxial cables: (**a**) *Short*, (**b**) *Medium*, and (**c**) *Long*, respectively, for the shape of the reference signal (left) sine and (right) square

## References


1. *SR4000 Data Sheet*. MESA Imaging AG, February 2011. http://www.adept.net.au/cameras/Mesa/pdf/SR4000.pdf.
2. *DS325 Data Sheet*. SoftKinetic, September 2012. https://www.sony-depthsensing.com/Portals/0/Download/WEB_20120907_SK_DS325_Datasheet_V2.1.pdf.
3. *Free Space Window Input Si Biased Detector*. Thorlabs, April 2017. https://www.thorlabs.com/thorproduct.cfm?partnumber=DET025A.
4. *Infiniium S-Series High-Definition Oscilloscopes – The New Standard for Superior Measurements*. Keysight Technologies, August 2018. https://literature.cdn.keysight.com/litweb/pdf/5991-3904EN.pdf?id=2447379.

5. *Agilent 8648A/B/C/D Signal Generators*. Agilent Technologies, November 2006. https://literature.cdn.keysight.com/litweb/pdf/5965-3432E.pdf?id=1000031128:epsg:dow.
6. Steven C Chapra and Raymond P Canale. *Numerical Methods for Engineers*, volume 7. McGraw-Hill New York, 2015.
7. *Agilent Technologies - Infiniium 54800 Series Oscilloscopes – Now with Mixed-Signal Oscilloscope, Models and MegaZoom, Deep Memory*. Agilent Technologies, November 2002. https://www.keysight.com/en/pd-1000004491%3Aepsg%3Apro-pn-54846B/infiniium-dso?cc=NZ&lc=eng#.

# Chapter 8
# Software-Defined Radio Technology for Jitter Extraction

**Abstract** In Chap. 7, the proposed algorithm to extract jitter in the light source of two AMCW ToF range imaging cameras without the reference clock was verified. There, the light source is measured by a medium speed (6 GHz) real-time digital oscilloscope, which is a very expensive instrument. Can we replace the oscilloscope with a much cheaper instrument? This chapter describes the possibility of re-purposing an emerging technology in current communication systems for jitter extraction in signals. We use the software-defined radio (SDR) technology with the proposed methodology to extract the jitter on the light sources of the same two AMCW ToF cameras used in Chap. 7. First, the background, some commercial SDR devices and the operational principle of the receiver side of the SDR technology are briefly reviewed. Then, the experimental setup is described for the ToF cameras with a currently available SDR receiver connected to a computer, coupled with the methodology described in Sect. 7.1. Next, the emitted light source from the SwissRanger 4000 and DepthSense 325 with known jitter is used as the signal with jitter to be tested. Finally, the comparison of the result for the signal acquired with the SDR receiver to the same signal acquired by the oscilloscope described in Chap. 7 is also included.

**Keywords** Software defined radio · Antenna · Tuner · re-programmable processors · Radio-frequency amplifier · Intermediate-frequency · Local oscillator · Analogue-digital converter · Digital-analogue converter · In-phase and quadrature components · Digital down converter · Digital signal processing · Down-conversion · Visible frequency · Inherent frequency errors

## 8.1 Background

It was once traditional in radios to process the received radio signal entirely with analogue circuitry. Over the last couple of decades, computer technology has become sufficiently powerful to largely perform the signal demodulation and process by mathematical calculations in software. This has led to the design of advanced radios to easily implement in software. In the late twentieth century, a new



G. Anthonys, *Timing Jitter in Time-of-Flight Range Imaging Cameras*,
https://doi.org/10.1007/978-3-030-94159-8_8

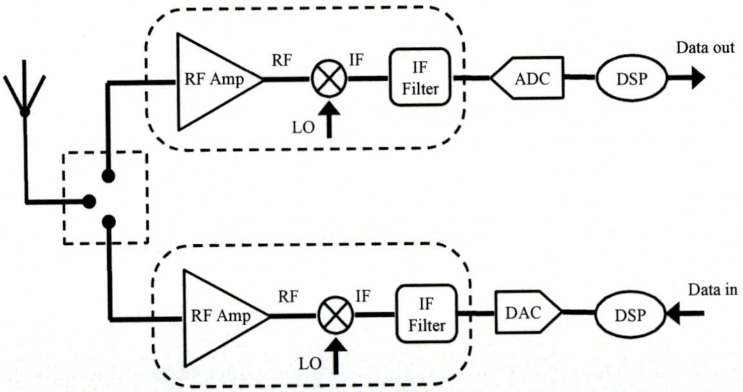


**Fig. 8.1** Possible structure of a SDR device

technology fusing software and digital radio called software radio became the choice for wireless communication. In 1991, the term "software-defined radio" (SDR) was coined by Joseph Mitola, who published the first paper on the topic in 1992 [1]. SDR is a type of re-configurable radio in which some or all of the physical layers of functionality are implemented in software and/or firmware [2]. SDR device consists of an analogue–digital converter (ADC) and a digital–analogue converter (DAC) with an antenna as shown in Fig. 8.1. The rest of the functionality is handled by re-programmable processors [1].

The hardware components of SDR are now available at affordable consumer prices, and software implementations are freely available. In SDR technology, the signal from the antenna is directly converted to the digital domain and all the further processing is under software control. The SDR can be totally configured by software so that a common platform can be used in many areas [3] such as in military applications in aviation [4, 5], mobile communication [6, 7], smart antennas [8, 9], cognitive radio research [10, 11], battery technologies [12, 13], satellite systems [14, 15], wildlife tracking [16, 17], and medical imaging [18, 19] to name some of them.

In certain markets, SDR is moving beyond innovators and early adopters. The adopters select a technology not because it is innovative but because it has been shown to successfully solve a problem within their specific market. One of the major advantages of SDR technology is that it can be configured to exactly meet the user requirements and is easy to implement. The same hardware platform can be used for several different radios. The reconfiguration and performance of the radios can be changed by updating the software are some other advantages of SDR technology. As processing power becomes cheaper to implement, so SDRs are being increasingly used for high-end applications and also increasingly they are moving into lower-end radios as well. The advances in these emerging technologies are making SDR technically and commercially realistic [20].

## 8.2 Commercial SDR Devices

Based on the first paper on the topic SDR [1], the first public software radio was developed as a US military project named SpeakEasy [4] from 1991 to 1995. The main goal of this project was to use programmable processing to emulate more than 10 existing military radios, operating in the frequency range (2–2000) MHz. As a result of this idea, it was able to easily incorporate new coding and modulation techniques in many areas during the last two decades. These days, new SDR products are popping up every few months due to the advantages of SDR technology described in Sect. 8.1. There are a number of commercial SDR devices available in the market [21]. Most of the products (RTL-SDR [22], Airspy [23, 24], ColibriNano [25], KiwiSDR [26, 27]) are SDR receivers, while for other products (LimeSDR [28], HackRF [29, 30]), the technology also includes the transmission capability. Table 8.1 shows the relevant parameters such as frequency range, bandwidth, resolution, and transmit capability, of the selected commercial SDR devices.

Most SDRs generate raw in-phase (I) and quadrature (Q) components for the data. They need to be connected directly to a personal computer running with operating system specific, installed software. There are network SDR devices (e.g., KiwiSDR) which are standalone devices that attach to the user local network and are optionally accessed through the internet. A browser is used to connect to the

**Table 8.1** Specifications of the selected commercial SDR devices

| Commercial name | Frequency range [MHz] | Bandwidth [MHz] or [MSa/s] | Resolution [in bits] | | Transmit capability |
|---|---|---|---|---|---|
| | | | in ADC | in DAC | |
| RTL-SDR 2831 | 24–1766 | 3.2 | 8 | N/A | No |
| RTL-SDR V3 | 0.5–1766 | 2.4–3.2 | 8 | | |
| Airspy HF+ | 0.009–31 | 0.768 | 18 | N/A | No |
| | 60–260 | | | | |
| Airspy R2 | 24–1700 | 10 | 12 | | |
| Airspy Mini | 24–1800 | 6 | | | |
| FUNcube Pro | 64–1700 | 0.096 | 16 | N/A | No |
| FUNcube Pro+ | 0.15–260 | 0.192 | | | |
| | 410–2050 | | | | |
| ColibriDDC | 0.01–62.5 | 0.038–0.312 | 14 | N/A | No |
| ColibriNANO | 0.1–55 | 0.768 | 24 | | |
| | 0.1–500 | 3 | 16 | | |
| KiwiSDR | 0.01–30 | 30 | N/A | N/A | No |
| LimeSDR | 0.1–3800 | 61.44–80 | 12 | N/A | Yes |
| LimeSDR Mini | 10–3500 | 30.72 | | | |
| HackRF | 30–600 | 20 | 8 | 8 | Yes |
| HackRF One | 1–6000 | | | | |
| BladeRF | 300–3800 | 40 | 12 | 12 | Yes |

N/A stands for Not Applicable

user interface. This type of SDR is more beneficial nowadays. High frequency communication reception quality may be improved using remote SDR receivers. As an example, Gudipati et al. have proposed a software-defined centralized control plane for radio access networks that abstracts all base stations in a local geographical area as a virtual big-base station comprised of a central controller and radio elements [31]. Listening to specific bands in remote locations can be useful for many applications [32–34].

SDR technologies have become increasingly accessible these days with low-cost devices and offering varying levels of performance. At present, the price of SDR receivers is starting from US $20, while for SDR transmitters it is from US $80 in the market [35].

## 8.3 Operational Principle of the SDR Receiver

SDR properly considered consists of a transmitter and a receiver. Since we need to capture the signals for jitter investigation, we are primarily interested in the operational principle of the receiver side of the SDR. In a SDR receiver, the signal from the antenna passes through wideband low-noise amplification, down-conversion to a lower frequency, before being digitally sampled. The operational principle of the SDR receiver will be briefly explained. However, note that there is a lot more to the theory and design of SDR systems than what is described below. One of the RTL-SDR receivers that is a cheap product in the current market is considered for the experiments in this chapter. Therefore, the detailed information from now is mainly focused on that product, but the general architecture of the SDR technology is common to other commercial products as well. Let us consider the hardware and software structures of the SDR receiver as follows.

### *8.3.1 SDR Hardware*

The basic structure of the SDR receiver is shown in Fig. 8.2. We use an RTL-SDR USB receiver [36] (with 3.2 MSa/s) that consists of the Rafael R820T2 tuner [37], which comprises the RF amp through to IF filter stages (see the figure) coupled to an RTL2832U demodulator [38], which houses the ADC and digital signal processing (DSP) blocks. The DSP block consists of a digital down converter (DDC) block with a DSP engine, which is under software control. The DDC block includes in-phase (I) and quadrature (Q) mixers, digital low-pass filters and down samplers. This DSP block provides the connection to the outside via the USB port.

The radio signal enters via the antenna and is amplified by a radio frequency (RF) tuner and then is downshifted by mixing with the local oscillator (LO) frequency to the intermediate frequency (IF). The IF signal is bandlimited (IF filter) before being sampled by the ADC at a fixed rate of 28.8 MHz, then passed through the DDC,

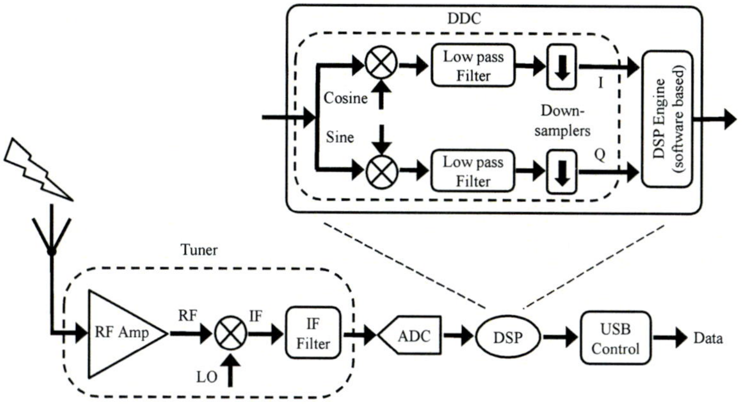


**Fig. 8.2** Structure of a SDR receiver side

a low-pass filter, and into a re-sampler (which then down-samples to the desired sample rate, 3.2 MHz at maximum). Therefore, from this SDR receiver, one can only measure frequencies below 1.6 MHz due to Nyquist, after down-conversion. In addition, the effective frequency for aliasing is much higher than this value (i.e., the rate 14.4 MHz).

Inside the DSP block, it uses separate ADC and low-pass filters each for I and Q components for the data. When the receiver uses a zero-IF tuner like the E4000 [39] or a low-IF tuner like the R820T, then the DSP block uses both I and Q ADCs; otherwise, it uses only one of them. Therefore, the DDC only activates for low-IF tuners or the direct sampling mode, and it shifts and filters out the desired part of the spectrum, generating I/Q samples from real samples [40].

In practice, an ADC will only work up to a specific frequency. Thus, to digitize the signals with the higher frequency we need a mixer stage to convert all received frequencies down to the frequencies that the ADC can use. So, this is the job of the tuner chip such as R820T2 and E4000.

### 8.3.2 SDR Software

In order to operate the SDR dongle from a computer, it is necessary for complementary dedicated drivers and software to enable the interaction of the dongle as a software-defined radio. Figure 8.3 shows the typical main screen of the RTL-SDR dongle used in this chapter.

In addition, one can set parameters such as the *Sampling Mode* and *Device Sample Rate* in the configuration window as shown in Fig. 8.4. The quadrature sampling mode is used and maximum sample rate ($f_s$ = 3.2 MHz at the DDC) for all experiments from this dongle. In order to get the maximum performance from the device, the other parameters are chosen as follows.

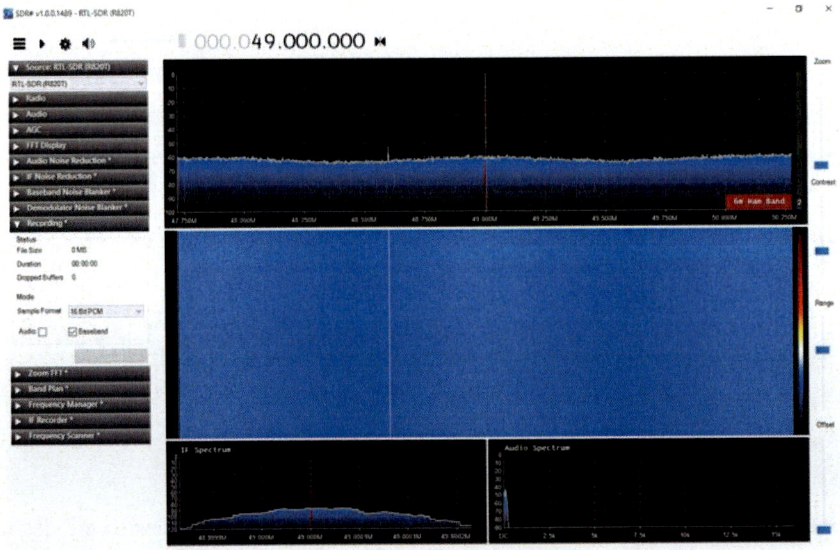


**Fig. 8.3** An example of the main menu of the RTL-SDR dongle

**Fig. 8.4** An example of the configuration window of RTL-SDR dongle

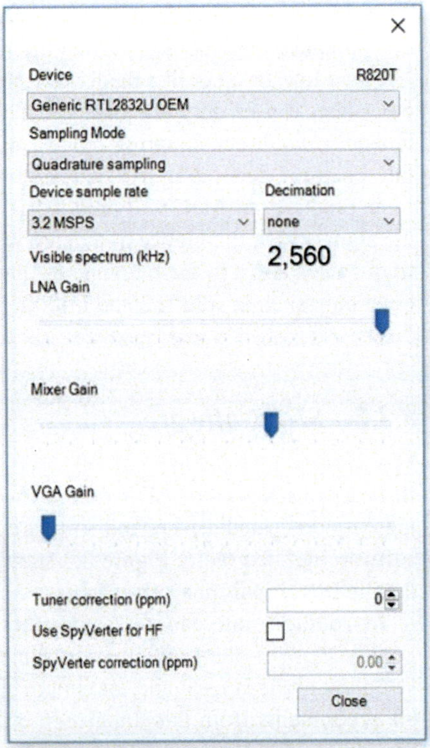

We set the "Decimation" tab to *none*, which results in the range of the visible spectrum as 2560 kHz. In real life, a rule of thumb is to use the 80% relationship:

$$\text{Visible frequency} = 0.8 \times f_s = 0.8 \times 3.2\,\text{MHz} = 2560\,\text{kHz}. \quad (8.1)$$

That is, the reduction can be applied up to an extra 20% without significantly affecting the quality of the result [41]. This 80% relationship can be seen by the values given in the configuration settings of the RTL-SDR dongle as tabulated in Table 8.2 and they are illustrated as shown in Fig. 8.5.

Next, we amplify the signal while minimizing additional noise by setting the low-noise amplifier "LNA Gain" to its maximum in the scale. The "Mixer Gain" tab is set to approximately 2/3 of its maximum. Then, the variable-gain amplifier "VGA Gain" is set to as much of a low value in the scale [42]. Finally, the corresponding signal of the light source is captured from this RTL-SDR receiver. The captured signal was recorded on the computer in the format of wave audio files with *.wav* extension.

**Table 8.2** Sample rate with its visible frequency of the RTL-SDR dongle

| Sample rate (MSa/s) | Visible frequency (MHz) |
|---|---|
| 0.001 | 0.0008 |
| 0.25 | 0.2 |
| 0.900001 | 0.720001 |
| 1.024 | 0.8192 |
| 1.4 | 1.120 |
| 1.8 | 1.440 |
| 1.92 | 1.536 |
| 2.048 | 1.6384 |
| 2.4 | 1.920 |
| 2.56 | 2.048 |
| 2.8 | 2.240 |
| 3.2 | 2.560 |

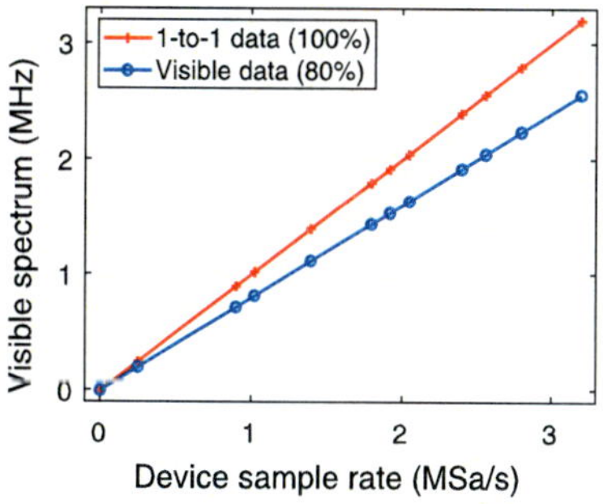


**Fig. 8.5** Visible spectrum versus sample rate of the RTL-SDR dongle is drawn using the data in Table 8.2

## 8.4 Experiment with MESA SR4000

As before, we captured the emitted light signal of the SR4000 camera using the same photodetector DET025A. Here, we used the SDR USB dongle, namely, RTL2832U R820T2 for the measurements. The experimental and connection setups are shown in Fig. 8.6, separately.

The range imaging camera was set to produce a 30 MHz modulated signal, which was characterized in Sect. 7.2.1. For the control experiment, we generated 30 MHz sine waves with the same signal generator (HP Agilent 8648B). Both signal sources were analyzed by the SDR USB dongle. We used the same two coaxial cables (as *Medium* and *Long*) that were experimented in Sect. 7.2.1, for

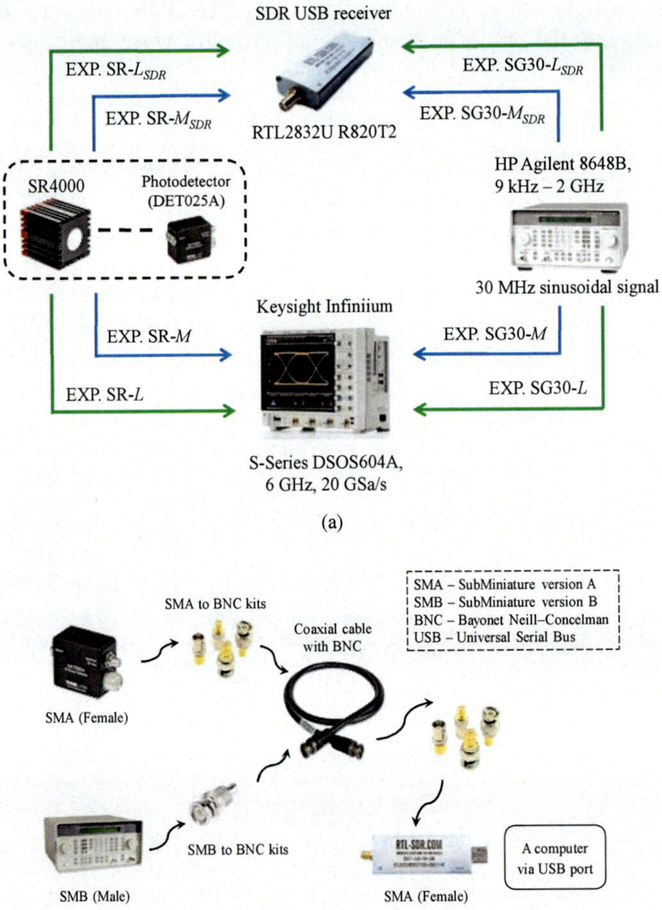


**Fig. 8.6** Setup of **(a)** components and **(b)** connections, for the experiment with the MESA Imaging SwissRanger 4000 by using SDR USB receiver

**Table 8.3** Parameters of the components and setup with RTL-SDR

| Device | Measuring instrument | | No. of cycles $N_{\Delta t}^{cal}$ | No. of cycles $N_{\Delta t}$ | EXP. # | |
|---|---|---|---|---|---|---|
| SR4000 with photodetector | SDR setup | SDR receiver 3.2 MSa/s (RTL2832U-R820T2) | 1345 | 1344 | *Medium* and *Long* cables | SR-$M_{SDR}$ |
| | | | | | | SR-$L_{SDR}$ |
| Signal Generator (30 MHz signal) | | | | | | SG30-$M_{SDR}$ |
| | | | | | | SG30-$L_{SDR}$ |
| SR4000 with photodetector | OSC setup | Oscilloscope 20 GSa/s (S-Series DSOS604A) | 15,000 | 14,998 | | SR-$M$ |
| | | | | | | SR-$L$ |
| Signal Generator (30 MHz signal) | | | | | | SG30-$M$ |
| | | | | | | SG30-$L$ |

all experiments in here. In addition, the corresponding results from the previous experiment with the oscilloscope in Sect. 7.2.2 are also used. The SDR dongle and the photodetector connect via SubMiniature version A (SMA) with the coaxial cable and BNC connectors. Also, the SDR dongle and the signal generator connect via SubMiniature version B (SMB) with the coaxial cable and BNC connectors; see Fig. 8.6b. The parameters of the experimental setup are shown in Table 8.3.

From Chap. 7, we found that the first component of the PJ for SR4000 is at 0.12 MHz (see Table 7.4) and 0.064 MHz (see Table 7.3) by using low speed and medium speed oscilloscopes, respectively. However, to satisfy both amounts, 0.12 MHz is selected as the first component of the PJ for SR4000. Then, for the full spectrum of the DFT, the upper limit of the frequency should be at least 0.24 MHz. However, with a safe margin, 0.4 MHz is chosen as the IF for the SDR receiver. Therefore, for the SDR setup, the emitted light signal (30 MHz) of the camera was down-mixed with 29.6 MHz as the RF tuner LO frequency to give a 400 kHz IF signal at the ADC. The number of jitter samples $N_{\Delta t}^{cal}$ were found by using (5.11) for the SDR with $f_m = 400$ kHz, $N_s \approx 10760$, and $F_s = 3.2$ MSa/s. In addition, $\lfloor N_{\Delta t}^{cal} \rfloor$ is calculated by using (5.12) for each measurement setup. These are tabulated in column four and five of Table 8.3, respectively.

In order to reduce the uncertainty of the result, 17,000 data sets for each experiment with the SDR setup are captured. However, about 1% of the data sets were found to contain errors and were discarded from the analysis. It appears that the SDR (or its device driver) occasionally drops samples [42]. For the comparison, the 100 acquisitions for each experiment with the oscilloscope from Chap. 7 are also used.

## 8.5 Results and Discussion

As before, for step ① of the proposed methodology, the data signal from the light source of the camera from the RTL-SDR receiver is captured. Note that the SDR dongle captures the data in quadrature form (i.e., real and imaginary parts,

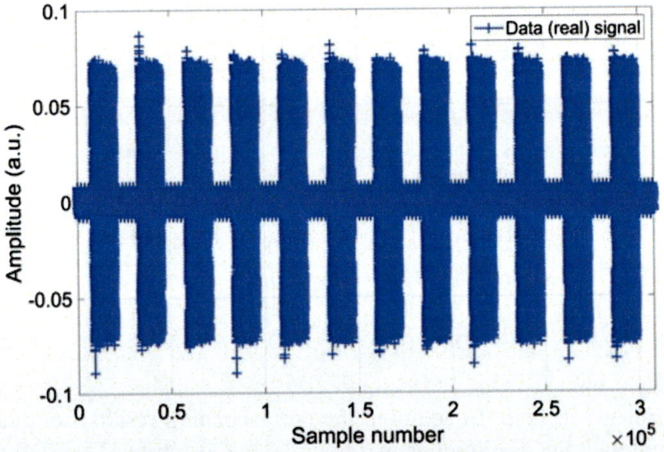


**Fig. 8.7** Real portion of the data signal captured from the SDR setup for a single acquisition of the light source of SR4000, experiment with *Medium* cable denoted as SR-$M_{SDR}$ in Table 8.3

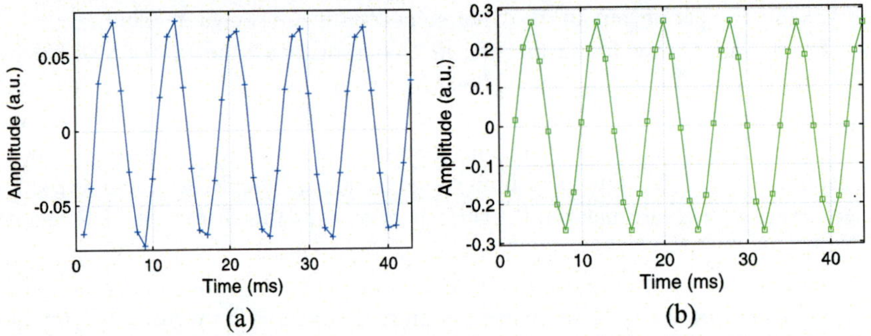


**Fig. 8.8** Data signals from the SDR setup for a single acquisition from (**a**) the light source of SR4000 and (**b**) signal generator, with the *Medium* cable denoted as SR-$M_{SDR}$ and SG30-$M_{SDR}$ in Table 8.3, respectively

separately) and here we concern the real portion of the light signal (imaginary portion also gives the same result) throughout the experiments. Figure 8.7 shows the real portion of the light signal of the SR4000 as step ①.

It shows the set of pulses since the camera shutter is at the "on" and "off" stages during the integration period. Now, we need to separate each section as follows. By comparison, the threshold value (here it is chosen as 0.01 a.u.) with the amplitude, we obtain the first sample positions that increase the amplitude of sections by adding a fixed amount greater than the number of samples $N_s$ of the section (i.e., we choose 11,000 since $N_s \approx 10{,}700$) to each section. So, now, the starting position (leftmost) of each section is known. Then in order to find the last position (rightmost) of the section, the same value $N_s$ is added to the leftmost position of each section. Finally, a couple of cycles at the end of the section are removed. In addition, Fig. 8.8 shows the corresponding expansion versions of the real portion of the data signals of the

**Table 8.4** Experimental parameters for setup with SDR and oscilloscope

| EXP. # | | Sampling time $1/F_s$ | Number of Sa/cycle | Frequency of the reference signal |
|---|---|---|---|---|
| SDR | SR-$M_{SDR}$ | 312.5 ns | 8 | $399.566 \pm 0.003$ kHz |
| | SG30-$M_{SDR}$ | | | $399.504 \pm 0.007$ kHz |
| | SR-$L_{SDR}$ | | | $399.546 \pm 0.001$ kHz |
| | SG30-$L_{SDR}$ | | | $399.508 \pm 0.008$ kHz |
| OSC | SR-$M$ | 50 ps | 2000/3 | $29.999645 \pm 0.000003$ MHz |
| | SG30-$M$ | | | $30.00001 \pm 0.00006$ MHz |
| | SR-$L$ | | | $29.999668 \pm 0.000003$ MHz |
| | SG30-$L$ | | | $30.00001 \pm 0.00006$ MHz |

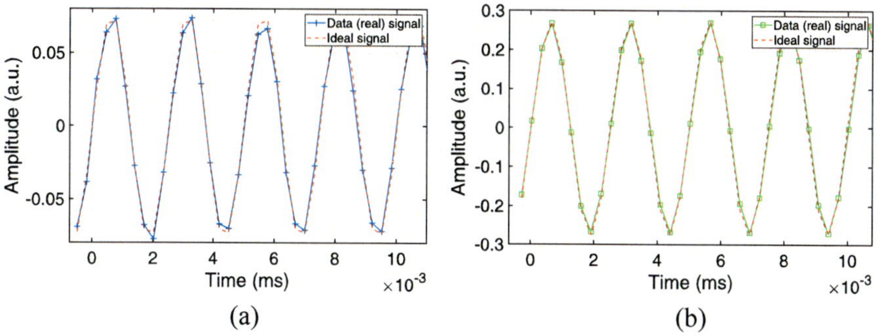


**Fig. 8.9** Aligned the same data signal and ideal signal for a single acquisition from (**a**) the SR4000 and (**b**) signal generator, using the SDR setup

SR4000 and signal generator (30 MHz signal), separately, in the experiment with *Medium* cable.

Then, by applying the process of sub-① in Fig. 7.1, we generate the corresponding ideal signal for each case (for step ②). Table 8.4 tabulates the experimental parameters for each setup and the fourth column represents the number of samples per cycle. The last column is the corresponding frequency with the standard deviation (STD) of the ideal signal that is extracted from the data, for each experimental setup. Each of these STDs is based on a large number of data sets (i.e., 100 sets with $\approx 10^7$ samples each for OSC setup and 17,000 sets with $\approx 10^4$ samples each for SDR setup). The frequencies of the signals are listed in Table 8.4 and determined very accurately. This is used to generate the reference signal.

Figure 8.9 shows the corresponding result for step ③ for the same acquisition of SR4000 and the signal generator from the SDR setup, separately. The data and ideal signals are aligned with respect to the first zero crossing point in each case. For the SDR setup, the number of samples per cycle 8 can be verified from Fig. 8.9a as well. This value is relatively very small, and the representation of the signal is not a great view (i.e., the number of samples per cycle is low). This likely increases the error in determining the zero crossings. By using the number of samples per cycle

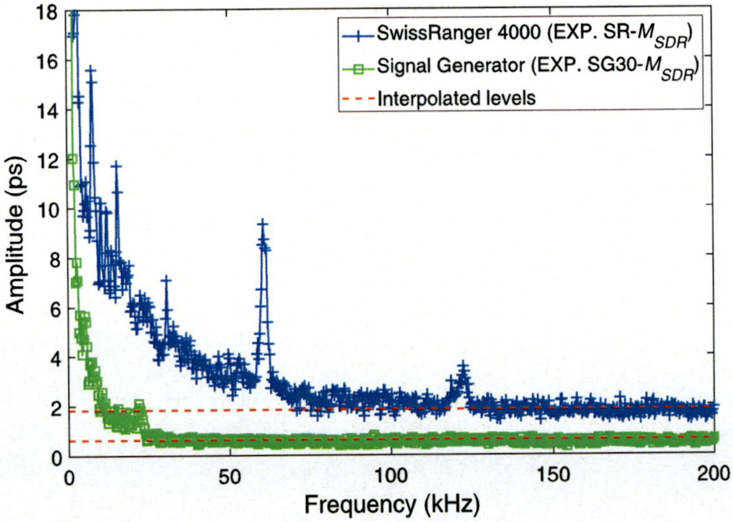


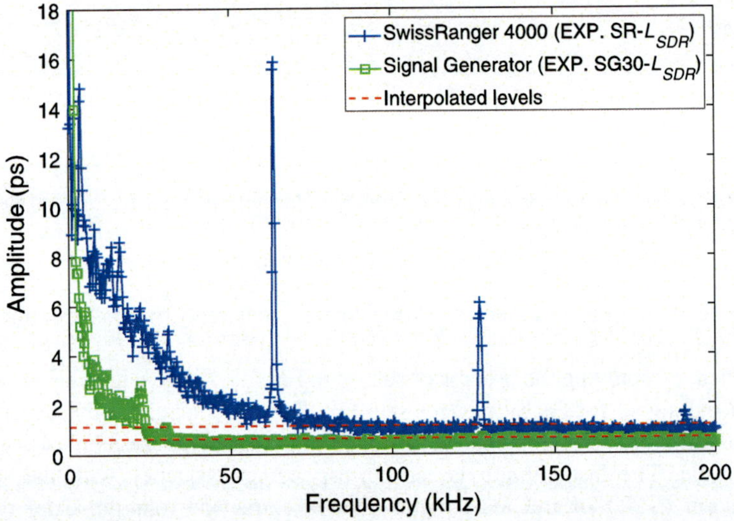


**Fig. 8.10** Jitter components for the SR4000 and signal generator from the experiments with the SDR receiver for (**a**) *Medium* (**b**) *Long* coaxial cables denoted as SDR setup in Table 8.3

(8 Sa/cycle) and the sampling rate of the dongle (3.2 MSa/s), we can compute the frequency of the ideal signal as 400 kHz. This can be verified from the last column of Table 8.4.

The jitter spectrum for each acquisition is obtained by using steps ① to ⑥ in the proposed methodology shown in Fig. 7.1. After step ⑥, jitter spectra for MESA SR4000 and 30 MHz signal for each cable are separately obtained, as shown in Fig. 8.10. The jitter components and the interpolated level between periodic and random jitter for each signal is marked in the figure. The frequency range for the SDR setup and oscilloscope setup is 0–200 kHz (since the IF signal is 400 kHz) and 0–15 MHz, respectively. In order to compare the results from the two instruments, we focus on the expansion version of the first 200 kHz from the oscilloscope results that were previously obtained as shown in Figs. 7.10 and 7.11 in Chap. 7. So, these figures are rescaled as shown in Fig. 8.11.

The curvature at low frequencies of the spectra obtained from SDR setup is higher than the same from oscilloscope (compare Figs. 8.10 and 8.11). We expect this may be due to the noise in the SDR USB dongle. Then this may significantly leak the energy to adjacent bins of the spectra. Note that the results from the SDR setup are compensated by the ratio 1:250/3 (see the fourth column of Table 8.4) since the number of samples per cycle has influenced the energy of the signal. As before, from Fig. 8.10, it is straightforward to identify the high peak on the jitter signal of SR4000 from the experiments with two cables, separately. They are at 61 and 64 kHz with the amplitudes 61 and 66 ps, respectively, for the experiment with *Medium* and *Long* cables. In addition, the jitter parameters for the first harmonic of this periodic jitter are 122 and 127 kHz with amplitudes of 22 and 29 ps, respectively, for the two cables. There is no PJ in the signal generator for both cases.

Furthermore, as referred in Chap. 7, the frequencies are 64 and 64 kHz with the amplitudes 71 and 72 ps, respectively, for the two cables were obtained from the oscilloscope. In addition, the parameters for the first harmonic of the PJ are 128 and 126 kHz with 29 and 28 ps amplitudes, respectively, for the two cables. There is a broad and flat spectrum being the RJ in the camera. By considering the full spectrum for each experiment, jitter parameters are calculated as shown in Table 8.5.

The PJ at $(62.5 \pm 1.1)$ kHz with the amplitude $(63.5 \pm 1.8)$ ps is represented by the SDR receiver, while it is 64 kHz from the oscilloscope. In addition, if the first harmonic of this PJ is considered, then they are $(124.5 \pm 1.8)$ kHz with the amplitude $(25.5 \pm 2.5)$ ps for the SDR setup while $(127.0 \pm 0.7)$ kHz with $(28.5 \pm 0.3)$ ps for the OSC setup. The RJ with RMS of $(70.0 \pm 0.2)$ ps and $(159.6 \pm 0.1)$ ps are obtained from SDR and OSC setup, respectively. The measurements of the RJ in the signal generator from the SDR and OSC setup are found as $(30.8 \pm 0.1)$ ps and $(35.6 \pm 0.1)$ ps, respectively.

Note that the RJ amounts for both camera and signal generator from the SDR setup are relatively less than the same from the oscilloscope setup. This is because the SDR receiver can measure only up to half of the IF (i.e., 200 kHz). There are relatively larger slopes at low frequencies with both *Medium* and *Long* cables for the signal generator from the SDR setup than with OSC setup (compare the signal generator curves of Figs. 8.10 and 8.11). This may be due to the lower quality of the SDR receiver.

As discussed in Sect. 5.3.2, the corresponding log–log plots for the jitter spectra of the SR4000 and signal generator for both experimental setup (relevant to Fig. 8.10) are shown in Fig. 8.12. In addition, the fitted curves with their characteristic equations for each case are stated in the figure.

The slopes at low frequencies of the total jitter for the SR4000 are higher than those at the rest of the frequencies (see Fig. 8.12) in each case. At low frequencies of

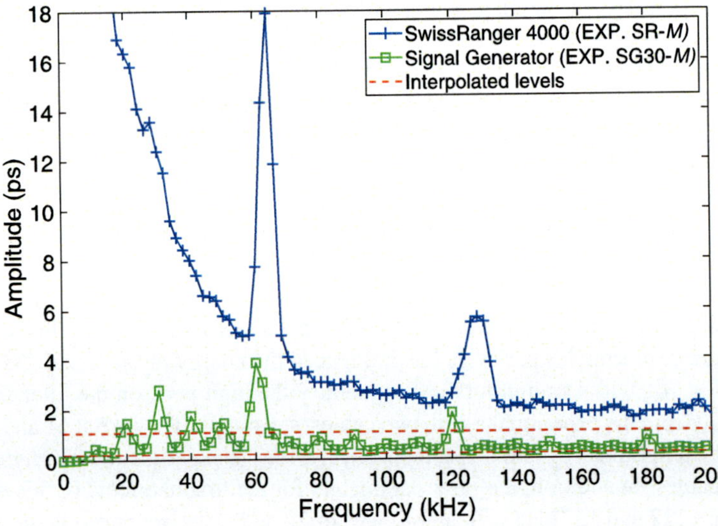


(a)

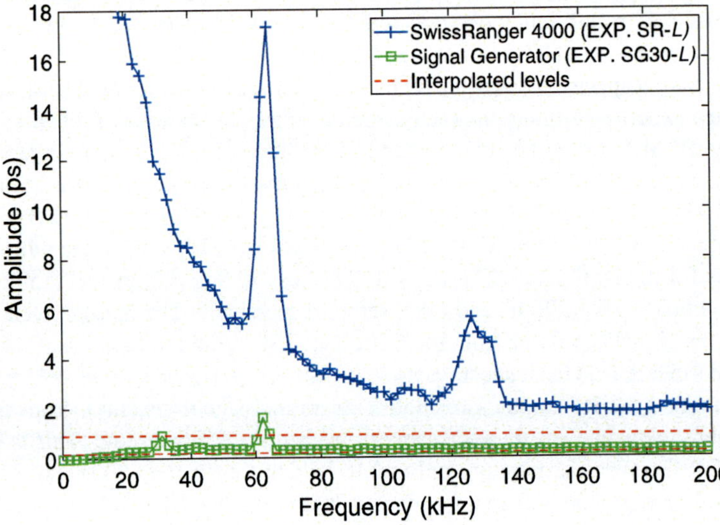


(b)

**Fig. 8.11** Expansion version of the first 200 kHz of the jitter components for the SR4000 and signal generator from the oscilloscope setup (corresponding to the Figs. 7.10 and 7.11 in Chap. 7) that experimented with (**a**) *Medium* and (**b**) *Long* coaxial cables, denoted as OSC setup in Table 8.3

**Table 8.5** Extracted results from measurements

| EXP. setup | | Periodic jitter $[f_{PJ}^{(p)}$ (MHz), $A_{PJ}^{(p)}$ (ps)] | Random jitter $\sigma_{RJ}$ (ps) |
|---|---|---|---|
| SDR | SR-$M_{SDR}$ | [0.061, 61] and [0.122, 22] | $85.8 \pm 0.2$ |
| | SG30-$M_{SDR}$ | nil | $30.8 \pm 0.2$ |
| | SR-$L_{SDR}$ | [0.064, 66] and [0.127, 29] | $54.3 \pm 0.3$ |
| | SG30-$L_{SDR}$ | nil | $30.8 \pm 0.1$ |
| OSC | SR-$M$ | [0.064, 71], [0.128, 29] and [5.001, 12] | $173.6 \pm 0.1$ |
| | SG30-$M$ | nil | $36.9 \pm 0.1$ |
| | SR-$L$ | [0.064, 72], [0.126, 28] and [5.173, 16] | $145.7 \pm 0.1$ |
| | SG30-$L$ | nil | $34.3 \pm 0.1$ |

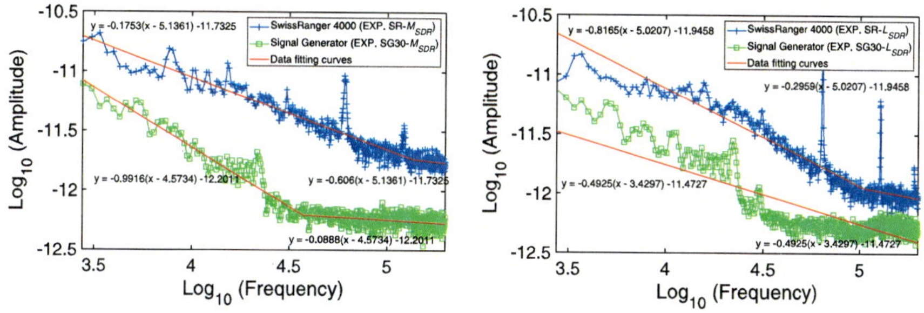


(a) (b)

**Fig. 8.12** Log–log scale with fitted curves for jitter spectra shown in Fig. 8.10 for the SR4000 and signal generator for the experiments with (**a**) *Medium* and (**b**) *Long* cables, denoted as SDR setup in Table 8.3

SR4000, the experiment with *Long* cable (see Fig. 8.12b) shows the slope $f^{-0.8165}$, while from the *Medium* cable (see Fig. 8.12a) it is $f^{-0.1753}$. This is because in the latter case the PJ is more spread out for neighbor bins than the experiments with *Long* cable.

## 8.6 Experiment with SoftKinetic DS325

We repeat the experiment with the DS325 camera that has the first component of the PJ at $(0.399 \pm 0.003)$ MHz using the SDR USB dongle, for the measurements. For the full spectrum of the DFT, the upper limit of the frequency should be at least 0.8 MHz. However, with a safe margin, we chose 1 MHz as the IF for the SDR receiver. Therefore, we down-mixed the emitted light signal (50 MHz) of the camera with 49 MHz as the RF tuner LO frequency to give a 1 MHz IF signal at the ADC. The captured light signal of the DS325 with *Long* cable is shown in Fig. 8.13.

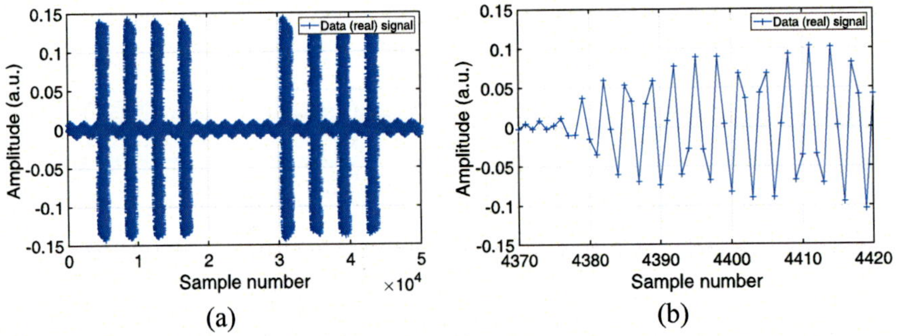


**Fig. 8.13** Data (real part) signal from the SDR setup for a single acquisition of (**a**) the light source of DS325 and (**b**) its expansion version, with the *Long* coaxial cable

Since the DS325 camera uses four phase steps in its correlation, Fig. 8.13a shows the corresponding portions for $0, \pi/2, \pi, 3\pi/2$ radians. Figure 8.13b indicates that the number of samples per cycle is a very low value (3 or 4). Therefore the representation of the signal is poor and it will give incorrect jitter amounts; hence, we do not further investigate the analysis of jitter of the light signal of the DS325 camera using this RTL-SDR dongle.

Both the oscilloscopes and SDR dongles have inherent frequency errors; however, the reasonable uncertainty of the measurements occurs from 200 MHz (about 0.2%) in oscilloscopes [43]. Since the modulation frequency of each camera is much less than 200 MHz, we believe the inherent frequency errors in the oscilloscope has not affected jitter measurement. We were unable to find any references about the inherent frequency errors in SDR dongles and there may be errors that effect the jitter measurement.

## 8.7 Chapter Remarks

The main aim of this chapter was to use the SDR technology (as an alternative measurement instrument for oscilloscope) to measure the jitter in signals. We experimented with a cheap SDR USB dongle with the proposed method to measure jitter on the emitted light signal of MESA SR4000 and SoftKinetic DS325 depth cameras. For SR4000, we downshifted the signal to a low frequency (400 kHz) and were able to find the periodic jitter at lower frequencies (less than 1.6 MHz, which is the Nyquist frequency of this particular dongle). By performing the proposed algorithm in Chap. 5, we extracted some of the jitter parameters for the camera SR4000.

We believe the high sample rates and high quality of SDRs can be used to explore periodic jitter at least at lower frequencies in an RF signal. We need to know the ratio of the number of samples per cycle between the two instruments: oscilloscope and SDR receiver for the correction of the amplitude calculations of the periodic and random jitter. From our proposed method, it is straightforward to find the corresponding frequency (only for less than the Nyquist frequency of the dongle) of the periodic jitter in the light source of the camera. However, jitter extraction was not a success for the other camera DS325, since the number of samples per cycle in the light source of the camera captured from the SDR receiver is very less it causes the representation of the signal to be incorrect.

Hence, we conclude the SDR USB dongle, which is a much cheaper instrument (respect to the oscilloscope), can be used to measure jitter in a signal, but there are limitations: (1) the periodic and random jitter only up to half of the intermediate frequency can be obtained from the downshift of the amplified radio frequency with the local oscillator, which is less than the Nyquist frequency of the dongle and (2) when the sampling frequency is less (i.e., if the number of sample points is not sufficient in a cycle), then this SDR dongle cannot be used for jitter measurement since the signal is not properly (smoothly) represented. Therefore, it has given incorrect jitter amounts.

## References

1. Joseph Mitola. Software radios: Survey, critical evaluation and future directions. *IEEE Aerospace and Electronic Systems Magazine*, 8(4): 25–36, 1993.
2. Walter Tuttlebee. *Software defined radio: origins, drivers and international perspectives*. Wiley, first edition, 2002.
3. José Raúl Machado-Fernández. Software defined radio: Basic principles and applications. *Facultad de Ingeniería*, 24(38): 79–96, 2015.
4. RI Lackey and Donald W Upmal. Speakeasy: the military software radio. *IEEE Communications Magazine*, 33(5): 56–61, 1995.
5. Walter HW Tuttlebee. *Software defined radio: enabling technologies*. John Wiley & Sons, 2003.
6. David Michael Davenport. Wireless communication with a mobile asset employing dynamic configuration of a software defined radio, August 30 2005. US Patent 6,937,877.
7. Pangan Ting, Hui-Ming Wang, and Nan-Sheng Huang. Software defined radio (SDR) architecture for wireless digital communication systems, December 19 2006. US Patent 7,151,925.
8. Namkyu Ryu, Yusuk Yun, Seungwon Choi, Ramesh Chembil Palat, and Jeffrey H Reed. Smart antenna base station open architecture for SDR networks. *IEEE Wireless Communications*, 13(3): 58–69, 2006.
9. Peter S Hall, Peter Gardner, and Antonio Faraone. Antenna requirements for software defined and cognitive radios. *Proceedings of the IEEE*, 100(7):2262–2270, 2012.
10. Friedrich K Jondral. Software-defined radio: basics and evolution to cognitive radio. *EURASIP journal on wireless communications and networking*, 2005(3): 275–283, 2005.
11. Rajeshree D Raut and Kishore D Kulat. SDR design for cognitive radio. In *2011 Fourth International Conference on Modeling, Simulation and Applied Optimization*, pages 1–8. IEEE, 2011.
12. Walter HW Tuttlebee. Software-defined radio: facets of a developing technology. *IEEE Personal Communications*, 6(2): 38–44, 1999.
13. Christoph Sandner, Franz Dielacher, Werner Hollinger, and Michael Staber. Power supply generation and battery management for energy efficient SDR. In *2008 IEEE International Conference on Acoustics, Speech and Signal Processing*, pages 5400–5403. IEEE, 2008.

14. Mamatha R Maheshwarappa, Mark Bowyer, and Christopher P Bridges. Software defined radio (SDR) architecture to support multi-satellite communications. In *2015 IEEE Aerospace Conference*, pages 1–10. IEEE, 2015.
15. Mamatha Ramapura Maheshwarappa, Mark DJ Bowyer, and Christopher P Bridges. Improvements in CPU and FPGA Performance for Small Satellite SDR applications. *IEEE Transactions on Aerospace and Electronic Systems*, 53(1): 310–322, 2017.
16. Kurt VonEhr, Seth Hilaski, Bruce E Dunne, and Jeffrey Ward. Software defined radio for direction-finding in UAV wildlife tracking. In *2016 IEEE International Conference on Electro Information Technology (EIT)*, pages 0464–0469. IEEE, 2016.
17. Amir Torabi, Michael W Shafer, Gabriel S Vega, and Kellan M Rothfus. UAV-RT: an SDR based aerial platform for wildlife tracking. In *2018 IEEE 88th Vehicular Technology Conference (VTC-Fall)*, pages 1–6. IEEE, 2018.
18. Jayaseelan Marimuthu, Konstanty S Bialkowski, and Amin M Abbosh. Software-defined radar for medical imaging. *IEEE Transactions on Microwave Theory and Techniques*, 64(2): 643–652, 2016.
19. KS Bialkowski, J Marimuthu, and AM Abbosh. Biomedical imaging system using software defined radio. In *2015 IEEE International Symposium on Antennas and Propagation & USNC/URSI National Radio Science Meeting*, pages 542–543. IEEE, 2015.
20. What is software-defined-radio used for. https://www.onesdr.com/2020/01/15/what-is-software-defined-radio-used-for/, 2020a. [Accessed June 2021].
21. ROUNDUP OF SOFTWARE DEFINED RADIOS. https://www.rtl-sdr.com/roundup-software-defined-radios/, 2014. [Accessed June 2021].
22. ABOUT RTL-SDR. https://www.rtl-sdr.com/about-rtl-sdr/, 2021a. [Accessed June 2021].
23. The Best HF/VHF Performance in the Smallest form factor. https://airspy.com/airspy-hf-discovery/, 2021b. [Accessed June 2021].
24. OUR REVIEW OF THE AIRSPY HF+: COMPARED AGAINST COLIBRINANO, AIRSPY MINI, RSP2. https://www.rtl-sdr.com/our-review-of-the-airspy-hf-compared-against-colibrinano-airspy-mini-rsp2-rtl-sdr/, 2017a. [Accessed June 2021].
25. RUSSIAN YOUTUBER COMPARES THE RTL-SDR BLOG V3 AGAINST THE COLIBRINANO ON HF. https://www.rtl-sdr.com/tag/colibrinano/, 2020b. [Accessed June 2021].
26. KiwiSDR: Wide-band SDR + GPS cape for the BeagleBone Black. http://kiwisdr.com/, 2021c. [Accessed June 2021].
27. A REVIEW OF THE KIWISDR: 10 KHZ–30 MHz WIDEBAND NETWORK SDR. https://www.rtl-sdr.com/a-review-of-the-kiwisdr-10-khz-30-mhz-wideband-network-sdr/, 2017b. [Accessed June 2021].
28. LIMESDR UNBOXING AND INITIAL REVIEW. https://www.rtl-sdr.com/limesdr-unboxing-initial-review/, 2017c. [Accessed June 2021].
29. A REVIEW OF THE HACKRF PORTAPACK (WITH HAVOC FIRMWARE). https://www.rtl-sdr.com/a-review-of-the-hackrf-portapack-with-havok-firmware/, 2018. [Accessed June 2021].
30. REVIEW: AIRSPY VS. SDRPLAY RSP VS. HACKRF. https://www.rtl-sdr.com/review-airspy-vs-sdrplay-rsp-vs-hackrf/, 2016. [Accessed June 2021].
31. Aditya Gudipati, Daniel Perry, Li Erran Li, and Sachin Katti. SoftRAN: Software defined radio access network. In *Proceedings of the second ACM SIGCOMM workshop on Hot topics in software defined networking*, pages 25–30, 2013.
32. Markus Dillinger, Kambiz Madani, and Nancy Alonistioti. *Software defined radio: Architectures, systems and functions*. John Wiley & Sons, 2005.
33. Songlin Sun, Michel Kadoch, Liang Gong, and Bo Rong. Integrating network function virtualization with SDR and SDN for 4G/5G networks. *IEEE Network*, 29(3): 54–59, 2015.
34. Alexandre Marquet, Nicolas Montavont, and Georgios Z Papadopoulos. Towards an SDR implementation of LoRa: Reverse-engineering, demodulation strategies and assessment over Rayleigh channel. *Computer Communications*, 2020.
35. BUY RTL-SDR DONGLES (RTL2832U). https://www.rtl-sdr.com/buy-rtl-sdr-dvb-t-dongles/, 2019. Accessed June 2021.

36. *RTL-SDR Blog V3 Datasheet*. Agilent Technologies, 2019. https://www.rtl-sdr.com/wp-content/uploads/2018/02/RTL-SDR-Blog-V3-Datasheet.pdf.
37. R820T2 Register Description. https://www.rtl-sdr.com/wp-content/uploads/2016/12/R820T2_Register_Description.pdf, 2011. Accessed June 2021.
38. Realtek RTL2832U - The mystery chip at the heart of RTL-SDR. http://datasheetcafe.databank.netdna-cdn.com/wp-content/uploads/2015/09/RTL2832U.pdf, 2017. Accessed June 2021.
39. *MULTI-STANDARD CMOS TERRESTRIAL RF TUNER*. Elonics Ltd, 2010.
40. RTL2832 datasheet/deep info? https://www.nooelec.com/files/e4000datasheet.pdf, 2014.
41. Rodger H Hosking. *Software-Defined Radio Handbook*. Pentek Inc, 12 edition, 2016.
42. Carl Laufer and Edward J Hoffman. *The Hobbyist's Guide to the RTL-SDR: Really Cheap Software Defined Radio: A Guide to the RTL-SDR and Cheap Software Defined Radio by the Authors of the RTL-SDR.com Blog*. Carl Laufer, 2016.
43. Paul Roberts. *Understanding and Dealing with High Frequency Error Sources in Oscilloscope Calibration*. Fluke Precision Measurement Ltd.

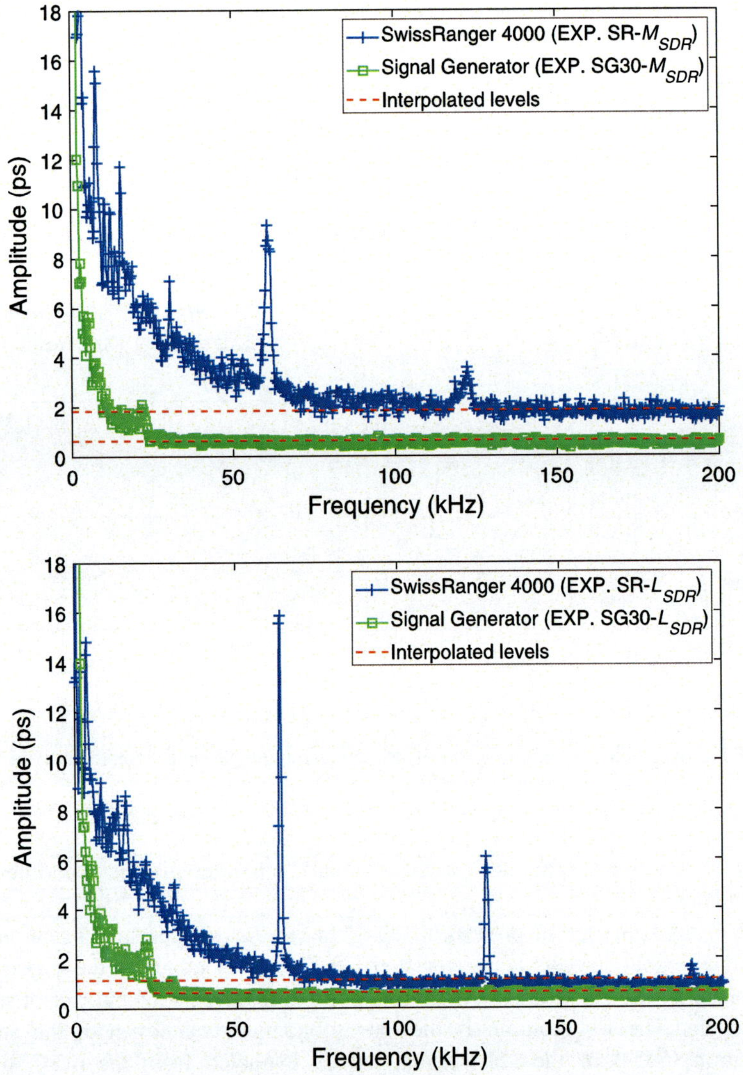


# Part III
# Jitter Influence on Range Measurements

From Chaps. 7 and 8, we clearly found the periodic and random jitter in the light signals of the two AMCW ToF ranging cameras and successfully measured by using a proposed methodology that can be used for any type of illumination signal of ranging systems. Even though those two types of jitter were measured, will it be important to the user and will it influence the range measurements? This interesting investigation will be inspected in the next two chapters, which is the last part of this book.

Part III comprises the inspection of the influence of two types of jitter: periodic and random, in AMCW ToF ranging systems. Chapter 9 is an investigation of the behavior of the periodic jitter on range measurements. The primary contribution is the derivation of an analytical model in the Fourier domain. Using Fourier analysis and standard statistical theorems, the benchmark stochastic model for the influence of random jitter in the AMCW ToF range imaging is constructed in Chap. 10.

# Chapter 9
# Influence of Periodic Jitter

**Abstract** This chapter presents a theoretical investigation of the influence of periodic jitter in AMCW ToF range imaging cameras. An analytical model with the help of Fourier analysis and standard mathematical functions is developed. First, the standard mathematical expressions with relevant properties of the Fourier transform are introduced. Then, the analytical model is explored to predict the effect of periodic jitter on range measurements in AMCW ToF cameras. Both homodyne and heterodyne operations of the camera are considered. In addition, two known numerical approaches: trapezoidal and Romberg integration methods are also supported to verify the analytical model obtained. For the numerical approaches, the four phase step computation method of the ToF range imaging cameras is considered. Next, the operation principle without the jitter in the emitted light source and shutter signals in AMCW lidar systems is investigated by deducting the analytical model obtained. The results and discussion for different configurations on simulation data with various parameters in AMCW lidar systems are presented. Finally, the approximate computation time for each configuration is pointed out.

**Keywords** Periodic jitter · Analytical model · Dirac delta · Rectangular function · sinc function · Bessel function · Fourier transform · Inverse Fourier transform · Scaling · Shifting · Convolution · Phase-step · Trapezoidal integration · Romberg integration · Phase and range errors

## 9.1 Background

As stated in Chap. 2, current modulation frequencies of AMCW ToF cameras are less than 100 MHz. All these cameras have different accuracy and precision levels (see Sect. 2.5). However, due to improvements of the technologies in camera sensors, it is not surprising the modulation frequencies of those cameras will be in GHz levels with the high frame rate in the future [1]. In general, the use of higher modulation frequencies guarantees higher range resolution yields on the higher noise and jitter levels as well [2]. Therefore, with the improvement of the technologies, it is important to consider the accuracy and precision of the depth


G. Anthonys, *Timing Jitter in Time-of-Flight Range Imaging Cameras*,
https://doi.org/10.1007/978-3-030-94159-8_9

measurements in AMCW ToF cameras. Accuracy is defined how closeness of the measurement to its actual value and precision is the distribution of random errors, refer Sect. 3.2.1.

Chapters 7 and 8 have demonstrated that periodic and random jitter are present in commercial AMCW ToF range imaging cameras, at least, in the emitted light signal. However, to date, there is no prior investigation of the impact of periodic jitter on range measurements in AMCW lidar systems. The investigation of jitter influence on the depth data is important because this time deviation causes errors on the phase difference (which is the ToF principle), hence on the range measurement. Once a model of jitter influence on the ToF cameras is developed, such a model can be used to correct for these effects on range imaging data. Not only that, the result of this investigation can be used in AMCW ToF camera manufacturing industries in future because the precision and accuracy are important for many applications.

A mathematical model is analytically obtained using some standard mathematical functions and the Fourier analysis for the correlation function in AMCW ToF range imaging cameras and evaluated with the numerical data for characteristic parameters of the camera and periodic jitter. In order to clarify the result, the numerical approach is used to solve the correlation function with the same parameters.

## 9.2 Notation and Definitions

Some definitions and notations that are needed for this investigation are described in this section. The standard mathematical functions and Fourier analysis are mainly reviewed. Let $i$ be the imaginary unit for complex $z = u + iv$, and let $\Re[z]$ be the real part of $z$, that is $\Re[z] = u$. Let $a$ and $b$ be non-zero real constant and a real constant, respectively. Also, let $g(t)$ and $h(t)$ be square-integrable functions, $*$ be the convolution operator, and $f$ be the frequency conjugate to time $t$.

### 9.2.1 *Mathematical Functions*

This section describes some standard mathematical functions with their useful properties and simplifies relationships for the advanced expressions used in this chapter.

#### 9.2.1.1 Definitions

Let $\delta(t)$ be the conventional Dirac delta function is defined by [3]

$$\delta(t) = \begin{cases} \infty & t = 0 \\ 0 & \text{otherwise}, \end{cases} \qquad \text{with} \qquad \int_{-\infty}^{\infty} \delta(t)\, dt = 1. \tag{9.1}$$

The value of $t$ can be controlled by substituting $t = t - t_0$ as

$$\delta(t - t_0) = \begin{cases} \infty & t = t_0 \\ 0 & \text{otherwise.} \end{cases} \tag{9.2}$$

The scaling property of $\delta(t)$ is

$$\delta(at) = \frac{\delta(t)}{|a|}. \tag{9.3}$$

Since the delta function is an even distribution, we have

$$\delta(-\alpha) = \delta(\alpha). \tag{9.4}$$

Let $\text{rect}_T(t)$ be the rectangular function of width $T$ defined by

$$\text{rect}_T(t) = \begin{cases} 1 & |t| < T/2 \\ 0 & \text{otherwise.} \end{cases} \tag{9.5}$$

The unnormalized sinc function (actually, the full name of the function is “sine cardinal”) be

$$\text{sinc}(t) = \begin{cases} \dfrac{\sin t}{t} & t \neq 0 \\ 1 & \text{otherwise} \end{cases} \tag{9.6}$$

and is shown in Fig. 9.1 with zero crossings denoted by small circles. Alternatively, the unnormalized sinc function is often called the *sampling function*.

#### 9.2.1.2 Relationships

The sinc function for any argument $\alpha \in \mathbb{C}$ has the relation when $T \to \infty$ [4]

$$\lim_{T\to\infty} T\, \text{sinc}(\alpha T) = \pi\, \delta(\alpha). \tag{9.7}$$

Let $J_m(x)$ be the Bessel function of the first kind of order $m$ given by [5]

$$J_m(x) = \frac{1}{2\pi}\int_{-\pi}^{\pi} e^{i(x\sin\theta - m\theta)}\, d\theta. \tag{9.8}$$

Note the expansion of the exponential in (9.8) is given by [6]

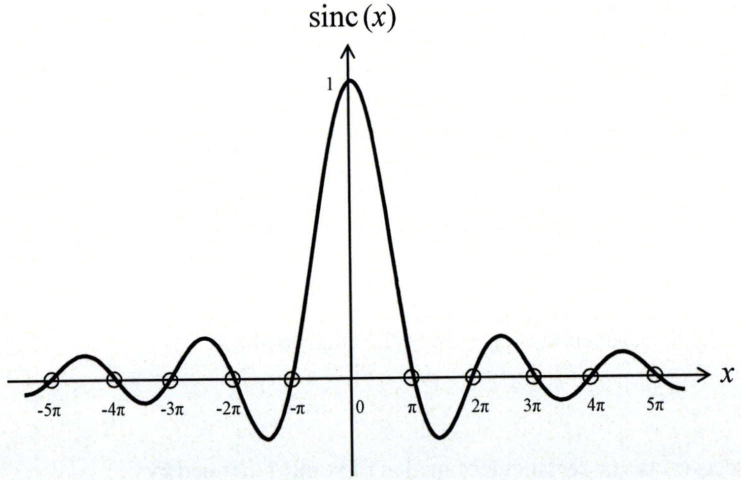


**Fig. 9.1** Indication of the zero crossings (denoted by ○) of a normalized sinc function

$$e^{ix\sin\theta} = \sum_{m=-\infty}^{\infty} J_m(x)\, e^{im\theta}, \tag{9.9}$$

and since the summation is linear, we have [6]

$$\Re\left[\sum_{m=-\infty}^{\infty} J_m(x)\, e^{im\theta}\right] = \sum_{m=-\infty}^{\infty} J_m(x)\ \cos(m\theta). \tag{9.10}$$

## 9.2.2 *Fourier Transform and Inverse Fourier Transform*

This section introduces the definitions with some identities of Fourier transform (FT) and inverse Fourier transform (IFT). In addition, the corresponding Fourier transforms of some functions are pointed out.

### 9.2.2.1 Definitions

Throughout this chapter, the following definitions for the FT and IFT are used. The Fourier transform of $g(t)$ is given by [7]

$$G(f) = \mathscr{F}[g(t)] = \int_{-\infty}^{\infty} g(t) e^{-i2\pi f t}\, dt, \tag{9.11}$$

with inverse Fourier transform [7],

$$g(t) = \mathscr{F}^{-1}[G(f)] = \mathscr{F}[g(-t)] = \mathscr{F}^{-1}[G(-f)] = \int_{-\infty}^{\infty} G(f)e^{i2\pi ft}\, df. \tag{9.12}$$

#### 9.2.2.2 Identities

The Fourier transform and its inverse satisfy the following well-known identities [7]: Scaling,

$$\mathscr{F}[g(at)] = \frac{1}{|a|} G\left(\frac{f}{a}\right), \tag{9.13}$$

Shifting,

$$\mathscr{F}[g(t \pm a)] = e^{\pm i2\pi af} G(f) \tag{9.14}$$

$$\mathscr{F}^{-1}[G(f \pm a)] = e^{\mp i2\pi at} g(t), \tag{9.15}$$

Convolution,

$$\mathscr{F}[gh] = \mathscr{F}[g] * \mathscr{F}[h] = G(f) * H(f) \tag{9.16}$$

$$\mathscr{F}^{-1}[G * H] = \mathscr{F}^{-1}[G] * \mathscr{F}^{-1}[H] = gh, \tag{9.17}$$

where $H(f)$ and $\mathscr{F}^{-1}[H]$ are the Fourier transform of $h(t)$ and inverse Fourier transform of $H(f)$, respectively.

#### 9.2.2.3 Fourier and Its Inverse Transforms of Some Functions

The following Fourier transforms are used in this chapter [7]:

$$\mathscr{F}[a] = a\,\delta(f) \tag{9.18}$$

$$\mathscr{F}[\mathrm{rect}_T(t)] = T\,\mathrm{sinc}(\pi f T) \tag{9.19}$$

$$\mathscr{F}[\cos(t)] = \frac{1}{2}\left[\delta\left(f - \frac{1}{2\pi}\right) + \delta\left(f + \frac{1}{2\pi}\right)\right] \tag{9.20}$$

$$\mathscr{F}[\cos(at + b)] = e^{i2\pi \frac{b}{a} f}\, \mathscr{F}[\cos(at)]. \tag{9.21}$$

Also, note that the sinc function is the inverse Fourier transform of the rectangular function in frequency centered at zero, with width $2\pi$ and unit height.

## 9.3 Correlation Model with Periodic Jitter

Re-consider the light and shutter relationship illustrated in Fig. 2.4 by setting the shutter signal with zero offset as shown in Fig. 9.2. Here, we slightly modified Fig. 2.4 to simplify the mathematical model to be obtained. The ToF is measured indirectly by correlating the reflected signal with a reference signal of the illumination and determining the phase shift $\phi$ due to the time of flight (refer Sect. 2.2.3).

Periodic jitter is defined as the repetition of the jitter at a certain period (or cycle) in a signal [8]. Let us introduce periodic jitter as a form of deterministic jitter into both the light and shutter sources to investigate its effect on reconstructing range. Introduce the periodic jitter $\Delta t^{\mathrm{PJ}}(t)$ of amplitude $A_{\mathrm{PJ}}^{(p)}$ and frequency $f_{\mathrm{PJ}}^{(p)}$ that are defined as (4.8) in Sect. 4.2.2, into both the light and shutter sources. Then, the modulated light signal is given by

$$l(t, \phi) = A \cos\left(2\pi f_l (t + \Delta t^{\mathrm{PJ}}(t)) - \phi\right) + B, \tag{9.22}$$

where $A$ is the amplitude of the illumination signal, $B$ is any offset due to background illumination, $f_l$ is the light modulation frequency, and $\phi$ is the ToF phase delay. Let $s(t, \theta)$ be the shutter signal given by

$$s(t, \theta) = \cos\left(2\pi f_s (t + \Delta t^{\mathrm{PJ}}(t)) + \theta\right), \tag{9.23}$$

where $f_s$ and $\theta$ are the modulation frequency of the shutter signal and an adjustable phase delay introduced to the shutter (the *phase step*, which is controlled by the camera), respectively. Note, it is assumed that the noise in the signals is zero and

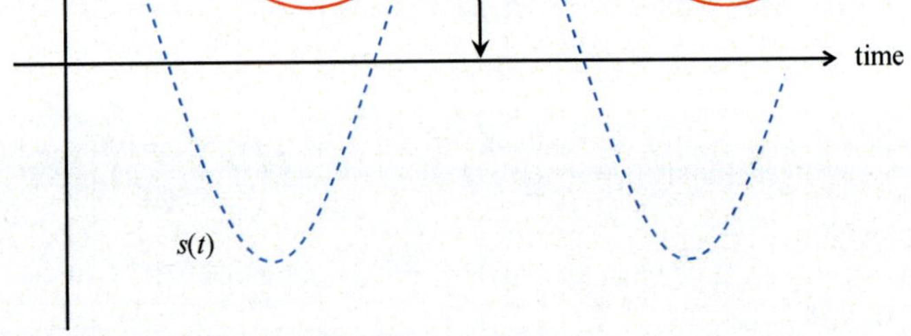


**Fig. 9.2** Light and shutter signals relationship in a ToF camera (slightly modified version of Fig. 2.4)

both the ToF phase shift and phase step are in units of radians by convention. In addition, we assume the offset background of the shutter signal to zero and its amplitude $a$ is embedded in $A$ and $B$ of the reflected light.

Here, we consider the periodic jitter at single frequency, that is, the superscript $^{(p)} = {}^{(1)}$ in (4.8). Then, substitute the PJ defined in (4.8) into (9.22) and (9.23), separately, which gives

$$l(t, \phi) = A\cos(2\pi f_l(t + A_{\mathrm{PJ}}\sin(2\pi f_{\mathrm{PJ}}t)) - \phi) + B, \tag{9.24}$$

and

$$s(t, \theta) = \cos(2\pi f_s(t + A_{\mathrm{PJ}}\sin(2\pi f_{\mathrm{PJ}}t + \phi^s_{\mathrm{PJ}})) + \theta), \tag{9.25}$$

where $\phi^s_{PJ}$ is the phase difference of the periodic jitter in shutter signal with respect to the light signal. Note that for the readability, all superscripts $^{(1)}$ in (9.24) and (9.25) have been ignored.

Now, let us consider how the correlation model with jitter can be solved numerically and analytically. However, the solution of the proposed analytical approach is mainly focused.

## 9.3.1 Numerical Approaches

In general, for computing an integral, most numerical methods split up the original integral into a sum of several integrals, each covering a small part of the original integration interval. Two types of integration methods, the trapezoidal and Romberg, for the approximate calculation of the definite integral are used [9]. Then, the integral can be found by

$$I(t) = \int_{-T/2}^{T/2} c(t)\, dt = \int_{-T/2}^{T/2} s(t, \theta)\, l(t, \phi)\, dt. \tag{9.26}$$

This integration in (9.26) is repeated for four phase frames $\theta = 0, \pi/2, \pi$ and $3\pi/2$ to compute the phase, $\phi^{\mathrm{cal}}_{\mathrm{num}}$ from (2.15a), and this will be used with the range measurement relation (2.12) to find the corresponding distance to the scene (refer Sect. 2.2.3).

### 9.3.1.1 Trapezoidal Integration

Let us consider the function $c(t)$ on the interval $[t_l\ t_u]$ as shown in Fig. 9.3. The integration in (9.26) can be represented as the area between the $t$-axis and the curve $c(t)$ of the integrand on the given interval. This is approximated with the sum of the

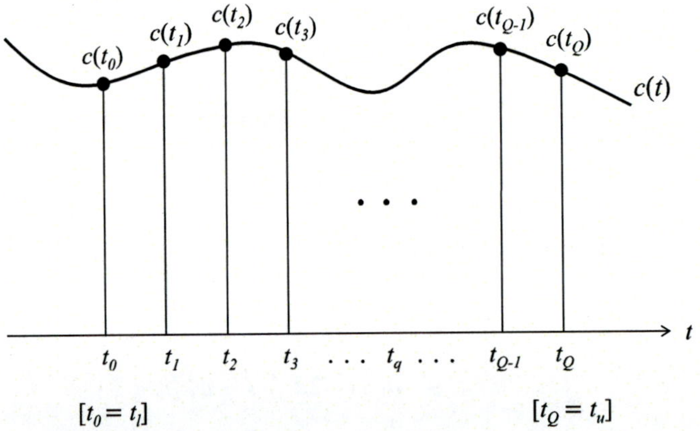


**Fig. 9.3** The trapezoidal integration for multiple segments on the function $c(t)$

areas of a series of parallel trapezoids as

$$I_{\text{trapz}} \approx \frac{1}{2} \sum_{q=0}^{Q-1} (t_{q+1} - t_q)\,(c(t_{q+1}) + c(t_q)), \quad Q \in \mathbb{Z}^+, \tag{9.27}$$

where $Q$ is the number of intervals and $t_q$ are the integration points that are distributed on the given interval $[t_0 = t_l \;\; t_Q = t_u]$. These points may, or may not, be evenly distributed. An even distribution simplifies the computation and is often sufficient. Then (9.27) becomes [10]

$$I_{\text{trapz}} \approx \frac{(t_u - t_l)}{2\,Q} \sum_{q=0}^{Q-1} \big(c(t_q) + c(t_{q+1})\big), \quad [t_l \; t_u], \tag{9.28}$$

where $(t_u - t_l)/Q$ is the gap on the $t$-axis between consecutive integration points. Here, the integration is computed under a straight-line segment to approximate the integral under a curve, and it is obvious that a substantial error can be incurred. If the function of the segment has a positive second derivative, then the integrand is concave-up; thus the error is negative and the trapezoidal integration overestimates the actual value.

A concave-down function yields an underestimate since the area is unaccounted for under the curve. Both of these situations are not considered in (9.28). Therefore, as usual, the error occurs in the trapezoidal method. The corresponding error bound formula for the trapezoidal rule is given by [10]

$$|Err_{\text{trapz}}| \leq \frac{K_{\text{trapz}}\,(t_u - t_l)^3}{12\,Q^2}, \tag{9.29}$$

where $K_{\text{trapz}}$ is a number such that $|c''(q)| \leq K_{\text{trapz}}$ for $\forall q \in [t_l \ t_u]$, and $c''(q)$ is the second derivative of the correlation function. Then, for a certain precision, we can obtain the sufficient number of subintervals as

$$Q \leq 0.5\sqrt{\frac{K_{\text{T}}(t_u - t_l)^3}{3\,|Err_{\text{trapz}}|}}. \tag{9.30}$$

However, from (9.28), it is clear that the number of intervals $Q$ increases and then the error decreases. Therefore, high accuracy can be attained by choosing a sufficiently large value for $Q$. But the computational work involved is roughly proportional to the number of points where $c(t)$ must be computed [11].

#### 9.3.1.2 Romberg Integration

When the interval size used in the trapezoidal rule is set to half by taking the number of subintervals to the power of two, these are called panels as shown in Fig. 9.4. Then by the recursive method, we can add the panels until the answer becomes to converge. This builds a form of a lower triangular matrix. The successive columns are computed by Richardson's extrapolation method, which is given by [9]

$$\mathcal{R}_{n+1,m+1} = \mathcal{R}_{n+1,m} + \frac{1}{4^m - 1}\left(\mathcal{R}_{n+1,m} - \mathcal{R}_{n,m}\right), \tag{9.31}$$

where $(n, m)$ is the $n$th row and $m$th column of the table.

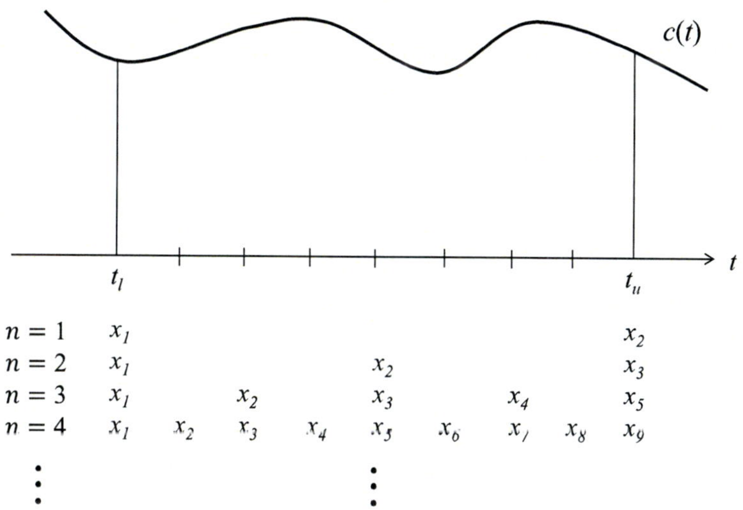


**Fig. 9.4** The recursive trapezoidal integration panels on the function $c(t)$

The first two columns are represented by the recursive trapezoidal and Simpson's rules, respectively. The success of repeated Richardson extrapolations is related to the behavior in the integration limit of the higher derivatives of the integrand. The most accurate estimate for the integrals is the value at the bottom right corner of the matrix as in (9.32).

$$[\mathcal{R}]_{M\times M} = \begin{pmatrix} \mathcal{R}_{1,1} & - & - & - & - \\ \mathcal{R}_{2,1} & \mathcal{R}_{2,2} & - & - & - \\ \mathcal{R}_{3,1} & \mathcal{R}_{3,2} & \mathcal{R}_{3,3} & - & - \\ \mathcal{R}_{4,1} & R_{4,2} & \mathcal{R}_{4,3} & \mathcal{R}_{4,4} & - \\ \vdots & \vdots & \vdots & \vdots & \ddots \end{pmatrix}, \quad M \in Z^{+}. \tag{9.32}$$

The error bound for Romberg's integration can be computed by [10]

$$Err_{\text{romb}} = 2^{m(1-2n+m)}(t_u - t_l)^{2m+1}\frac{|B_{2m}|}{(2m)!}c^{(2m)}(\xi), \qquad \xi(t_l, t_u), \tag{9.33}$$

where $B_{2m}$ is the Bernoulli numbers and $c^{(2m)}$ is the (2m)th derivative of $c(t)$ at $\xi$ that lies somewhere in the given interval.

Since the main purpose of this chapter is to explore an analytical approach, the investigation of error bounds in numerical approaches is not considered here. Those who are interested in such topics can refer [10].

## 9.3.2 *Analytical Approach*

The proposing analytical approach is the main contribution of this chapter and is systematically explained throughout this section. As stated before, the camera integrates a frame capture over an integration period of $T$ for a specific phase step $\theta$; hence, the received signal $I(\theta)$ is given by

$$I(\theta) = \int_{-T/2}^{T/2} s(t,\theta)\, l(t,\phi)\, dt. \tag{9.34}$$

In practice, it is typical to adjust $\theta$ stepwise between captures, but here we take the phase step $\theta$ to be continuous so that we can take the Fourier transform with respect to $\theta$, namely let

$$\begin{aligned}\widehat{I}(\vartheta) = \mathcal{F}[I(\theta)]_{\theta\to\vartheta} &= \mathcal{F}\left[\int_{-T/2}^{T/2} s(t,\theta)\, l(t,\phi)\, dt\right]_{\theta\to\vartheta} \\ &= \int_{-T/2}^{T/2} \mathcal{F}[s(t,\theta)]_{\theta\to\vartheta}\, l(t,\phi)\, dt,\end{aligned} \tag{9.35}$$

where $\vartheta$ is the angular frequency conjugate to $\theta$ in units of per radian (i.e., rad$^{-1}$). With (9.24) and (9.25), let us consider the correlation function in (9.35). By application of the Fourier theorems in (9.14) and (9.20), the Fourier transform of the shutter signal can be obtained, namely

$$\begin{aligned}\mathscr{F}[s(t,\theta)] &= \mathscr{F}\left[\cos\left(2\pi f_s(t + A_{\mathrm{PJ}}\sin\left(2\pi f_{\mathrm{PJ}}t + \phi^s_{\mathrm{PJ}}\right)) + \theta\right)\right]_{\theta\to\vartheta} \\ &= e^{i(2\pi)^2 f_s(t + A_{\mathrm{PJ}}\sin(2\pi f_{\mathrm{PJ}}t + \phi^s_{\mathrm{PJ}}))\vartheta}\,\mathscr{F}[\cos(\theta)] \\ &= \frac{1}{2}e^{i(2\pi)^2 f_s(t + A_{\mathrm{PJ}}\sin(2\pi f_{\mathrm{PJ}}t + \phi^s_{\mathrm{PJ}}))\vartheta}\left[\delta\left(\vartheta - \frac{1}{2\pi}\right) + \delta\left(\vartheta + \frac{1}{2\pi}\right)\right].\end{aligned} \tag{9.36}$$

Now, let $\Theta = 2\pi\vartheta$ be the "frequency" conjugate to $\theta$, and with the scaling property of delta function (9.3), substituting into (9.36), we have

$$\begin{aligned}\mathscr{F}[s(t,\theta)] &= \frac{e^{i2\pi f_s\Theta(t + A_{\mathrm{PJ}}\sin(2\pi f_{\mathrm{PJ}}t + \phi^s_{\mathrm{PJ}}))}}{2}\left[\delta\left(\frac{\Theta-1}{2\pi}\right) + \delta\left(\frac{\Theta+1}{2\pi}\right)\right] \\ &= \frac{e^{i2\pi f_s\Theta(t + A_{\mathrm{PJ}}\sin(2\pi f_{\mathrm{PJ}}t + \phi^s_{\mathrm{PJ}}))}}{2}\left[\frac{\delta(\Theta-1)}{|1/2\pi|} + \frac{\delta(\Theta+1)}{|1/2\pi|}\right], \quad \because (9.3) \\ &= \pi e^{i2\pi f_s\Theta(t + A_{\mathrm{PJ}}\sin(2\pi f_{\mathrm{PJ}}t + \phi^s_{\mathrm{PJ}}))}\left[\delta(\Theta-1) + \delta(\Theta+1)\right];\end{aligned} \tag{9.37}$$

hence, (9.35) becomes, with (9.24) gives,

$$\begin{aligned}\widehat{I}(\Theta) = \pi[\delta(\Theta-1) + \delta(\Theta+1)]\int_{-T/2}^{T/2} e^{i2\pi f_s\Theta(t + A_{\mathrm{PJ}}\sin(2\pi f_{\mathrm{PJ}}t + \phi^s_{\mathrm{PJ}}))} \\ \times\left\{A\cos\left(2\pi f_l(t + A_{\mathrm{PJ}}\sin(2\pi f_{\mathrm{PJ}}t)) - \phi\right) + B\right\}dt.\end{aligned} \tag{9.38}$$

Let $\beta_s = 2\pi f_s A_{\mathrm{PJ}}$ and $\beta_l = 2\pi f_l A_{\mathrm{PJ}}$, and rearrangement gives

$$\begin{aligned}\widehat{I}(\Theta) &= \pi[\delta(\Theta-1) + \delta(\Theta+1)]\int_{-T/2}^{T/2} e^{i2\pi f_s\Theta t}\, e^{i\beta_s\Theta\sin(2\pi f_{\mathrm{PJ}}t + \phi^s_{\mathrm{PJ}})} \\ &\quad\times\left\{A\cos\left(2\pi f_l t - \phi + \beta_l\sin(2\pi f_{\mathrm{PJ}}t)\right) + B\right\}dt \\ &= \pi\left[\delta(\Theta-1) + \delta(\Theta+1)\right]\left\{A\int_{-T/2}^{T/2} e^{i2\pi f_s\Theta t}\, e^{i\beta_s\Theta\sin(2\pi f_{\mathrm{PJ}}t + \phi^s_{\mathrm{PJ}})}\right. \\ &\quad\times\cos\left(2\pi f_l t - \phi + \beta_l\sin(2\pi f_{\mathrm{PJ}}t)\right)dt \\ &\quad\left. + B\int_{-T/2}^{T/2} e^{i2\pi f_s\Theta t}\, e^{i\beta_s\Theta\sin(2\pi f_{\mathrm{PJ}}t + \phi^s_{\mathrm{PJ}})}\,dt\right\}\end{aligned}$$

$$= \pi\big[\delta(\Theta-1)+\delta(\Theta+1)\big]\Big\{A\int_{-T/2}^{T/2} e^{i2\pi f_s\Theta t}\, e^{i\beta_s\Theta\sin(2\pi f_{\mathrm{PJ}}t+\phi^s_{\mathrm{PJ}})}$$

$$\times\,\Re\Big[e^{i(2\pi f_l t-\phi)}\, e^{i\beta_l\sin(2\pi f_{\mathrm{PJ}}t)}\Big]\,dt$$

$$+\,B\int_{-T/2}^{T/2} e^{i2\pi f_s\Theta t}\, e^{i\beta_s\Theta\sin(2\pi f_{\mathrm{PJ}}t+\phi^s_{\mathrm{PJ}})}\,dt\Big\}$$

$$= \pi\big[\delta(\Theta-1)+\delta(\Theta+1)\big]\Big\{A\,\mathcal{I}_{\mathcal{A}}+B\,\mathcal{I}_{\mathcal{B}}\Big\}, \tag{9.39}$$

where $\mathcal{I}_{\mathcal{A}}$ and $\mathcal{I}_{\mathcal{B}}$ are denoted for the corresponding amplitude and background integrals, respectively. Now, let us simplify the two integrals in (9.39), separately.

#### 9.3.2.1 Solution of the Amplitude Portion

Consider the integral denoted by $\mathcal{I}_{\mathcal{A}}$ in (9.39). Substitute (9.9) and evaluate the real function by using (9.10), and then separating time dependent terms gives

$$\mathcal{I}_{\mathcal{A}} = \int_{-T/2}^{T/2} e^{i2\pi f_s\Theta t}\, e^{i\beta_s\Theta\sin(2\pi f_{\mathrm{PJ}}t+\phi^s_{\mathrm{PJ}})}\,\Re\Big[e^{i(2\pi f_l t-\phi)}\, e^{i\beta_l\sin(2\pi f_{\mathrm{PJ}}t)}\Big]\,dt$$

$$= \int_{-T/2}^{T/2} e^{i2\pi f_s\Theta t}\sum_{m=-\infty}^{\infty} J_m(\beta_s\Theta)\, e^{im(2\pi f_{\mathrm{PJ}}t+\phi^s_{\mathrm{PJ}})}dt$$

$$\times\,\Re\left[e^{i(2\pi f_l t-\phi)}\sum_{n=-\infty}^{\infty} J_n(\beta_l)\, e^{in(2\pi f_{\mathrm{PJ}}t)}\right]dt, \quad \because (9.9)$$

$$= \int_{-T/2}^{T/2}\sum_{m=-\infty}^{\infty} J_m(\beta_s\Theta)\, e^{im\phi^s_{\mathrm{PJ}}}\, e^{i2\pi(f_s\Theta+mf_{\mathrm{PJ}})t}$$

$$\times\,\Re\left[\sum_{n=-\infty}^{\infty} J_n(\beta_l)\, e^{i(2\pi(f_l+nf_{\mathrm{PJ}})t-\phi)}\right]dt$$

$$= \sum_{m=-\infty}^{\infty} J_m(\beta_s\Theta)\, e^{im\phi^s_{\mathrm{PJ}}}\sum_{n=-\infty}^{\infty} J_n(\beta_l)$$

$$\times\int_{-T/2}^{T/2} e^{i2\pi(f_s\Theta+mf_{\mathrm{PJ}})t}\cos\big(2\pi(f_l+nf_{\mathrm{PJ}})t-\phi\big)dt, \quad \because (9.10). \tag{9.40}$$

The integral in (9.40) is found by

$$\int_{-T/2}^{T/2} e^{i2\pi(f_s\Theta + mf_{\mathrm{PJ}})t} \cos\big(2\pi(f_l + nf_{\mathrm{PJ}})t - \phi\big)dt$$

$$= \int_{-\infty}^{\infty} \mathrm{rect}_T(t) \cos\big(2\pi(f_l + nf_{\mathrm{PJ}})t - \phi\big)\, e^{i2\pi(f_s\Theta + mf_{\mathrm{PJ}})t}\, dt$$

$$= \mathscr{F}^{-1}\left[\mathrm{rect}_T(t)\, \cos\big(2\pi(f_l + nf_{\mathrm{PJ}})t - \phi\big)\right]_{t \to f_s\Theta + mf_{\mathrm{PJ}}}$$

$$= \mathscr{F}^{-1}\left[\mathrm{rect}_T(t)\right] * \mathscr{F}^{-1}\left[\cos\big(2\pi(f_l + nf_{\mathrm{PJ}})t - \phi\big)\right]$$

$$= T\ \mathrm{sinc}(\pi(f_s\Theta + mf_{\mathrm{PJ}})T) * \left(\exp\left(\frac{i2\pi(f_s\Theta + mf_{\mathrm{PJ}})\,\phi}{2\pi(f_l + nf_{\mathrm{PJ}})}\right)\right.$$

$$\left. \times \mathscr{F}^{-1}\big[\cos\big(2\pi(f_l + nf_{\mathrm{PJ}})t\big)\big]\right), \quad \because (9.15)$$

$$= T\ \mathrm{sinc}(\pi(f_s\Theta + mf_{\mathrm{PJ}})T) * \left(\exp\left(\frac{i2\pi(f_s\Theta + mf_{\mathrm{PJ}})\,\phi}{2\pi(f_l + nf_{\mathrm{PJ}})}\right)\right.$$

$$\left. \times \frac{1}{2}\big[\delta(f_s\Theta + mf_{\mathrm{PJ}} - f_l - nf_{\mathrm{PJ}}) + \delta(f_s\Theta + mf_{\mathrm{PJ}} + f_l + nf_{\mathrm{PJ}})\big]\right), \quad \because (9.20)$$

$$= \frac{T}{2} \exp\left(i\left(\frac{f_s\Theta + mf_{\mathrm{PJ}}}{f_l + nf_{\mathrm{PJ}}}\right)\phi\right)\Big[\mathrm{sinc}(\pi(f_s\Theta + mf_{\mathrm{PJ}} - f_l - nf_{\mathrm{PJ}})T)$$

$$+ \mathrm{sinc}(\pi(f_s\Theta + mf_{\mathrm{PJ}} + f_l + nf_{\mathrm{PJ}})T)\Big], \tag{9.41}$$

where we have used the fact that the sinc(·) and delta(·) functions are both even, and thus there is no difference between the forward and inverse transforms. We used inverse Fourier shift theorem (9.15) and the Fourier transform of the cosine (9.20). Hence, from (9.40), the integral $\mathcal{I}_{\mathcal{A}}$ is given by

$$\mathcal{I}_{\mathcal{A}} = \frac{T}{2} \sum_{m=-\infty}^{\infty} J_m(\beta_s\Theta)\, e^{im\phi^s_{\mathrm{PJ}}} \sum_{n=-\infty}^{\infty} J_n(\beta_l)\ \exp\left(i\left(\frac{f_s\Theta + mf_{\mathrm{PJ}}}{f_l + nf_{\mathrm{PJ}}}\right)\phi\right)$$

$$\times \big[\mathrm{sinc}(\pi(f_s\Theta + mf_{\mathrm{PJ}} - f_l - nf_{\mathrm{PJ}})T)$$

$$+ \mathrm{sinc}(\pi(f_s\Theta + mf_{\mathrm{PJ}} + f_l + nf_{\mathrm{PJ}})T)\big]. \tag{9.42}$$

Note that when $m = n$ and at the $|\Theta| = f_l/f_s$ frequency, the phase shift $\phi$ in (9.42) due to ToF appears as the phase of the signal scaled by $(f_s\Theta + mf_{\mathrm{PJ}})/(f_l + nf_{\mathrm{PJ}})$. It is therefore typical to operate a camera in homodyne ($f_l = f_s$) and capture a sequence of $N$ frames with $\theta = k/N$ for $k = 0, \ldots, N-1$, and then the phase can be recovered from the fundamental frequency of an $N$-point DFT of the captured frames.

#### 9.3.2.2 Solution of the Background Portion

In practice, the background lighting $B$ degrades SNR because it reduces dynamic range at the sensor [12]. Consider the integral denoted by $\mathcal{I}_{\mathcal{B}}$ in (9.39) that corresponds to the background illumination as

$$\begin{aligned}
\mathcal{I}_{\mathcal{B}} &= \int_{-T/2}^{T/2} e^{i2\pi f_s \Theta t}\, e^{i\beta_s \Theta \sin(2\pi f_{\mathrm{PJ}} t + \phi^s_{\mathrm{PJ}})}\, dt \\
&= \int_{-T/2}^{T/2} e^{i2\pi f_s \Theta t} \sum_{w=-\infty}^{\infty} J_w(\beta_s\Theta)\, e^{iw(2\pi f_{\mathrm{PJ}} t + \phi^s_{\mathrm{PJ}})}\, dt \\
&= \int_{-T/2}^{T/2} \sum_{w=-\infty}^{\infty} J_w(\beta_s\Theta)\, e^{iw\phi^s_{\mathrm{PJ}}}\, e^{i2\pi(f_s\Theta + wf_{\mathrm{PJ}})t}\, dt \\
&= \sum_{w=-\infty}^{\infty} J_w(\beta_s\Theta) \exp\left(iw\phi^s_{\mathrm{PJ}}\right) \int_{-\infty}^{\infty} \mathrm{rect}_T(t)\, e^{i2\pi(f_s\Theta + wf_{\mathrm{PJ}})t}\, dt \\
&= \sum_{w=-\infty}^{\infty} J_w(\beta_s\Theta) \exp\left(iw\phi^s_{\mathrm{PJ}}\right) \mathscr{F}^{-1}\left[\mathrm{rect}_T(t)\right]_{t\to(f_s\Theta + wf_{\mathrm{PJ}})} \\
&= T \sum_{w=-\infty}^{\infty} J_w(\beta_s\Theta) \exp\left(iw\phi^s_{\mathrm{PJ}}\right) \mathrm{sinc}(\pi(f_s\Theta + wf_{\mathrm{PJ}})T), \quad \because (9.19).
\end{aligned} \tag{9.43}$$

According to (9.43), the background light $B$ is canceled (as $B\,\mathcal{I}_{\mathcal{B}}$ in (9.39)) only if the frame integration time $T$ is an exact integer multiple of the modulation period of the term "$f_s\Theta + wf_{\mathrm{PJ}}$." In current cameras this term contributes an unmeasurably small perturbation to the phase measurement. These zero crossings are located at integer values of argument $x$ (see Fig. 9.1). Now, substituting (9.42) and (9.43) into (9.39) gives

$$\begin{aligned}
\widehat{I}(\Theta) = \frac{\pi T}{2}[\delta(\Theta-1) + \delta(\Theta+1)]\Bigg\{ A \sum_{m=-\infty}^{\infty}\sum_{n=-\infty}^{\infty} J_m(\beta_s\Theta)\, J_n(\beta_l) \\
\times \exp\left(i\left\{\left(\frac{f_s\Theta + mf_{\mathrm{PJ}}}{f_l + nf_{\mathrm{PJ}}}\right)\phi + m\phi^s_{\mathrm{PJ}}\right\}\right)\left[\mathrm{sinc}(\pi(f_s\Theta + mf_{\mathrm{PJ}} - f_l - nf_{\mathrm{PJ}})T)\right. \\
\left. + \,\mathrm{sinc}(\pi(f_s\Theta + mf_{\mathrm{PJ}} + f_l + nf_{\mathrm{PJ}})T)\right] \\
+ 2B \sum_{w=-\infty}^{\infty} J_w(\beta_s\Theta) \exp\left(iw\phi^s_{\mathrm{PJ}}\right) \mathrm{sinc}(\pi(f_s\Theta + wf_{\mathrm{PJ}})T)\Bigg\}.
\end{aligned} \tag{9.44}$$

Considering the fundamental frequency ($\Theta = 1$) in (9.44) is given by

$$\widehat{I}(1) = \frac{\pi T}{2}\Bigg\{A \sum_{m=-\infty}^{\infty} \sum_{n=-\infty}^{\infty} J_m(\beta_s) J_n(\beta_l) \exp\left(i \left\{\left(\frac{f_s + m f_{\mathrm{PJ}}}{f_l + n f_{\mathrm{PJ}}}\right)\phi + m\phi^s_{\mathrm{PJ}}\right\}\right)$$
$$\times \left[\operatorname{sinc}(\pi(f_s - f_l + (m-n) f_{\mathrm{PJ}})T) + \operatorname{sinc}(\pi(f_s + f_l + (m+n) f_{\mathrm{PJ}})T)\right]$$
$$+ 2B \sum_{w=-\infty}^{\infty} J_w(\beta_s) \exp\left(i w \phi^s_{\mathrm{PJ}}\right) \operatorname{sinc}(\pi(f_s + w f_{\mathrm{PJ}})T)\Bigg\}. \tag{9.45}$$

Note that since the amplitude and background portions are independent, the index $w$ can be replaced by either $m$ or $n$. This replacement is more effective when numerical computations are performed since the number of variables has to be minimized. In (9.45), the amplitude portion consists of two complex series with the product of common Bessel functions and two sinc terms, while the background portion has a complex series with a Bessel function and a sinc term. These series affect to the ToF phase as well as the measured magnitude of $\widehat{I}(1)$. Let us thoroughly analyze the behavior of these terms and some characteristics of (9.45) as follows.

#### 9.3.2.3 Long Integration Period

In the current implementations the integration period of the frame capture $T$ is very long compared to the period of the modulation cycle, and thus, in the limit of long integration time (i.e., $T \gg 1/f_s, 1/f_l$), substituting (9.7) and (9.3) into (9.45) gives

$$\lim_{T\to\infty} \widehat{I}(1) = \frac{\pi}{2}\Bigg\{A \sum_{m=-\infty}^{\infty} \sum_{n=-\infty}^{\infty} J_m(\beta_s) J_n(\beta_l) \exp\left(i \left\{\left(\frac{f_s + m f_{\mathrm{PJ}}}{f_l + n f_{\mathrm{PJ}}}\right)\phi + m\phi^s_{\mathrm{PJ}}\right\}\right)$$
$$\times \left[\delta(f_s - f_l + (m-n) f_{\mathrm{PJ}}) + \delta(f_s + f_l + (m+n) f_{\mathrm{PJ}})\right]$$
$$+ 2B \sum_{w=-\infty}^{\infty} J_w(\beta_s) \exp\left(i w \phi^s_{\mathrm{PJ}}\right) \delta(f_s + w f_{\mathrm{PJ}})\Bigg\}. \tag{9.46}$$

With the long integration period, the sinc functions become delta functions. From the definition, the delta function produces a value when its argument is zero only. The frequency of the periodic jitter and the frequencies in both the illumination and shutter modulation signals are contributed to the arguments of delta functions. So, it is needed to find the behavior of those frequencies in three delta functions obtained in (9.46).

#### 9.3.2.4 Jitter Frequency Is a Non-factor of Modulation Frequency

If $f_{\mathrm{PJ}}$ is not a factor of $f_s - f_l$ in (9.46), then the first delta function in the double summation is only non-zero if both $f_s = f_l$ and $m = n$. The second delta function is only ever potentially non-zero for large (and negative) $m$ and $n$ at which point the Bessel functions are effectively zero. Similarly, in the single summation (i.e., background part) when $f_{\mathrm{PJ}}$ is not a factor of $f_s$ then the delta function is only non-zero for large $w$ which the Bessel functions are approaching to zero. Hence, the background portion is vanished. In this case, let $\beta_s = \beta_l = \beta$ and then (9.46) gives

$$\begin{aligned}\lim_{T\to\infty} \widehat{I}(1) &= \frac{\pi A}{2} \sum_{n=-\infty}^{\infty} J_n^2(\beta)\, e^{i(\phi + n\phi^s_{\mathrm{PJ}})} \\ &= \frac{\pi A}{2} e^{i\phi} \left[ J_0^2(\beta) + \sum_{n=1}^{\infty} \left( J_n^2(\beta)\, e^{in\phi^s_{\mathrm{PJ}}} + J_{-n}^2(\beta)\, e^{-in\phi^s_{\mathrm{PJ}}} \right) \right] \\ &= \frac{\pi A}{2} e^{i\phi} \left[ J_0^2(\beta) + 2\sum_{n=1}^{\infty} J_n^2(\beta) \cos(n\phi^s_{\mathrm{PJ}}) \right], \end{aligned} \tag{9.47}$$

where we have used a property of the Bessel function $J_{-n}(\beta) = (-1)^n J_n(\beta)$. The last line shows that while there may be distortions in received amplitude (assuming sufficiently large $\beta$), there is nevertheless no error in the ToF phase.

### 9.3.2.5 Jitter Frequency Is a Factor of Modulation Frequency

If on the other hand $f_{\mathrm{PJ}}$ is a factor of $f_s - f_l$, then the first delta function of the amplitude portion in (9.46) can be non-zero when $m \neq n$. Also, if $f_{\mathrm{PJ}}$ is a factor of $f_s$ in the background part of (9.46), the delta function can be non-zero only when $w = -f_s/f_{\mathrm{PJ}}$. These will lead to phases of the form $((f_s + mf_{\mathrm{PJ}})/(f_l + nf_{\mathrm{PJ}}))\phi$ and $w\phi^s_{\mathrm{PJ}}$ appearing in the exponentials in summations of the amplitude and background portions in (9.46), respectively, thus distorting the phase from the ToF phase. Whether these terms make a significant contribution depends on the value of $\beta$. Table 9.1 shows first five terms of $J_n(\beta)$ when $A_{\mathrm{PJ}} = 100$ ps with $f = 2^k \times 50\,\mathrm{MHz}$, $k = 0, 1, 2, 3$.

**Table 9.1** First five values of Bessel function of first kind of argument $\beta$ when $f = 2^k \times 50\,\mathrm{MHz}$, $k = 0, 1, 2, 3$, and $A_{\mathrm{PJ}} = 100$ ps

| | $\beta = 2\pi f A_{\mathrm{PJ}}$ and $n = 0, 1, 2, 3, 4$ | | | |
|---|---|---|---|---|
| $J_n(\beta)$ | 0.0314 | 0.0628 | 0.1257 | 0.2513 |
| $J_0(\beta)$ | $9.997 \times 10^{-1}$ | $9.990 \times 10^{-1}$ | $9.960 \times 10^{-1}$ | $9.843 \times 10^{-1}$ |
| $J_1(\beta)$ | $1.571 \times 10^{-2}$ | $3.140 \times 10^{-2}$ | $6.271 \times 10^{-2}$ | $1.247 \times 10^{-1}$ |
| $J_2(\beta)$ | $1.233 \times 10^{-4}$ | $4.933 \times 10^{-4}$ | $1.971 \times 10^{-3}$ | $7.854 \times 10^{-3}$ |
| $J_3(\beta)$ | $6.459 \times 10^{-7}$ | $5.166 \times 10^{-6}$ | $4.130 \times 10^{-5}$ | $3.294 \times 10^{-4}$ |
| $J_4(\beta)$ | $2.536 \times 10^{-9}$ | $4.058 \times 10^{-8}$ | $6.489 \times 10^{-7}$ | $1.036 \times 10^{-5}$ |

It seems that the first three values: $J_0(\beta)$, $J_1(\beta)$, and $J_2(\beta)$ for each $\beta$ in the table are reasonable amounts. Thus, it is conceivable that particular combinations of $f_s$, $f_l$ and $f_{\text{PJ}}$ might lead to detectable errors in measuring the ToF phase.

#### 9.3.2.6 Correlation Model in Homodyne Operation

It is normal to operate a system in homodyne operation, that is $f_s = f_l = f$ and $\beta = 2\pi f A_{\text{PJ}} = \beta_s = \beta_l$, then (9.45) becomes

$$\begin{aligned}\widehat{I}(1) = \frac{\pi T}{2}\Bigg\{ & A \sum_{m=-\infty}^{\infty} \sum_{n=-\infty}^{\infty} J_m(\beta) J_n(\beta) \exp\left( i \left\{ \left( \frac{f + m f_{\text{PJ}}}{f + n f_{\text{PJ}}} \right) \phi + m \phi_{\text{PJ}}^{s} \right\} \right) \\ & \times \left[ \operatorname{sinc}(\pi(m-n) f_{\text{PJ}} T) + \operatorname{sinc}(\pi(2f + (m+n) f_{\text{PJ}})T) \right] \\ & + 2B \sum_{w=-\infty}^{\infty} J_w(\beta) \exp\left( i w \phi_{\text{PJ}}^{s} \right) \operatorname{sinc}(\pi(f + w f_{\text{PJ}})T) \Bigg\},\end{aligned} \tag{9.48}$$

and hence, the calculated phase from the analytical approach is given by

$$\phi_{\text{ana}}^{\text{cal}} = \angle \widehat{I}(1), \tag{9.49}$$

where $\angle$ is the angle of the resulted function.

Multiples of the phase relationship of the jitter, both $\phi$ and $\phi_{\text{PJ}}^{s}$, appear in the amplitude portion, while only $\phi_{\text{PJ}}^{s}$ appears in the background portion of the correlation model. Thus, these can contaminate the overall phase of the result. The amount of contamination of the phase depends on the frame integration time ($T$) and modulation frequency ($f$) of the camera, the amplitude of the periodic jitter ($A_{\text{PJ}}$), and the relationship of the periodic jitter frequency ($f_{\text{PJ}}$) to the modulation frequency.

Even without the phase difference of the periodic jitter between the shutter and light signals (that is $\phi_{\text{PJ}}^{s} = 0$ in (9.48)), the periodic jitter influences range measurements. In this case, the contribution from the amplitude portion of (9.48) is still complex series due to the presence of the term $(f + m f_{\text{PJ}})/(f + n f_{\text{PJ}})$, while the background portion contributes real part only. Thus, the overall phase of the result can be decreased than the same when $\phi_{\text{PJ}}^{s} \neq 0$.

When the frame integration is not a factor of the modulation cycle period, the first and second sinc functions in the amplitude portion in (9.48) are corresponding to the integration for a complete number of cycles and the remaining fraction of the last cycle during the modulation, respectively. If $f_{\text{PJ}} T$ is an integer, then the first sinc function in the amplitude portion is zero for all $m \neq n$ (see Fig. 9.1). In addition, both $f_{\text{PJ}} T$ and $f T$ are integers, the sinc function in the background portion is also zero, and hence the contribution of the background portion in (9.48) has vanished.

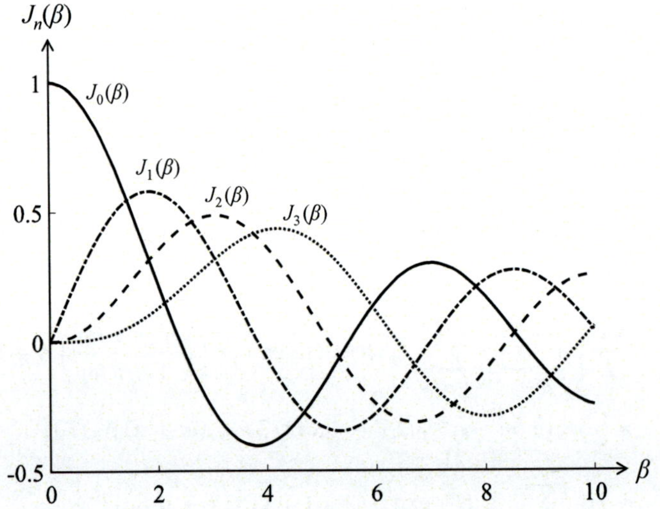


**Fig. 9.5** Bessel function of the first kind, $J_n(\beta)$, for integer orders $n = 0, 1, 2,$ and 3

On the other hand, the Bessel functions of the first kind of argument $\beta$ are finite at the origin ($\beta = 0$) for any order (integer or non-integer positive) and diverge as $x$ approaches zero for negative orders [5]. The Bessel functions oscillate but are not periodic; however, in the limit as $\beta \to \infty$, the zeros become equidistant. The amplitude of $J_n(\beta)$ is not constant but decreases asymptotically as $1/\sqrt{\beta}$. The results for $J_0$, $J_1$, $J_2$, and $J_3$ are shown in Fig. 9.5. Here, we are interested only in integer orders of Bessel functions with positive arguments in (9.48) since the order $n$ represents the harmonic number in the Fourier series expansion and the $\beta$ is always greater than or equal to zero, i.e., $\beta = 2\pi f A_{PJ} \geq 0$.

### 9.3.3 The Phase Error Due to the Periodic Jitter

The phase error $\phi_{err}$ due to the periodic jitter in the correlation model can be computed from the two approaches, separately as follows:

$$\phi_{err} = \begin{cases} \left|\phi - \phi_{num}^{cal}\right|, & \text{by numercial approach with (9.26) and (2.12),} \\ \left|\phi - \phi_{ana}^{cal}\right|, & \text{by analytical approach with (9.48).} \end{cases} \tag{9.50}$$

Now, let us analyze the phase error computed from analytical approach in (9.50) with (9.49), which gives

$$\phi_{\text{err}} = \left| \phi - \angle \widehat{I}(1) \right|$$
$$= \Bigg| \phi - \angle \Bigg\{ \frac{\pi T}{2} \Bigg\{ A \sum_{m=-\infty}^{\infty} \sum_{n=-\infty}^{\infty} J_m(\beta) J_n(\beta) \exp\left( i \left\{ \left( \frac{f + m f_{\text{PJ}}}{f + n f_{\text{PJ}}} \right) \phi + m \phi_{\text{PJ}}^s \right\} \right)$$
$$\times \left[ \operatorname{sinc}(\pi(m-n) f_{\text{PJ}} T) + \operatorname{sinc}(\pi(2f + (m+n) f_{\text{PJ}}) T) \right]$$
$$+ 2B \sum_{n=-\infty}^{\infty} J_n(\beta) \exp\left( i n \phi_{\text{PJ}}^s \right) \operatorname{sinc}(\pi(f + n f_{\text{PJ}}) T) \Bigg\} \Bigg\} \Bigg|. \tag{9.51}$$

Note that as stated before, the index $w$ in (9.48) has been replaced by $n$. The scalar factor $\pi T/2$ in the phase error (9.51) does not affect the phase. It is clear that the ToF phase shift $\phi$ is always added to the double summation (i.e., amplitude portion) in the analytical model obtained; therefore, some portion of the phase shift contributes to the phase error in the system. For the fundamental frequency, when the indices are the same (i.e., $m = n$), the $\phi$ is independent of the double summation, and hence, the phase error is independent of $\phi$. However, when the indices are $m \neq n \neq 0$, then the phase error is dependent on $\phi$ for those terms. When $m = n = 0$, the term of the series is more contributed to the ToF phase. Because the Bessel functions decrease to zero quickly with increasing of the order, the series converges very fast for small arguments, especially $\beta < 1$ (see Fig. 9.5). In addition, even though the ToF phase shift is zero (i.e., $\phi = 0$), the rest of the parameters are contributed to the phase error.

Thus, we have realized that the periodic jitter on the modulation signals introduces new frequencies that can interact together via the sinc and Bessel functions to contaminate the phase of the cosine, i.e., the correlation model.

## 9.4 Correlation Model Without Periodic Jitter

Most of the results and implications of this section are well-known [12, 13] and are described in Sect. 2.2; however, this section analyzes the ToF operation in the frequency domain. It is straightforward, we can immediately note that the limit of no jitter ($A_{\text{PJ}} = f_{\text{PJ}} = \phi_{\text{PJ}}^s = 0$) results for the fundamental frequency in (9.45) gives

$$\widehat{I}(1) = \frac{\pi T}{2} \Bigg\{ A \exp\left( i \frac{f_s}{f_l} \phi \right) \left[ \operatorname{sinc}(\pi(f_s - f_l) T) + \operatorname{sinc}(\pi(f_s + f_l) T) \right]$$
$$+ 2B \operatorname{sinc}(\pi f_s T) \Bigg\}. \tag{9.52}$$

In the limit of $T$ large, which with applications of (9.7) and (9.3) is found to be,

$$\lim_{T \to \infty} \widehat{I}(1) = \frac{\pi A}{2} \exp\left( i \frac{f_s}{f_l} \phi \right) \left[ \delta(f_s - f_l) + \delta(f_s + f_l) \right]. \tag{9.53}$$

It is clear that only frequencies that are present in both the illumination and shutter modulation signals contribute to the measurement of the ToF phase. This is a true correlation in the integral and is standard wisdom in AMCW ToF cameras, but we note that, while accurate for current hardware, as modulation frequencies increase and frame capture times decrease the assumption will eventually be violated in hardware.

Consider the homodyne operation, that is $f_s = f_l = f$. Returning to a finite integration time, (9.52), is then

$$\widehat{I}(1) = \frac{\pi T}{2}\left\{Ae^{i\phi}\left[1 + \text{sinc}(2\pi f T)\right] + 2B\,\text{sinc}(\pi f T)\right\}. \tag{9.54}$$

The ToF phase is recovered by taking the arctangent of (9.54). Both the amplitude and background portions affected the ToF phase. Now, first, let us analyze the behavior of the terms in the amplitude portion in (9.54). The term 1 in the square brackets is corresponding to the integration across a whole number of modulation cycles, while the term $\text{sinc}(2\pi f T)$ is due to the integration of the remaining fraction of a cycle. Moreover, for sufficiently large $T$ the sinc term is very small compared to 1 and thus taking the phase of the fundamental frequency of the DFT of a sequence of frame captures gives the ToF phase. It is instructive to ask when the sinc term might be significant.

Next, let us consider the final sinc term corresponding to the background portion in (9.54). As stated in Sect. 9.3.2.6, this term affects the ToF phase as well as the measured magnitude of $\widehat{I}(1)$. If the background lighting $B$ is sufficiently large and $T$ sufficiently small, then the background light not canceled out in the final part cycle of frame integration may introduce an error to the measurement of the ToF phase.

Returning to the general solution (9.52) note that if the modulation signals are not pure and the integration period is not sufficiently large, then it is possible that the side-lobes of the sinc functions for modulation frequencies $|f_s| \neq |f_l|$ may contribute to the result, and if they do, then the ToF phase is distorted because of the $f_s/f_l$ multiplier in the argument of the exponential. Retrieving the ToF phase by DFT will return a value with some error. That harmonic content on the modulation signals leads to measurement error is well-known [14], and many effective mitigation methods have been proposed [15–17].

## 9.5 Simulation Setup

The homodyne operation is considered for various modulation frequencies (201 values) span in $f = 0$ to 1000 MHz by 5 MHz steps. For jitter frequency $f_{\text{PJ}}$, the two fixed values are chosen for the simulations, one is 5 MHz that exactly divides the modulation frequencies used (i.e., $10^6$ times of the greatest common divisor (GCD) among $f$). The second value is 4.9261 MHz (less than and closer to the above

**Table 9.2** Parameters of the simulation setup

| ToF camera | Periodic jitter |
|---|---|
| Modulation frequency (MHz) | Frequency (MHz) |
| $f = 5(k-1),\ k = 1, \ldots, 201$ | $f_{PJ} = 4.9261$ and 5 |
| Integration period (ms) | Amplitude (ps) |
| $T = 1, 0.1, 0.01$ and 0.001 | $A_{PJ} = 10(k-1),\ k = 1, \ldots, 101$ |
| ToF phase shift (rad) | Phase shift (rad) |
| $\phi = 0$ to $5\pi/6$ by $\pi/6$ steps | $\phi_{PJ}^{s} = 0$ to $\pi$ by $\pi/19$ rad steps |
| Argument of Bessel function $\beta = 2\pi f A_{PJ} = 0$ to $2\pi$ | |
| For the purpose of analysis | |
| Choose four phase error criteria $\phi_{err} = 10^{-1}, 10^{-2}, 10^{-3}$, and $10^{-4}$ rad | |
| Compute the maximum (worst) and mean phase errors observed among the 20 phase shifts, $\phi_{PJ}^{s}$ | |

GCD) that does not exactly divide modulation frequencies. For jitter amplitude, 101 values that span in $A_{PJ} = 0$ to 1000 ps by 10 ps steps are selected throughout the simulations. In addition, for the phase difference of the jitter in shutter with respect to the light signal, we use multiple phases that are non-integer factors of each other such as $\phi_{PJ}^{s} = 0$ to $\pi$ by $\pi/19$ rad steps, for each evaluation, and then select the mean and maximum (worst) phase error among them, separately. Table 9.2 shows all corresponding parameters of the simulation setup.

The four precisions (i.e., phase errors, $\phi_{err} = 10^{-1}, 10^{-2}, 10^{-3}$, and $10^{-4}$ rad) of range measurements on simulated signals from the analytical approach are evaluated. These values are the corresponding phase errors for depth errors of approximately $d_{err} = 100, 10, 1$, and 0.1 mm respectively, for the MESA SR4000 when $f = 30$ MHz with non-ambiguity range of 5.0 m (but actual calculated phase errors are approximately $\phi_{err} = 0.12566, 0.012566, 0.0012566$, and 0.00012566 rad). In addition, three different integration periods ($T = 1, 0.1$, and 0.01 ms) on the phase error are also performed for each evaluation. These values are chosen to investigate the influence of PJ on range measurements in current ToF cameras and predicting for the future cameras. As an example, the integration period in the MESA Imaging SR4000 [18] can be selectable in the range of 0.3 to 25.8 ms by 0.1 ms steps.

## 9.6 Results and Discussion

In order to use the analytical model (9.51), first, we need to find the optimal value of the order that converges the Bessel functions. With the maximum amounts of $f$ and $A_{PJ}$ (see Table 9.2), the corresponding maximum argument of Bessel functions is given by $\beta = 2\pi f A_{PJ} = 2\pi$. For the verification of the optimal value, we consider the order $n = \{10, 30, 50\}$ for the evaluation purpose. Then, the behavior of the analytical model is performed when the frequency of the periodic jitter $f_{PJ} = 4.9261$ and 5 MHz, separately, as in the following two sections.

### 9.6.1 Jitter Frequency Is a Non-factor of Modulation Frequency

Tables 9.3 and 9.4 show the corresponding maximum phase errors with fifteen significant digits due to periodic jitter at the frequency $f_{PJ} = 4.9261$ MHz, for a selected ToF phase shift (i.e., $\phi = \pi/6$) with different order values of the Bessel function with two integration periods $T = 1$ and 0.01 ms, respectively. This analysis

**Table 9.3** Calculated maximum phase error (worst) with five different orders $n$ of Bessel functions: $T = 1$ ms, $fA_{PJ} = 0.5$, $\phi = \pi/6$, $f_{PJ} = 4.9261$ MHz, $\beta = (0 : \pi/10 : \pi)$ and $A = B = 1$ in (9.51)

| # | $\beta = (0 : \pi)$ | $n = 10$ | $n = 30$ | $n = 50$ |
|---|---|---|---|---|
| 1 | 0 | −0.00000008975334 | −0.00000008975334 | −0.00000008975334 |
| 2 | 0.3142 | −0.00000006689951 | −0.00000006689951 | −0.00000006689951 |
| 3 | 0.6283 | −0.00000004833030 | −0.00000004833030 | −0.00000004833030 |
| 4 | 0.9425 | −0.00000001444109 | −0.00000001444109 | −0.00000001444109 |
| 5 | 1.2566 | 2.09439543051489 | 2.09439543051489 | 2.09439543051489 |
| 6 | 1.5708 | 2.09439521539121 | 2.09439521539121 | 2.09439521539121 |
| 7 | 1.8850 | 2.09439521306326 | 2.09439521306326 | 2.09439521306326 |
| 8 | 2.1991 | 2.09439524517782 | 2.09439524517784 | 2.09439524517784 |
| 9 | 2.5133 | 2.09439539253153 | 2.09439539253162 | 2.09439539253162 |
| 10 | 2.8274 | 2.09440131786666 | 2.09440131769860 | 2.09440131769860 |
| 11 | 3.1416 | 2.09441037827703 | 2.09441037791601 | 2.09441037791601 |

**Table 9.4** Calculated maximum phase error (worst) with five different orders $n$ of Bessel functions: $T = 0.01$ ms, $fA_{PJ} = 0.5$, $\phi = \pi/6$, $f_{PJ} = 4.9261$ MHz, $\beta = (0 : \pi/10 : \pi)$ and $A = B = 1$ in (9.51)

| # | $\beta = (0 : \pi)$ | $n = 10$ | $n = 30$ | $n = 50$ |
|---|---|---|---|---|
| 1 | 0 | 0.00000231774490 | 2.00000231774490 | 2.00000231774490 |
| 2 | 0.3142 | 0.00046048028555 | 0.00046048028555 | 0.00046048028555 |
| 3 | 0.6283 | 0.00102133985067 | 0.00102133985067 | 0.00102133985067 |
| 4 | 0.9425 | 0.00205495468332 | 0.00205495468332 | 0.00205495468332 |
| 5 | 1.2566 | 2.13634287435225 | 2.13634287435049 | 2.13634287435049 |
| 6 | 1.5708 | 2.32934755951315 | 2.32934755958461 | 2.32934755958461 |
| 7 | 1.8850 | 2.22457014100934 | 2.22457014112242 | 2.22457014112242 |
| 8 | 2.1991 | 2.11462995757400 | 2.11462995773640 | 2.11462995773640 |
| 9 | 2.5133 | 2.11651801824188 | 2.11651801914674 | 2.11651801914674 |
| 10 | 2.8274 | 2.09942774124086 | 2.09942774253111 | 2.09942774253111 |
| 11 | 3.1416 | 2.10338740981628 | 2.10338741536859 | 2.10338741536859 |

is also valid for $T = 0.1$ ms. The last row of both tables corresponds to half of the maximum argument $\beta = \pi$ for the Bessel function with the parameters chosen ($f A_{PJ} = 0.5$).

The thick border line in each table indicates the corresponding last row for each order $n$ that satisfies the precision level of fifteen significant digits. Thus, in order to get this precision, it is sufficient to select the order $n = 30$ for those chosen parameters. But, to be safe, the order of the Bessel function has been selected to be 40. Thus, the summation of (9.51) is truncated for both $m$ and $n$ to be $-40$ to 40 each, which is well sufficient for the fundamental frequency. Then, there are $81 \times 81 = 6561$ terms in the double summation in (9.51).

As stated at the beginning of this chapter, the current commercial ToF cameras are less than 100 MHz modulation frequency. First, let us consider an analysis when the modulation frequencies of $f = 30, 50$, and 100 MHz before the main results from the analytical model. The amount of the periodic jitter in each ToF camera experimented is found less than 100 ps (see Chap. 7). Let us take the double of this amount as a range $A_{PJ} = 0{:}200$ ps, and then the maximum corresponding $\beta = 2\pi f A_{PJ} \approx 0.04$ and 0.12 for the two frequencies $f = 30$ and 100 MHz, respectively. Later, we will analyze the behavior for the full range of amounts of the parameters described in Table 9.2.

Figures 9.6, 9.7, and 9.8 show the maximum phase error (worst) due to the periodic jitter at the frequency $f_{PJ} = 4.9261$ MHz for six ToF phase shifts $\phi = 0$ to $5\pi/6$ by $\pi/6$ steps, separately, for the three integration periods with the argument range of $\beta = 0$ to 0.04 for each evaluation for 30, 50, and 100 MHZ signals, respectively. It is clear that the phase error increases when the integration period $T$ is shorter and this error is further worst when the modulation frequency $f$ increases. That is, for both lower modulation and integration periods, the phase errors are vanished for non-zero ToF phase shifts ($\phi \neq 0$) when $T = 0.01$ and 0.1 ms (compare Figs. 9.6, 9.7, and 9.8 each other).

For 100 MHz signal (see Fig. 9.8), among the six ToF phase shifts with the maximum argument of $\beta = 0.12$, the phase errors are approximately in the ranges of $(1.56\text{–}1.17)\times10^{-6}$, $(0.60\text{–}2.40)\times10^{-5}$, and $(1.89\text{–}3.41)\times10^{-4}$ rad for the integration periods $T = 1, 0.1$ and 0.01 ms, respectively. This seems when $T = 1$ ms, phase error is relatively very small (in microradians), which is insignificant to depth measurements. As examples, from (2.12) with $f = 100$ MHz, the corresponding range errors due to phase errors $\phi_{err} = 10^{-6}, 10^{-5}$, and $10^{-4}$ rad are given by $d_{err} = 0.00075/\pi, 0.0075/\pi$, and $0.075/\pi$ mm, respectively. These values are more insignificant when the modulation frequency $f$ increases. However, the phase error increases when the integration period decreases (short $T$) within the same frequency. That is, the frame rates of the camera increase.

In addition, the corresponding results without the background portion (i.e., $B = 0$ in (9.51)) in the analytical model for 30, 50, and 100 MHz signals are shown in Figs. 9.9, 9.10, and 9.11, respectively.

It seems the corresponding results for all four integration periods chosen have occurred with the presence of the background portion in the model (compare Figs. 9.6, 9.7, and 9.8, separately with Figs. 9.9, 9.10, and 9.11, respectively). For

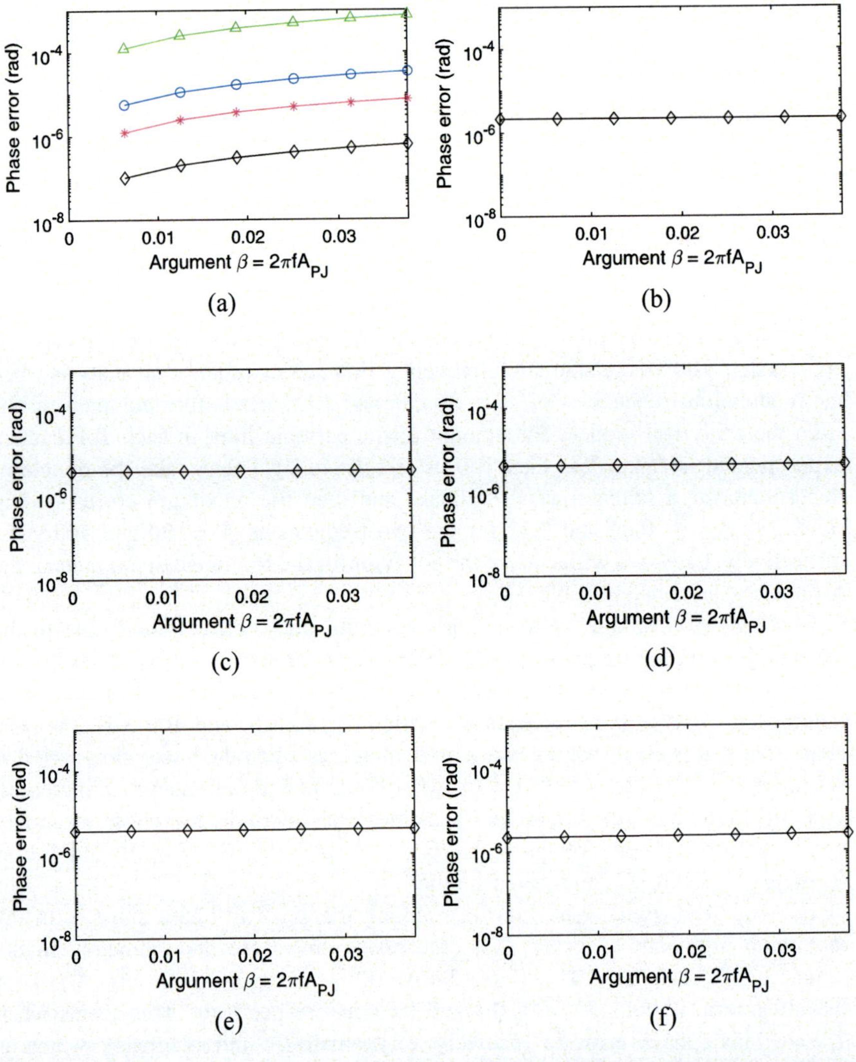


(a) (b) (c) (d) (e) (f)

**Fig. 9.6** Phase error (**a–f**) for the ToF phase shift $\phi = 0$ to $5\pi/6$ by $\pi/6$ steps, respectively, due to the PJ at $f_{PJ} = 4.9261$ MHz and $A_{PJ} = 0{:}200$ ps, when $f = 30$ MHz for the integration periods: 1 ms (solid line with open diamond symbol), 0.1 ms (magenta solid line with asterisk symbol), 0.01 ms (blue solid line with open circle symbol), and 0.001 ms (green solid line with open triangle symbol) with $A = B = 1$ and $m, n = -40{:}40$ each using the analytical model in (9.51)

100 MHz signal (see Fig. 9.11), among the six ToF phase shifts with the maximum argument of $\beta = 0.12$, the phase errors are approximately in the ranges of (0.07–1.40)$\times 10^{-6}$, (0.12–2.39)$\times 10^{-5}$ and (1.68–1.73)$\times 10^{-4}$ rad for the integration periods $T = 1, 0.1$, and 0.01 ms, respectively. As before, when $T = 1$ ms, the phase error is in microradians, which is insignificant to depth measurements. So,

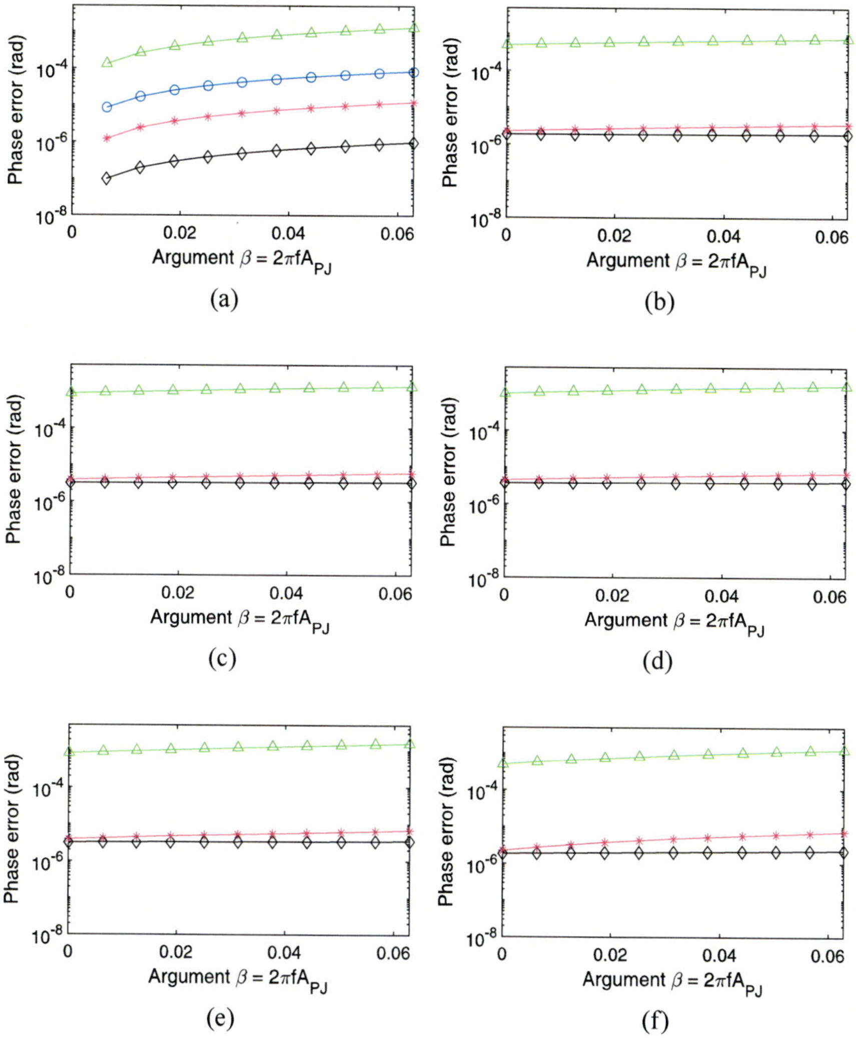


**Fig. 9.7** Phase errors (**a**–**f**) for the ToF phase shift $\phi = 0$ to $5\pi/6$ by $\pi/6$ steps, respectively, due to the PJ at $f_{PJ} = 4.9261$ MHz and $A_{PJ} = 0{:}200$ ps, when $f = 50$ MHz for the integration periods: 1 ms (solid line with open diamond symbol), 0.1 ms (magenta solid line with asterisk symbol), 0.01 ms (blue solid line with open circle symbol), and 0.001 ms (green solid line with open triangle symbol) with $A = B = 1$ and $m, n = -40{:}40$ each using the analytical model in (9.51)

the analysis with the presence of background is valid here as well. Note that here we fixed the amount of periodic jitter $A_{PJ} = 200$ ps for each modulation. But it is interesting to investigate whether these analyses are valid with the increasing of modulation frequencies and/or periodic jitter amplitudes.

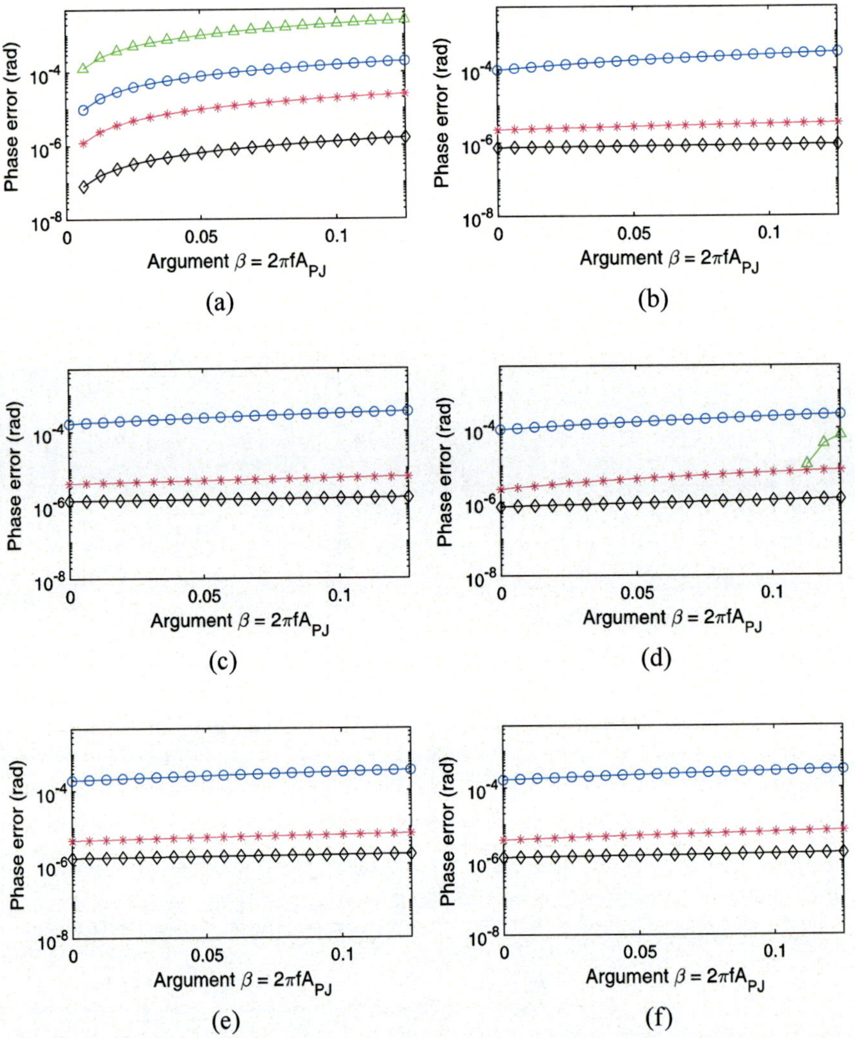


**Fig. 9.8** Phase errors **(a–f)** for the ToF phase shift $\phi = 0$ to $5\pi/6$ by $\pi/6$ steps, respectively, due to the PJ at $f_{\mathrm{PJ}} = 4.9261$ MHz and $A_{\mathrm{PJ}} = 0{:}200$ ps, when $f = 100$ MHz for the integration periods: 1 ms (solid line with open diamond symbol), 0.1 ms (magenta solid line with asterisk symbol), 0.01 ms (blue solid line with open circle symbol), and 0.001 ms (green solid line with open triangle symbol) with $A = B = 1$ and $m, n = -40{:}40$ each using the analytical model in (9.51)

Let us consider the worst behaviors of the model with all parameters. The evaluation is extended on the multiple set of characteristic parameters (see Table 9.2) such as the maximum (worst) and mean phase errors $\phi_{\mathrm{err}}$ with various modulation frequencies $f$ and periodic jitter amplitudes $A_{\mathrm{PJ}}$ for different integration periods. Figures 9.12, 9.13, 9.14, 9.15, and 9.16 show the contour diagrams for the chosen

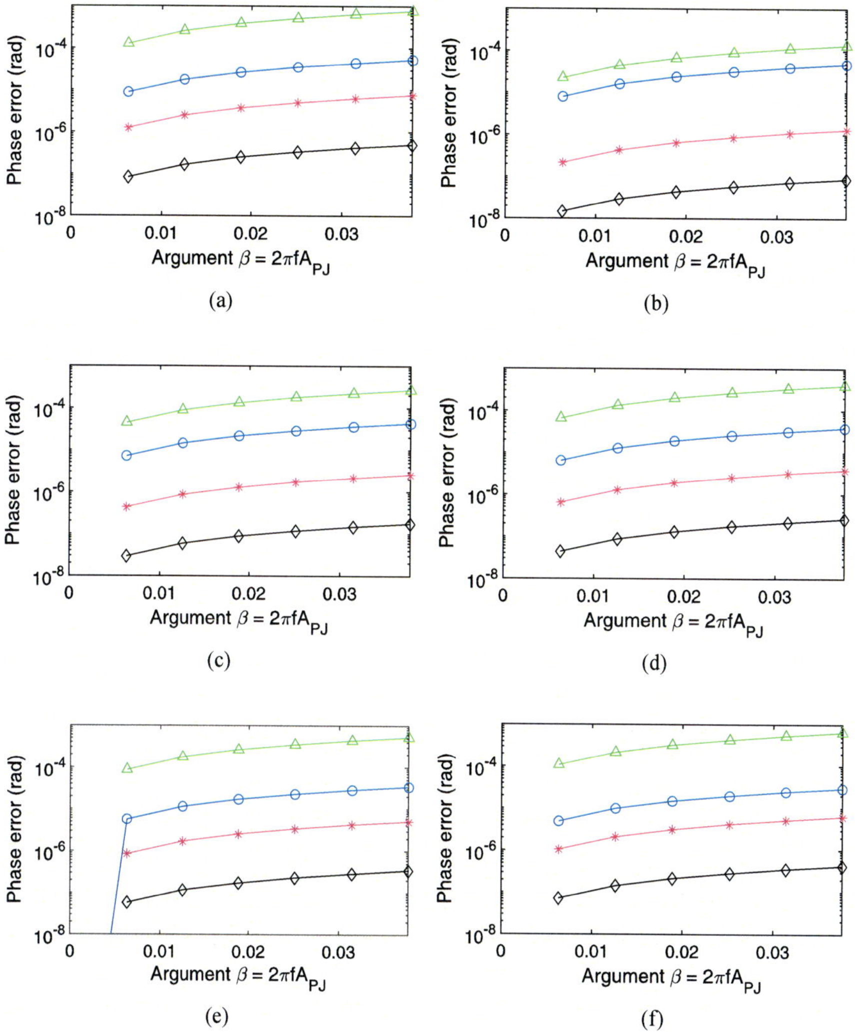


**Fig. 9.9** Phase errors (**a–f**) for the ToF phase shift $\phi = 0$ to $5\pi/6$ by $\pi/6$ steps, respectively, due to the PJ at $f_{\mathrm{PJ}} = 4.9261\,\mathrm{MHz}$, when $f = 30\,\mathrm{MHz}$ for the integration periods: 1 ms (solid line with open diamond symbol), 0.1 ms (magenta solid line with asterisk symbol), 0.01 ms (blue solid line with open circle symbol), and 0.001 ms (green solid line with open triangle symbol)with $A = 1$, $B = 0$ and $m, n = -40{:}40$ each using the analytical model in (9.51)

four phase error criteria against the modulation frequency and jitter amplitude at the frequency $f_{\mathrm{PJ}} = 4.9261\,\mathrm{MHz}$ for the ToF phase shift $\phi = \pi/6$ for the integration periods $T = 1, 0.1, 0.01$, and 0.001 ms, separately, with the various amounts of background portion in the analytical model in (9.51). Four amounts of the background amount such as one, ten, fifty, and hundred times of the unity amplitude are chosen, i.e., $B = 1$, 10, 50, and 100. Figure 9.12 shows the corresponding result

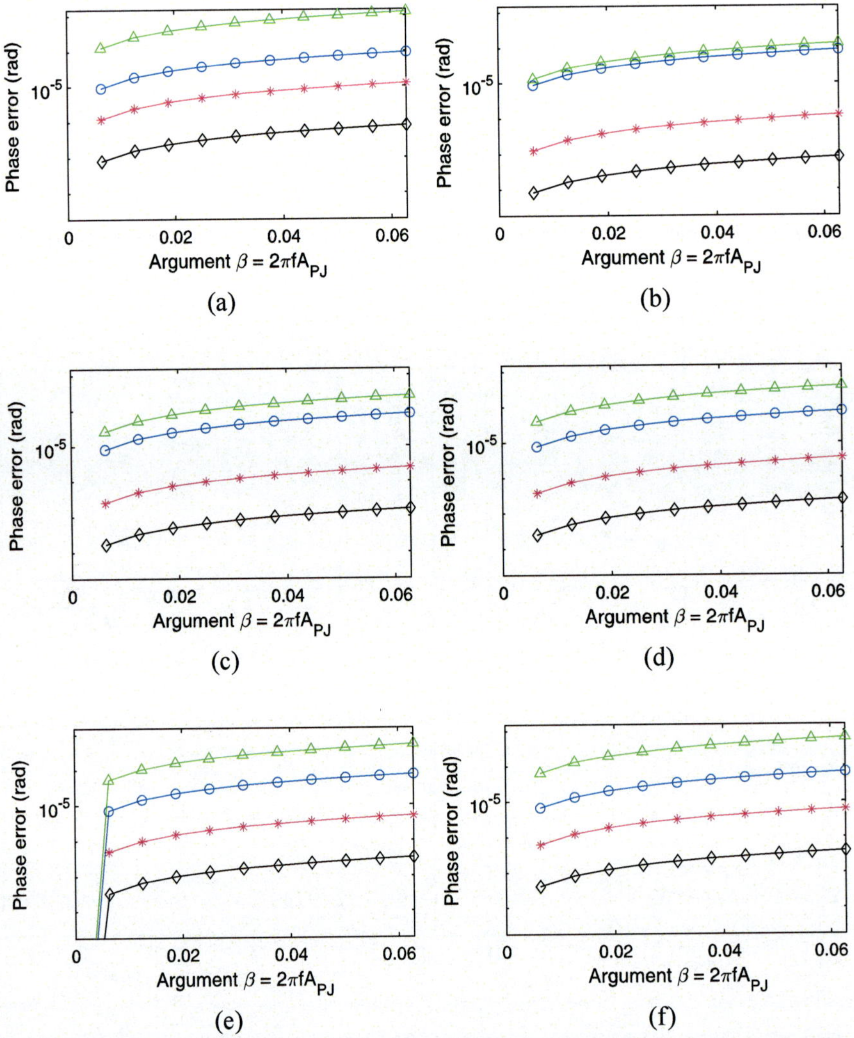


**Fig. 9.10** Phase errors **(a–f)** for the ToF phase shift $\phi = 0$ to $5\pi/6$ by $\pi/6$ steps, respectively, due to the PJ at $f_{PJ} = 4.9261$ MHz, when $f = 50$ MHz for the integration periods: 1 ms (solid line with open diamond symbol), 0.1 ms (magenta solid line with asterisk symbol), 0.01 ms (blue solid line with open circle symbol), and 0.001 ms (green solid line with open triangle symbol) with $A = 1$, $B = 0$ and $m, n = -40{:}40$ each using the analytical model in (9.51)

without the background portion in the model. It seems that the corresponding results are approximately similar and do not depend on the background portion in the model by comparing each relevant sub-figure in Fig. 9.12 with the sub-figure in Figs. 9.13, 9.14, 9.15, and 9.16. Also, it is noted that the values chosen for the indexes $m$ and $n$ are not sufficient to be stable of the analytic results when the background portion

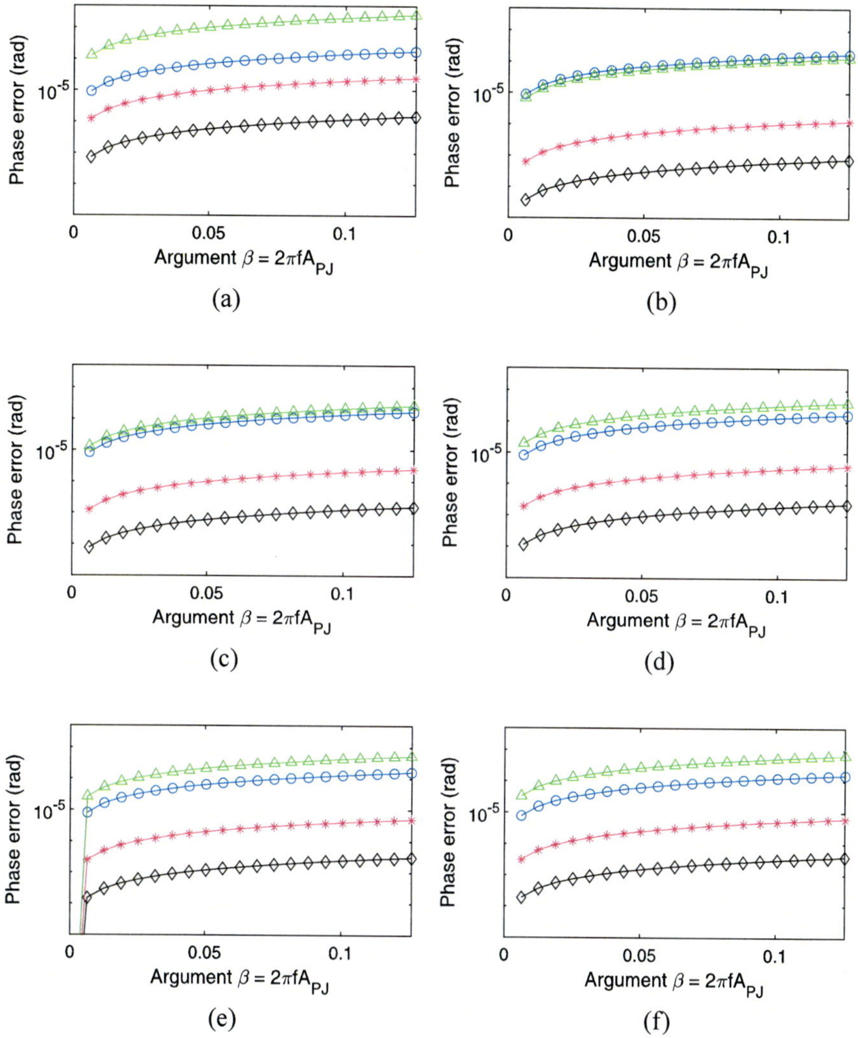


**Fig. 9.11** Phase errors (**a–f**) for the ToF phase shift $\phi = 0$ to $5\pi/6$ by $\pi/6$ steps, respectively, due to the PJ at $f_{PJ} = 4.9261$ MHz and $A_{PJ} = 0{:}200$ ps, when $f = 100$ MHz for the integration periods: 1 ms (solid line with open diamond symbol), 0.1 ms (magenta solid line with asterisk symbol), 0.01 ms (blue solid line with open circle symbol), and 0.001 ms (green solid line with open triangle symbol) with $A = 1$, $B = 0$ and $m, n = -40{:}40$ each using the analytical model in (9.51)

is increased and the integration period is decreased, i.e., $m, n = 50$ are not enough in (9.51). Compare each (d) in Figs. 9.13, 9.14, 9.15, and 9.16, which indicates the spanning of each curve (i.e., not smooth) in sub-figures.

Each of the figures seems to be approximately symmetric along the diagonal from origin to the maximum amounts of the scales, i.e., (0, 0) to (1000, 1000), because

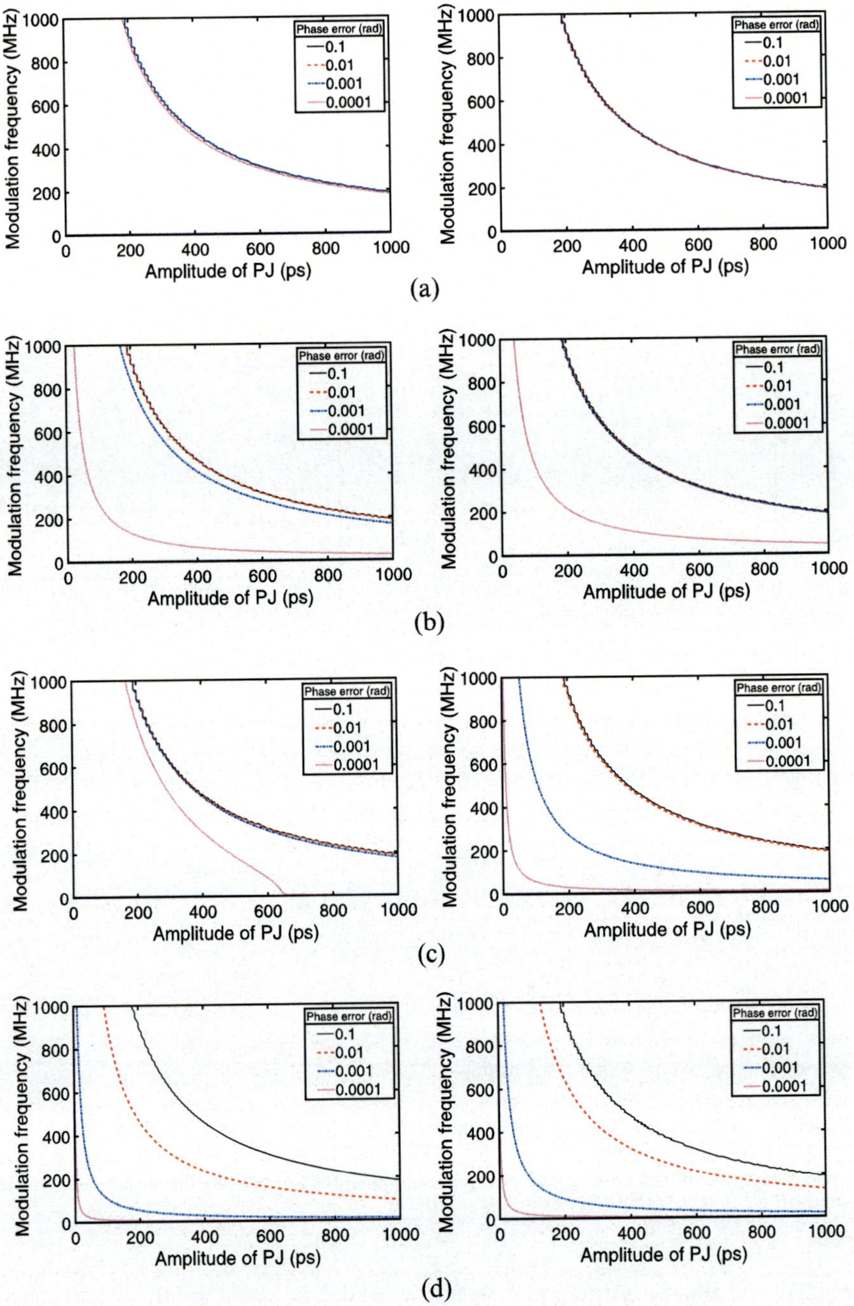


**Fig. 9.12** (Left) Maximum and (right) mean phase errors due to the periodic jitter at $f_{PJ} =$ 4.9261 MHz with various modulation frequencies against various jitter amplitudes by the analytical model with order $m, n = -40{:}40$ in (9.51), for the ToF phase shift $\phi = \pi/6$ and $B = 0$, when the integration period, $T$: (**a**) 1 ms, (**b**) 0.1 ms, (**c**) 0.01 ms, and (**d**) 0.001 ms

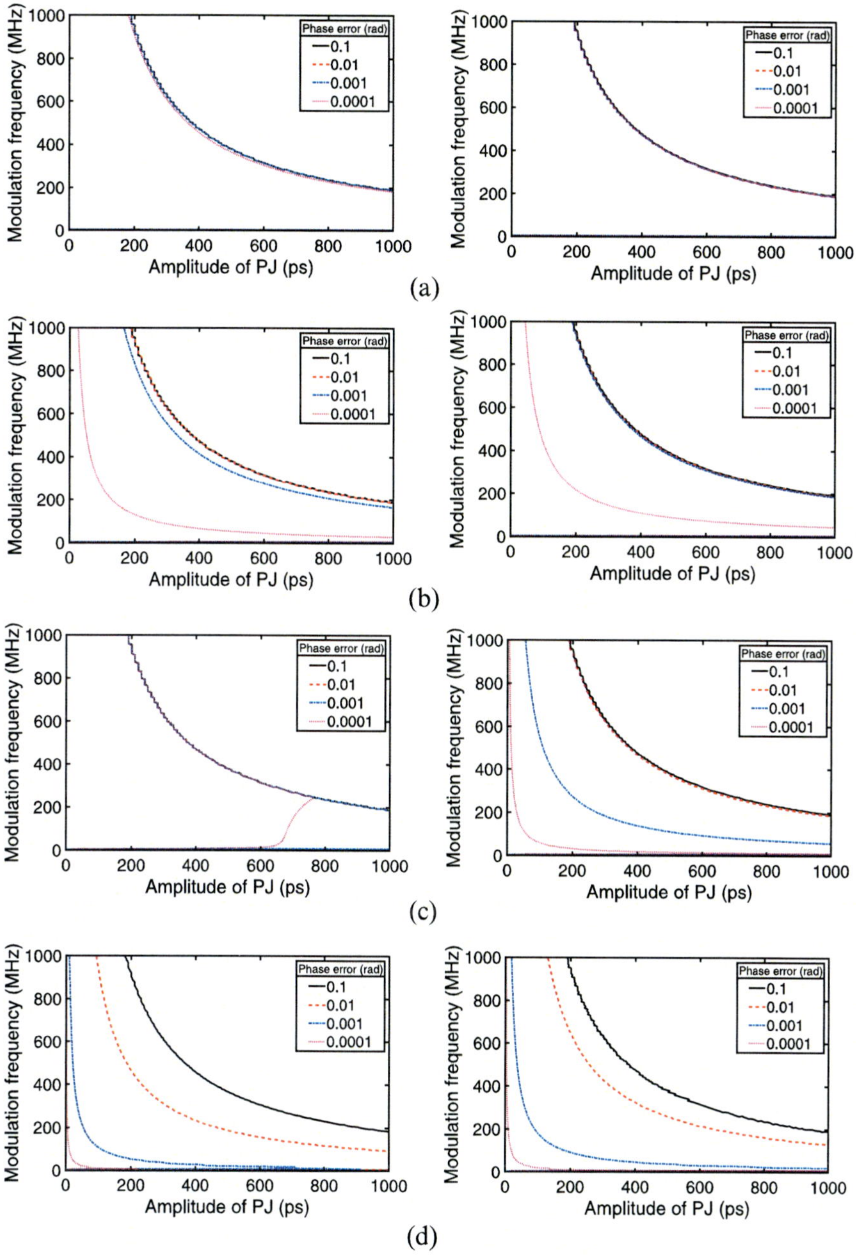


**Fig. 9.13** (Left) Maximum and (right) mean phase errors due to the periodic jitter at $f_{PJ}$ = 4.9261 MHz with various modulation frequencies against various jitter amplitudes by the analytical model with order $m, n = -40{:}40$ in (9.51), for the ToF phase shift $\phi = \pi/6$ and $B = 1$, when the integration period, $T$: **(a)** 1 ms, **(b)** 0.1 ms, **(c)** 0.01 ms, and **(d)** 0.001 ms

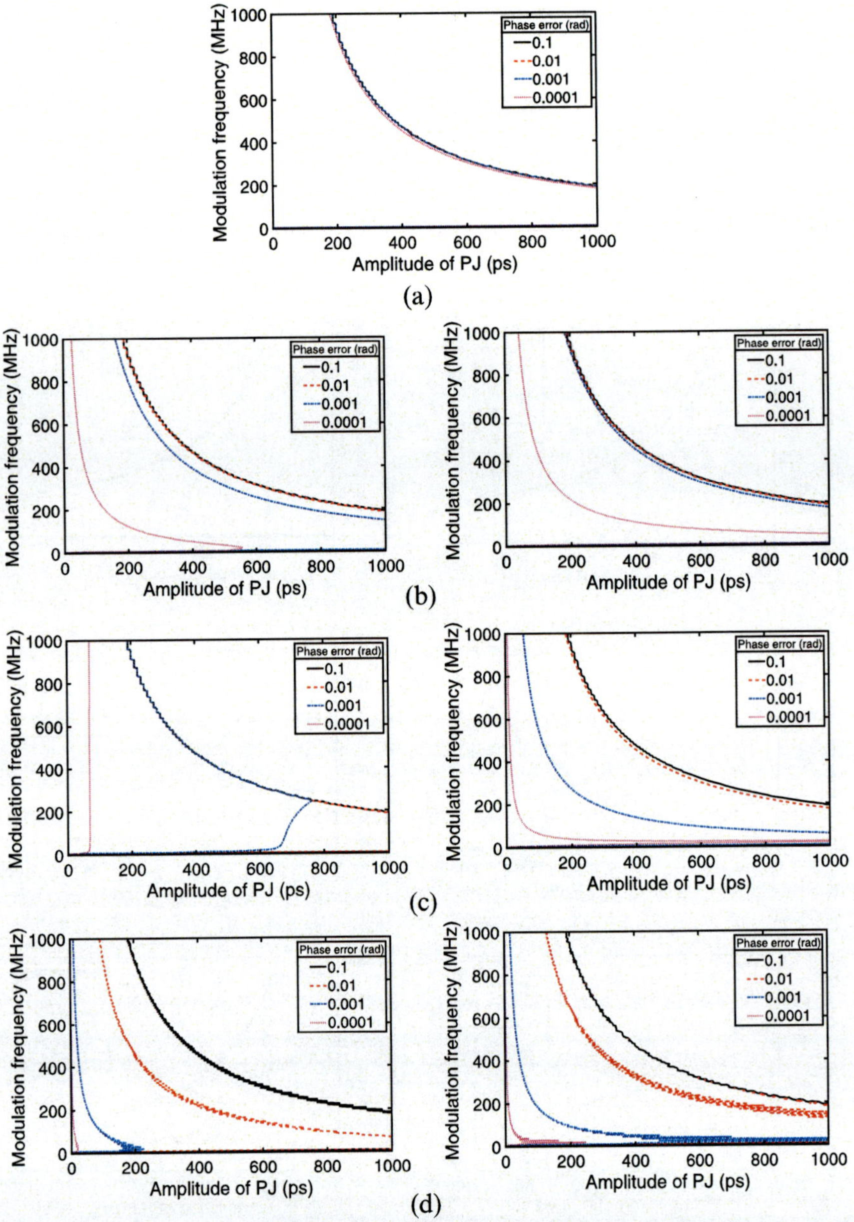


**Fig. 9.14** (Left) Maximum and (right) mean phase errors due to the periodic jitter at $f_{PJ}$ = 4.9261 MHz with various modulation frequencies against various jitter amplitudes by the analytical model with order $m, n = -40{:}40$ in (9.51), for the ToF phase shift $\phi = \pi/6$ and $B = 10$, when the integration period, $T$: **(a)** 1 ms, **(b)** 0.1 ms, **(c)** 0.01 ms, and **(d)** 0.001 ms

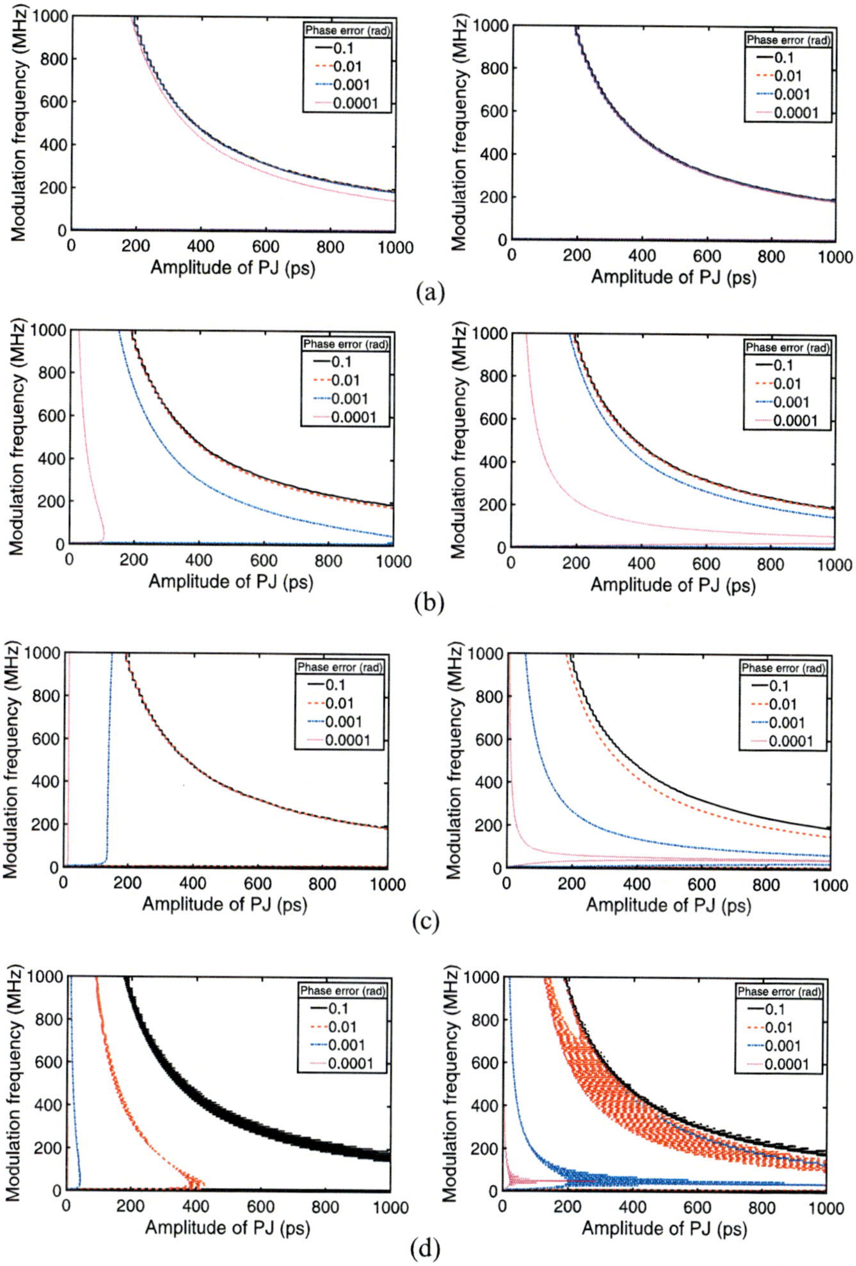


**Fig. 9.15** (Left) Maximum and (right) mean phase errors due to the periodic jitter at $f_{\mathrm{PJ}} = 4.9261$ MHz with various modulation frequencies against various jitter amplitudes by the analytical model with order $m, n = -40{:}40$ in (9.51), for the ToF phase shift $\phi = \pi/6$ and $B = 50$, when the integration period, $T$: (**a**) 1 ms, (**b**) 0.1 ms, (**c**) 0.01 ms, and (**d**) 0.001 ms

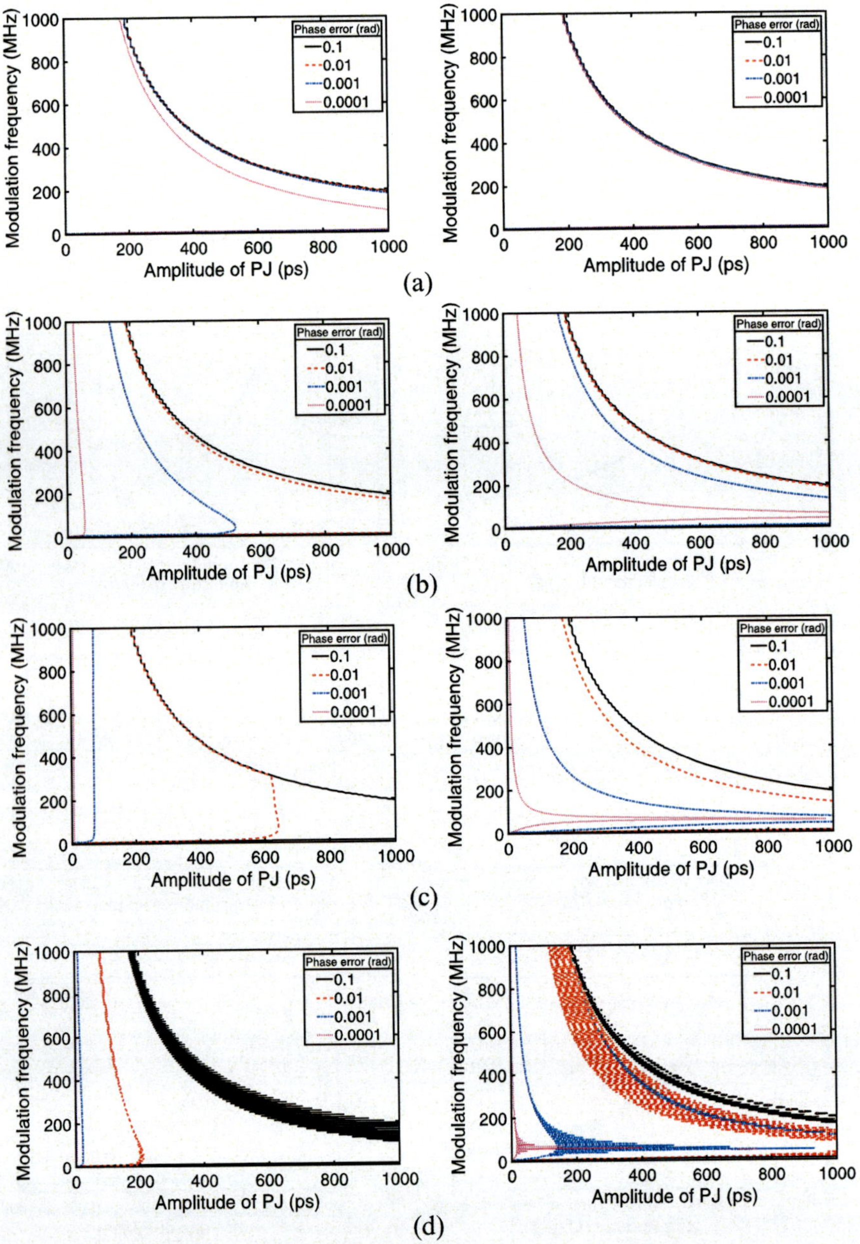


**Fig. 9.16** (Left) Maximum and (right) mean phase errors due to the periodic jitter at $f_{PJ}$ = 4.9261 MHz with various modulation frequencies against various jitter amplitudes by the analytical model with order $m, n = -40{:}40$ in (9.51), for the ToF phase shift $\phi = \pi/6$ and $B = 100$, when the integration period, $T$: **(a)** 1 ms, **(b)** 0.1 ms, **(c)** 0.01 ms, and **(d)** 0.001 ms

the model obtained is almost a reciprocal. The aforementioned four phase errors are systematically increased in the figures when $T$ becomes shorter. But, only for $T = 1$ ms, interestingly the four chosen contours land very close together; hence, the changes of the values cannot be visualized for the naked eyes, see each left and right sub-figure of (a) Figs. 9.12, 9.13, 9.14, 9.15, and 9.16. This indicates that there is a catastrophic decrease in accuracy when the periodic jitter exceeds a certain amount. This behavior is occurred for both worst and mean errors when $T = 1$ ms. The results for the worst phase error and the mean phase error behavior are approximately similar in each case. Compare the corresponding left and right sub-figures in Figs. 9.12, 9.13, 9.14, 9.15, and 9.16. In addition, the phase error curves occurred at around 5 MHz modulation frequency for all jitter amplitude amounts in each analytical result.

In addition, it is noted that the worst and mean phase error curves of the 0.1 rad criterion are similar for each integration period regardless of the background amounts. However, for the other three phase error criteria, curves moved toward the origin when the integration period is shorter. As examples, for the occurrence of the 0.1 mrad phase error when the integration period is 1 $\mu$s, even the very small amounts of 20 and 5 ps of periodic jitter are well sufficient for the modulation frequencies of 50 and 200 MHz, respectively (see the sub-figure (d) in Figs. 9.12, 9.13, 9.14, 9.15, and 9.16). For the same stated frequencies and the phase error criterion, the periodic jitter amounts are given 550 and 130 ps when the integration period is 0.1 ms, respectively (see the sub-figure (b) in Figs. 9.12, 9.13, 9.14, 9.15, and 9.16).

Now, we have evidence, the periodic jitter affects range measurement at least for the future ToF range imaging cameras that will be available with higher modulation frequencies and shorter integration periods. Next, let us analyze the result when the amplitude of the periodic jitter is a factor of the modulation frequency in the analytical model in (9.51).

## 9.6.2 Jitter Frequency Is a Factor of Modulation Frequency

The above analysis is repeated when the amplitude of the periodic jitter is a factor of the modulation frequency by selecting $f_{PJ} = 5$ MHz in the analytical model in (9.51). Tables 9.5 and 9.6 show the corresponding phase error with fifteen significant digits due to periodic jitter at an arbitrary chosen of the frequency $f_{PJ} = 5$ MHz, for a selected ToF phase shift (i.e., $\phi = \pi/6$) with different order values of the Bessel function with two integration periods $T = 1$ and 0.01 ms, respectively. As before, half of the maximum argument $\beta = \pi$ for the Bessel function with the parameters chosen as $fA_{PJ} = 0.5$ is located at the last row of each table.

As before, the precision level of fifteen significant digits for each order $n$ is indicated by the thick border line. Then, it is clear that even $n = 30$ is good enough to obtain the fifteen significant digits, but for safety, we choose the order of the Bessel function to be 40 that summed up the 6561 terms in (9.51). Now, let

**Table 9.5** Calculated maximum phase error (worst) with five different orders $n$ of Bessel functions: $T = 1$ ms, $fA_{PJ} = 0.5$, $\phi = \pi/6$, $f_{PJ} = 5$ MHz, $\beta = (0 : \pi/10 : \pi)$, and $A = B = 1$ in (9.51)

| # | $\beta = (0 : \pi)$ | $n = 10$ | $n = 30$ | $n = 50$ |
|---|---|---|---|---|
| 1 | 0 | −0.00000008975334 | −0.00000008975334 | −0.00000008975334 |
| 2 | 0.3142 | −0.00000009125246 | −0.00000009125246 | −0.00000009125246 |
| 3 | 0.6283 | −0.0000009262771 | −0.0000009262771 | −0.0000009262771 |
| 4 | 0.9425 | −0.00000009387768 | −0.00000009387768 | −0.00000009387768 |
| 5 | 1.2566 | 2.09439387908781 | 2.09439387908781 | 2.09439387908781 |
| 6 | 1.5708 | 2.09439490027607 | 2.09439490027607 | 2.09439490027607 |
| 7 | 1.8850 | 2.09439496371611 | 2.09439496371612 | 2.09439496371612 |
| 8 | 2.1991 | 2.09439495640765 | 2.09439495640771 | 2.09439495640771 |
| 9 | 2.5133 | 2.09439483990624 | 2.09439483990669 | 2.09439483990669 |
| 10 | 2.8274 | 2.09439355807616 | 2.09439355808376 | 2.09439355808376 |
| 11 | 3.1416 | 2.09439527688131 | 2.09439527690973 | 2.09439527690973 |

**Table 9.6** Calculated maximum phase error (worst) with five different orders $n$ of Bessel functions: $T = 0.01$ ms, $fA_{PJ} = 0.5$, $\phi = \pi/6$, $f_{PJ} = 5$ MHz, $\beta = (0 : \pi/10 : \pi)$, and $A = B = 1$ in (9.51)

| # | $\beta = (0 : \pi)$ | $n = 10$ | $n = 30$ | $n = 50$ |
|---|---|---|---|---|
| 1 | 0 | 0.00000231774490 | 0.00000231774490 | 0.00000231774490 |
| 2 | 0.3142 | 0.00000308184222 | 0.00000308184222 | 0.00000308184222 |
| 3 | 0.6283 | 0.00000382961341 | 0.00000382961341 | 0.00000382961341 |
| 4 | 0.9425 | 0.00000811709330 | 0.00000811709330 | 0.00000811709330 |
| 5 | 1.2566 | 2.09443712418130 | 2.09443712418130 | 2.09443712418130 |
| 6 | 1.5708 | 2.09440218267109 | 2.09440218267108 | 2.09440218267108 |
| 7 | 1.8850 | 2.09439992179043 | 2.09439992179039 | 2.09439992179039 |
| 8 | 2.1991 | 2.09439987621373 | 2.09439987621338 | 2.09439987621338 |
| 9 | 2.5133 | 2.09440245850361 | 2.09440245849943 | 2.09440245849943 |
| 10 | 2.8274 | 2.09399742708254 | 2.09399742752765 | 2.09399742752765 |
| 11 | 3.1416 | 2.09433849787919 | 2.09433849957900 | 2.09433849957900 |

us consider the behavior of the phase error due to the PJ at $f_{PJ} = 5$ MHz for the frequencies 30 and 100 MHz as shown in Figs. 9.17 and 9.18, respectively.

Similarly, the corresponding results without the background portion (i.e., $B = 0$ in (9.51)) when $f_{PJ} = 5$ MHz for the frequencies 30 and 100 MHz are shown in Figs. 9.19 and 9.20, respectively.

By comparing the results for 100 MHz signal with the complete model when $f_{PJ} = 4.9261$ and 5 MHz (see Figs. 9.8 and 9.18), the phase errors when $f_{PJ} =$ 5 MHz are extremely small (less than a microradian) when $T = 1$ ms, which is more insignificant to depth measurements.

It is clear that the phase error increases when the $T$ decreases. That is, the frame rates of the camera increase. Furthermore, for both fixed $T$ and $\beta$, the phase error decreases when the ToF phase shift increases. For examples, with $\beta = 0.1005$ and

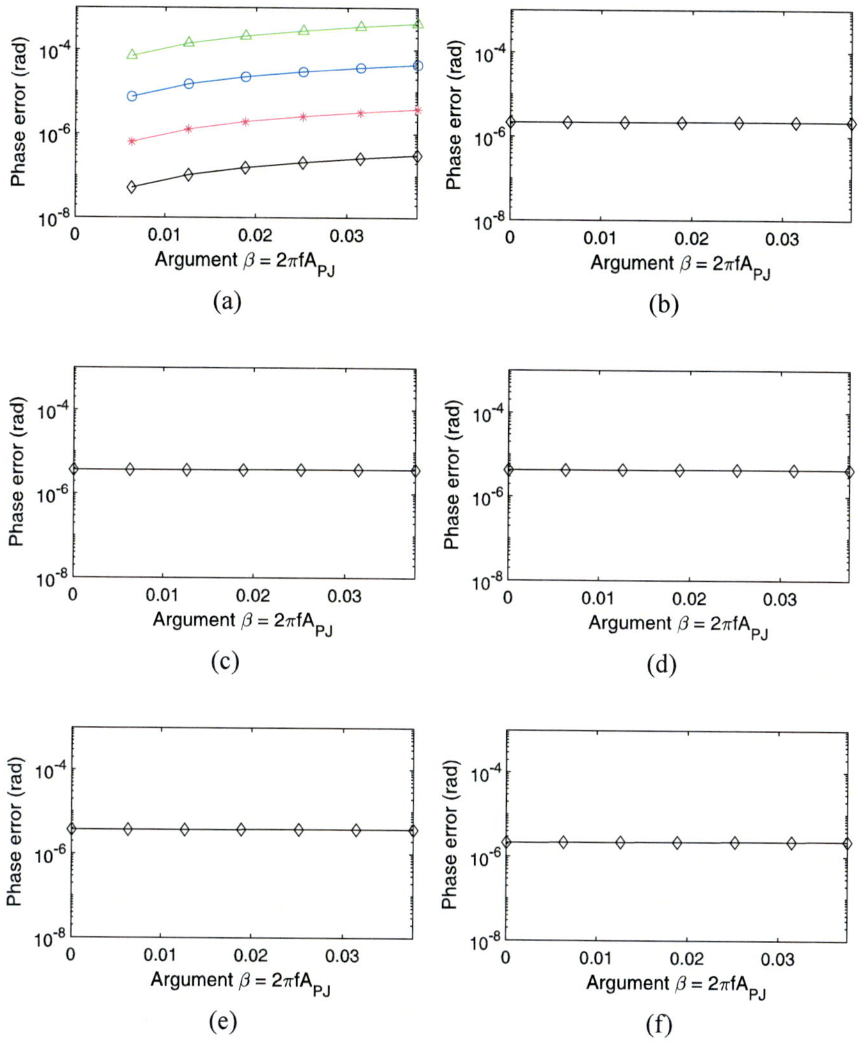


**Fig. 9.17** Phase errors (**a–f**) for the ToF phase shift $\phi = 0$ to $5\pi/6$ by $\pi/6$ steps, respectively, due to the PJ at $f_{PJ} = 5$ MHz and $A_{PJ} = 0$:200 ps, when $f = 30$ MHz for the integration periods: 1 ms (solid line with open diamond symbol), 0.1 ms (magenta solid line with asterisk symbol), 0.01 ms (blue solid line with open circle symbol), and 0.001 ms (green solid line with open triangle symbol) with $A = B = 1$ and $m, n = -40$:40 each using the analytical model in (9.51)

$T = 0.01$ ms, the phase errors are $\phi_{err} = 9.71, 28.72,$ and 46.55 microradians when $\phi = \pi/6, \pi/2,$ and $5\pi/6$, respectively. In addition, when comparing the results for 100 MHz signal without the background portion in the model for two $f_{PJ} = 4.9261$ and 5 MHz (see Figs. 9.11 and 9.20), the phase errors when $f_{PJ} =$ 5 MHz are extremely small (less than a microradian) when $T = 1$ ms, which is more insignificant to depth measurements.

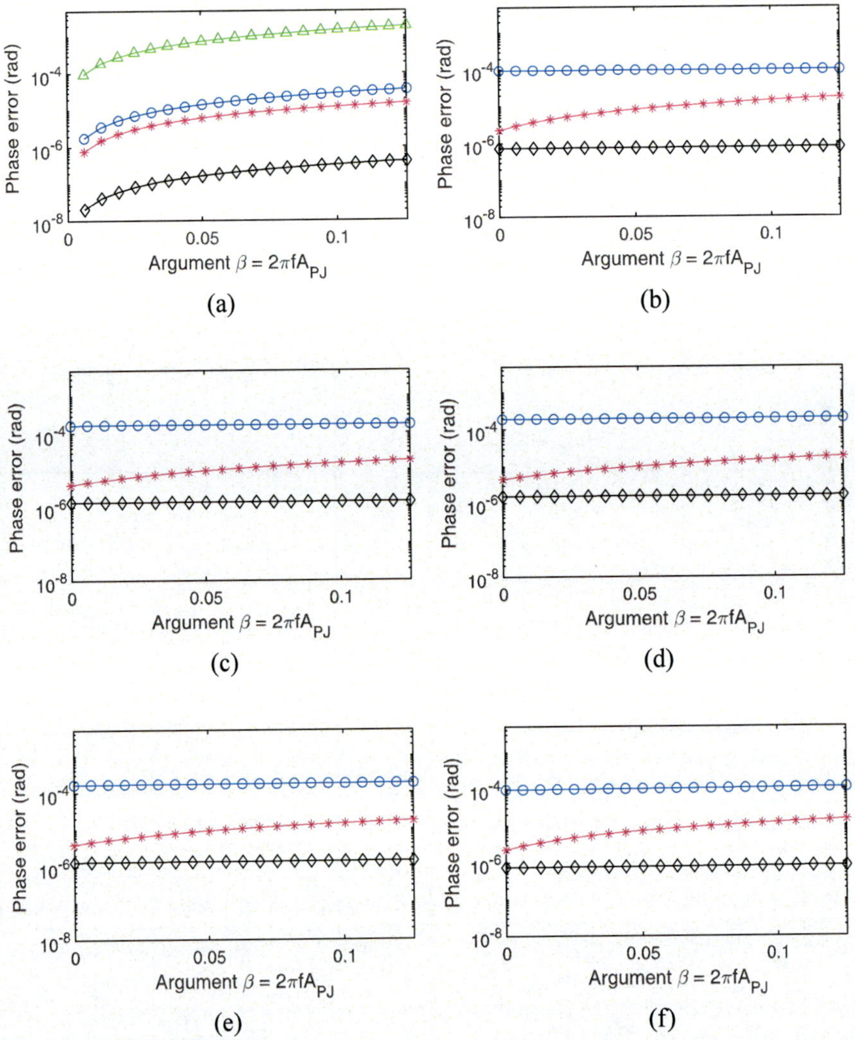


**Fig. 9.18** Phase errors (**a–f**) for the ToF phase shift $\phi = 0$ to $5\pi/6$ by $\pi/6$ steps, respectively, due to the PJ at $f_{PJ} = 5$ MHz and $A_{PJ} = 0{:}200$ ps, when $f = 100$ MHz for the integration periods: 1 ms (solid line with open diamond symbol), 0.1 ms (magenta solid line with asterisk symbol), 0.01 ms (blue solid line with open circle symbol), and 0.001 ms (green solid line with open triangle symbol) with $A = B = 1$ and $m, n = -40{:}40$ each using the analytical model in (9.51)

Let us extend the evaluation for all sets of parameters (see Table 9.2). Figures 9.21 and 9.25 show contour diagrams for the aforementioned phase error against the $f$ and $A_{PJ}$ when the frequency $f_{PJ} = 5$ MHz with $T = 1, 0.1$, and 0.01 ms, respectively, for $\phi = \pi/6$ with and without background portion in the analytical model in (9.51).

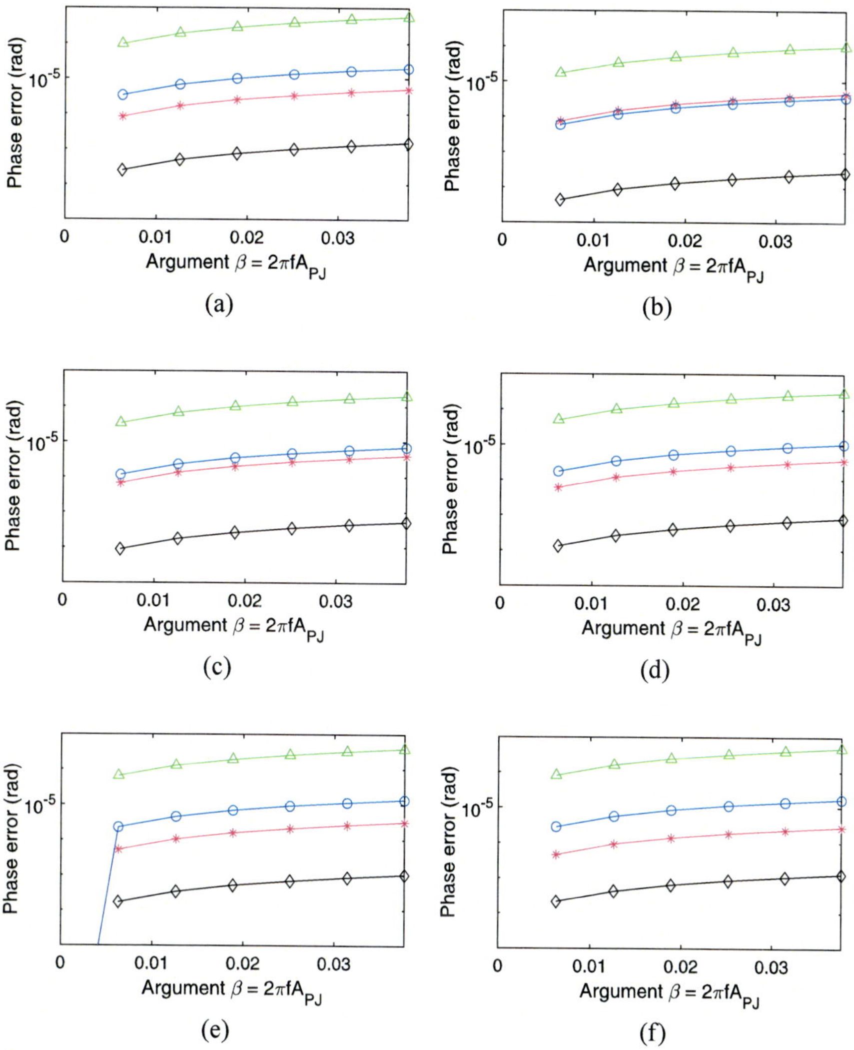


**Fig. 9.19** Phase errors (**a–f**) for the ToF phase shift $\phi = 0$ to $5\pi/6$ by $\pi/6$ steps, respectively, due to the PJ at $f_{\mathrm{PJ}} = 5\,\mathrm{MHz}$ and $A_{\mathrm{PJ}} = 0{:}200\,\mathrm{ps}$, when $f = 30\,\mathrm{MHz}$ and $T$ : 1 ms (solid line with open diamond symbol), 0.1 ms (magenta solid line with asterisk symbol), 0.01 ms (blue solid line with open circle symbol), and 0.001 ms (green solid line with open triangle symbol) with $A = 1$, $B = 0$ and $m, n = -40{:}40$ each using the analytical model in (9.51)

As before, when the integration period becomes longer, there is a trend to systematically vanish the large phase error (see the disappearing of the phase error $\phi_{err} = 0.1$ rad, which is indicating by ——) through Figs. 9.21, 9.22, 9.23, 9.24, and 9.25. It can be seen, the phase errors have occurred in a similar way. However, it is interesting to see, each of the sub-figures in Figs. 9.21, 9.22, 9.23, 9.24, and 9.25

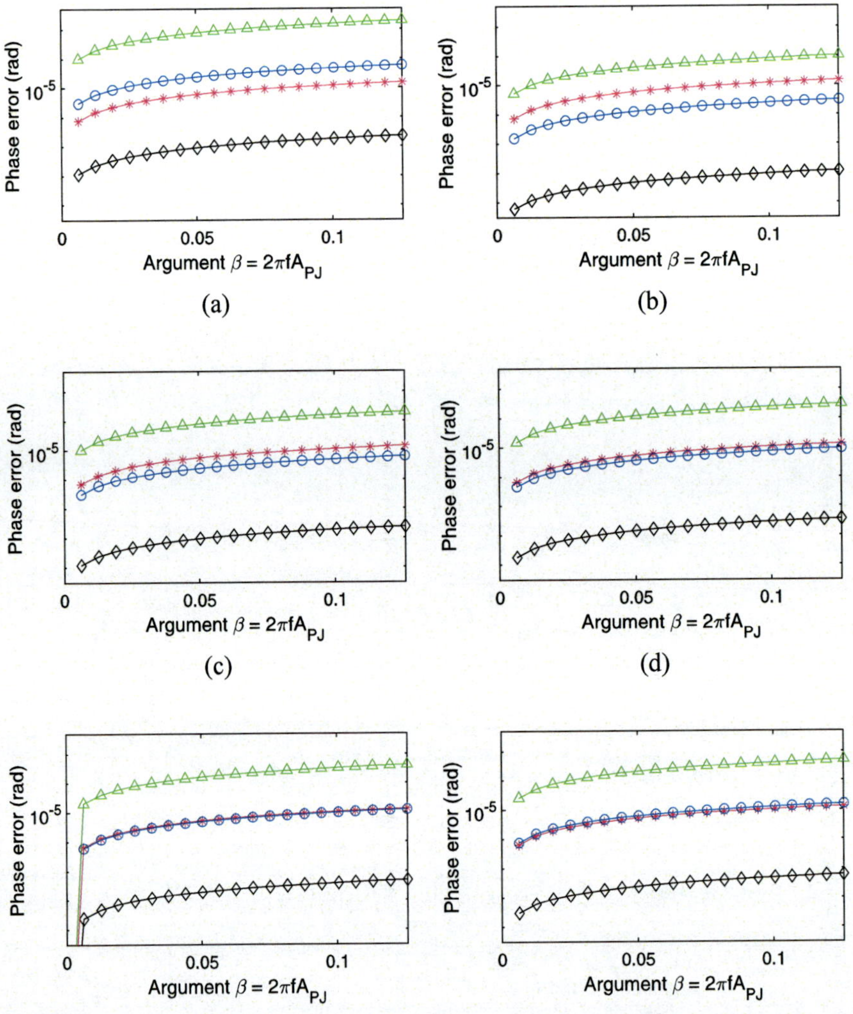


(a) (b)

(c) (d)

(e) (f)

**Fig. 9.20** Phase errors (**a–f**) for the ToF phase shift $\phi = 0$ to $5\pi/6$ by $\pi/6$ steps, respectively, due to the PJ at $f_{PJ} = 5$ MHz and $A_{PJ} = 0{:}200$ ps, when $f = 100$ MHz and $T$ : 1 ms (solid line with open diamond symbol), 0.1 ms (magenta solid line with asterisk symbol), 0.01 ms (blue solid line with open circle symbol), and 0.001 ms (green solid line with open triangle symbol) with $A = 1$, $B = 0$ and $m, n = -40{:}40$ each using the analytical model in (9.51)

gives the similar results regardless of the integration period, which is not agreed when the frequency of the periodic jitter is not a factor of modulation frequency, see Figs. 9.12, 9.13, 9.14, 9.15, and 9.16. The results are not dependent on the background illumination. In addition, by comparing each corresponding left and right sub-figure, it is clear that the worst and the mean phase error curves are similar.

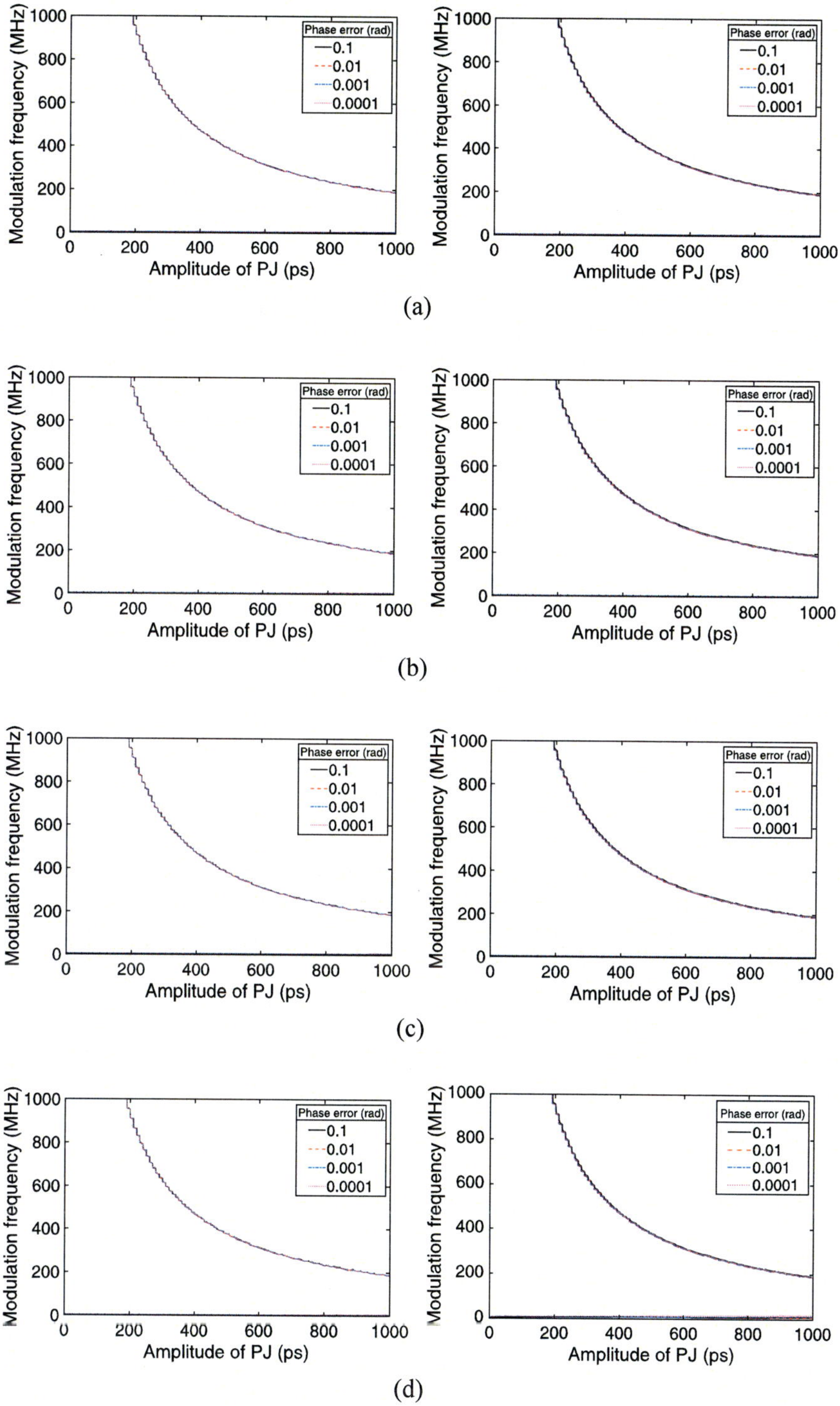


**Fig. 9.21** (Left) Maximum and (right) mean phase errors due to the periodic jitter at $f_{\text{PJ}} = 5\,\text{MHz}$ with various modulation frequencies against various jitter amplitudes by the analytical model with order $m, n = -40{:}40$ in (9.51), for the ToF phase shift $\phi = \pi/6$ and $B = 0$, when the integration period, $T$: (**a**) 1 ms, (**b**) 0.1 ms, (**c**) 0.01 ms, and (**d**) 0.001 ms

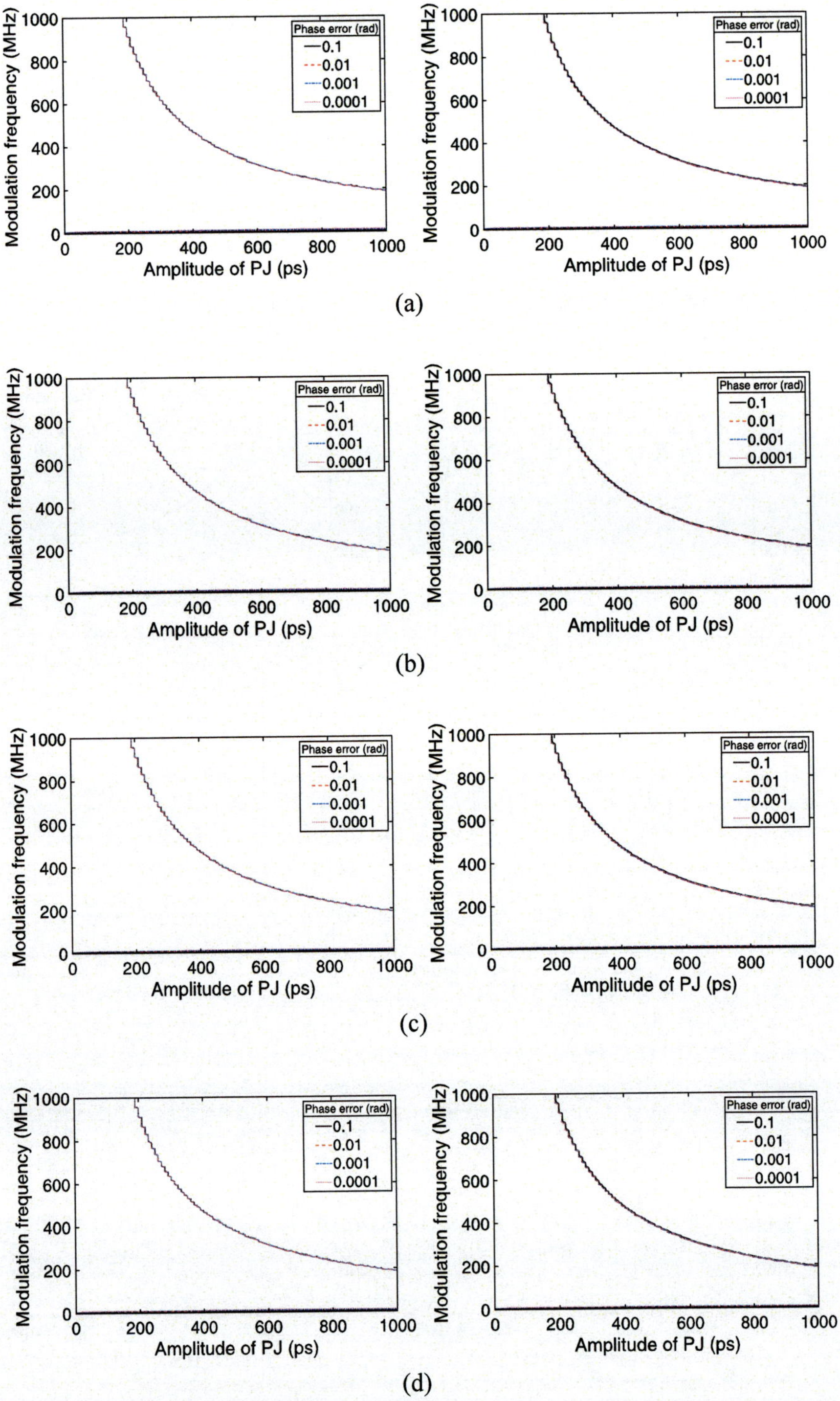


**Fig. 9.22** (Left) Maximum and (right) mean phase errors due to the periodic jitter at $f_{\mathrm{PJ}} = 5\,\mathrm{MHz}$ with various modulation frequencies against various jitter amplitudes by the analytical model with order $m, n = -40{:}40$ in (9.51), for the ToF phase shift $\phi = \pi/6$ and $B = 1$, when the integration period, $T$: (**a**) 1 ms, (**b**) 0.1 ms, (**c**) 0.01 ms, and (**d**) 0.001 ms

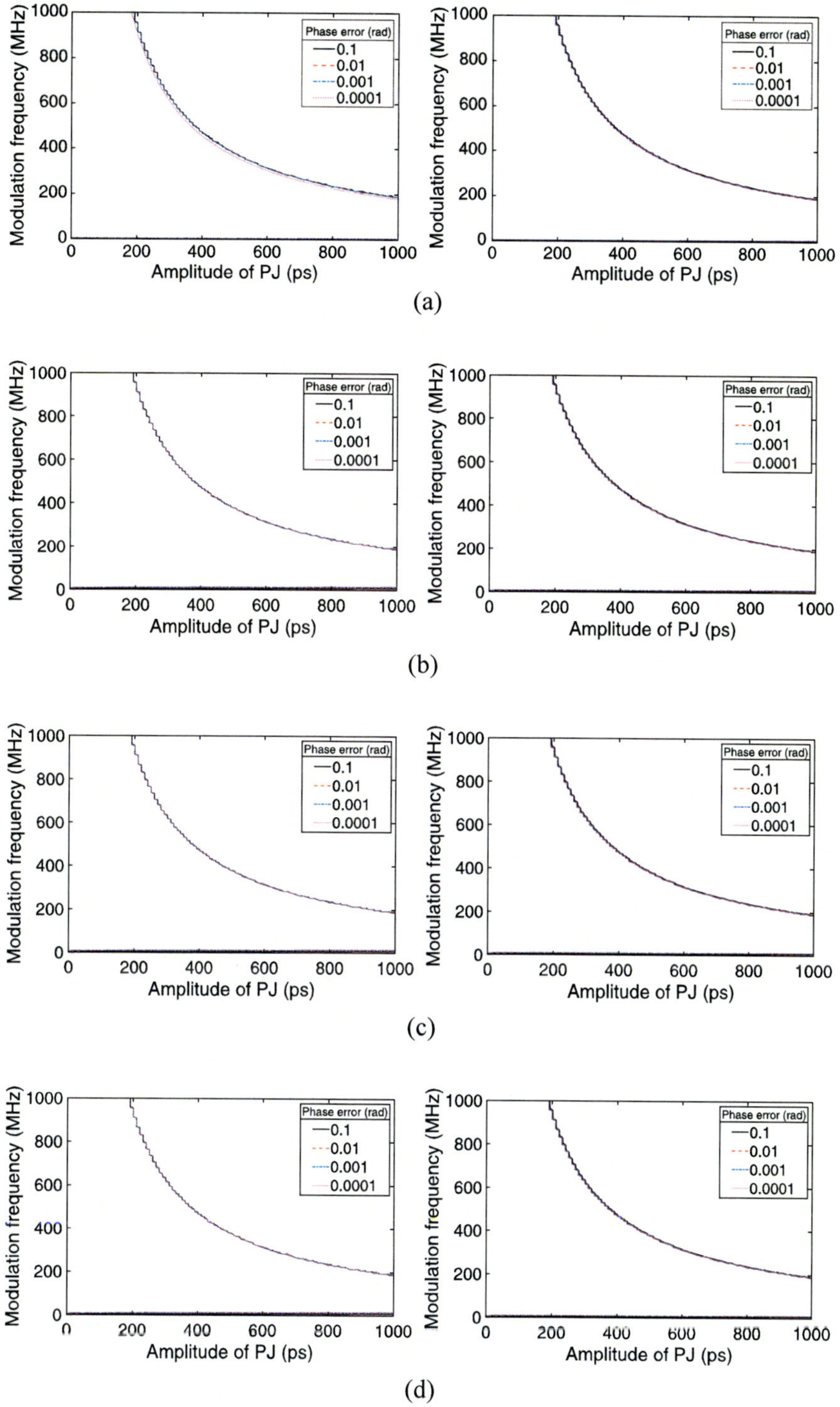


**Fig. 9.23** (Left) Maximum and (right) mean phase errors due to the periodic jitter at $f_{\mathrm{PJ}} = 5\,\mathrm{MHz}$ with various modulation frequencies against various jitter amplitudes by the analytical model with order $m, n = -40{:}40$ in (9.51), for the ToF phase shift $\phi = \pi/6$ and $B = 10$, when the integration period, $T$: (**a**) 1 ms, (**b**) 0.1 ms, (**c**) 0.01 ms, and (**d**) 0.001 ms

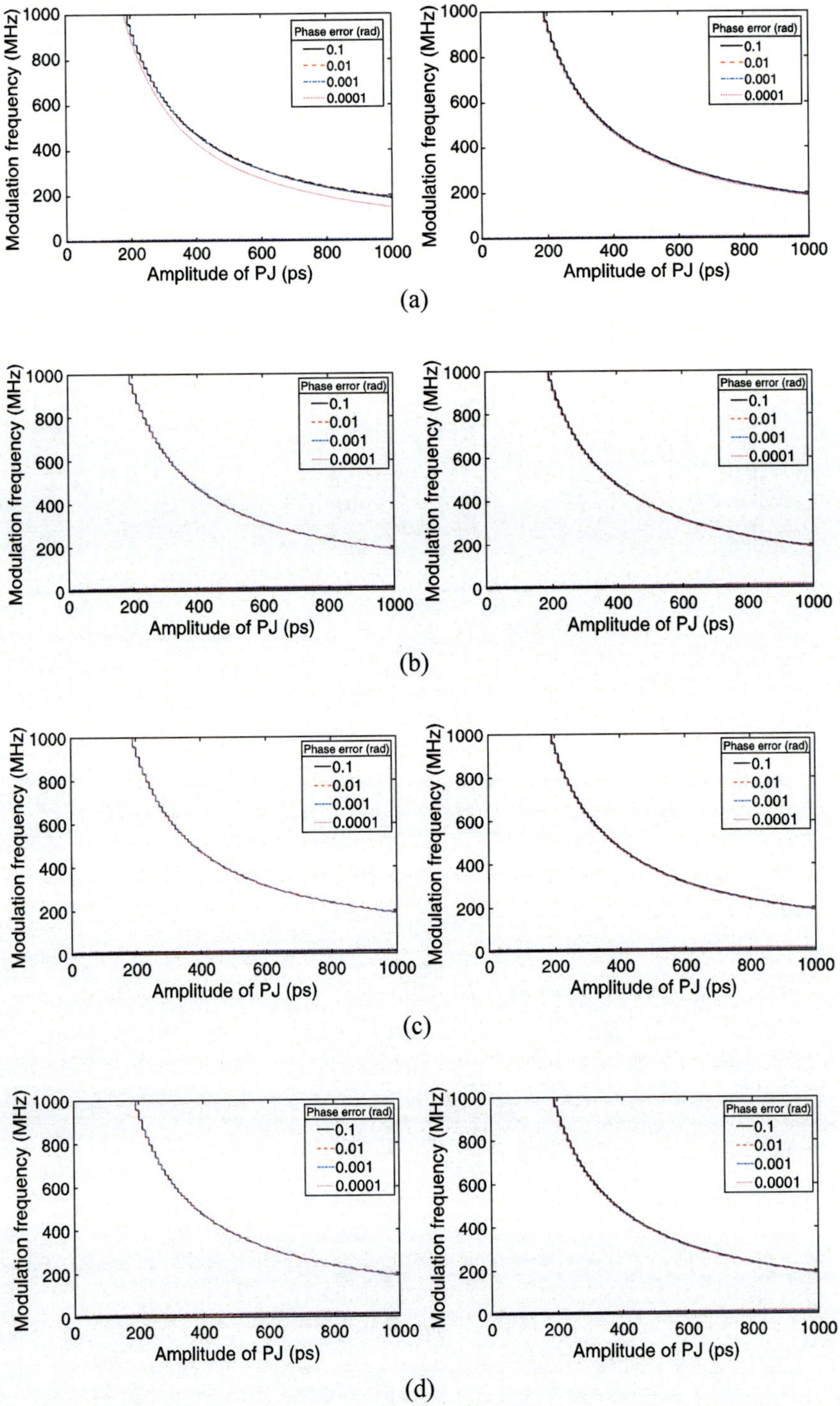


**Fig. 9.24** (Left) Maximum and (right) mean phase errors due to the periodic jitter at $f_{\mathrm{PJ}} = 5$ MHz with various modulation frequencies against various jitter amplitudes by the analytical model with order $m, n = -40{:}40$ in (9.51), for the ToF phase shift $\phi = \pi/6$ and $B = 50$, when the integration period, $T$: (**a**) 1 ms, (**b**) 0.1 ms, (**c**) 0.01 ms, and (**d**) 0.001 ms

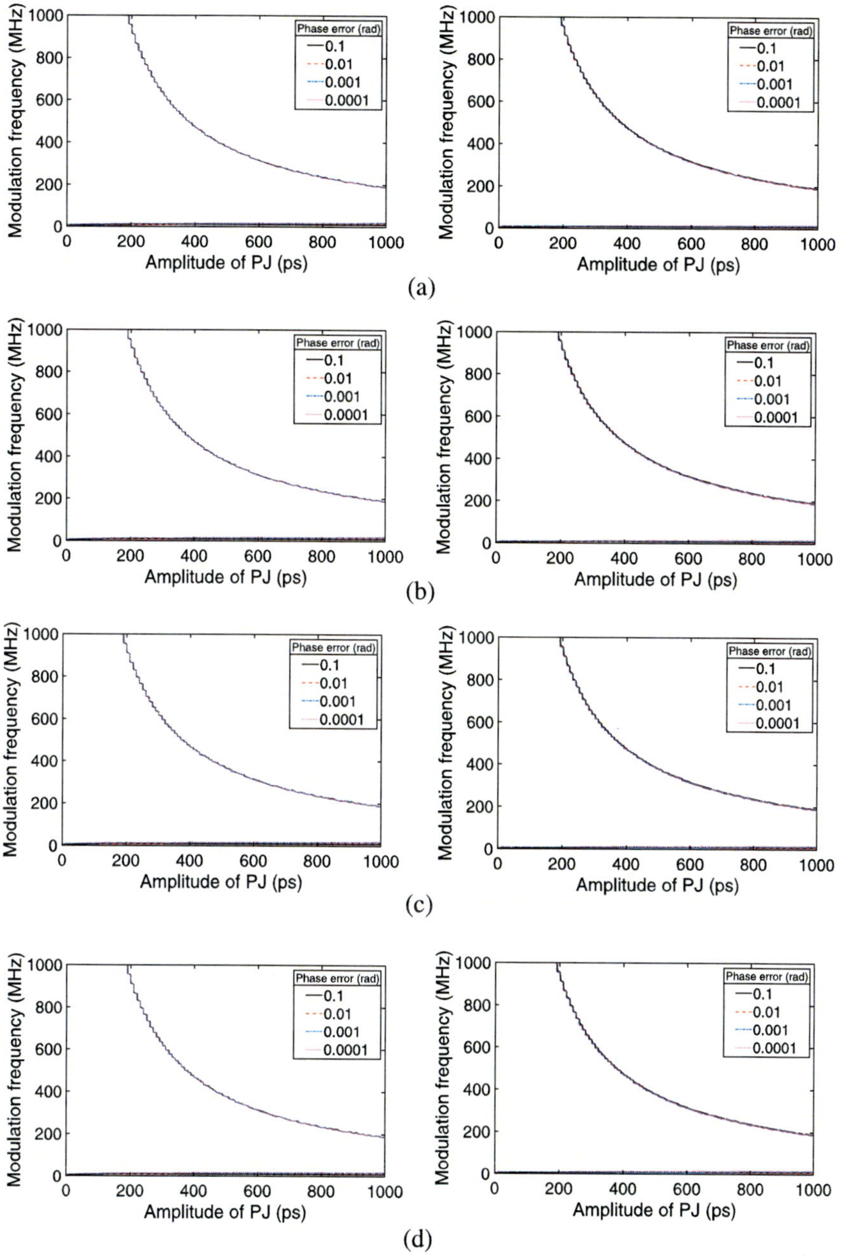


**Fig. 9.25** (Left) Maximum and (right) mean phase errors due to the periodic jitter at $f_{PJ} = 5$ MHz with various modulation frequencies against various jitter amplitudes by the analytical model with order $m, n = -40{:}40$ in (9.51), for the ToF phase shift $\phi = \pi/6$ and $B = 100$, when the integration period, $T$: (**a**) 1 ms, (**b**) 0.1 ms, (**c**) 0.01 ms, and (**d**) 0.001 ms

So, again it can be verified that the phase errors have occurred due to the periodic jitter regardless of the relation between the frequencies of the periodic jitter and the camera. Now, let us analyze the results from the two numerical methods.

### 9.6.3 Comparison of the Numerical and Analytical Approaches

For the simplicity, we consider zero background in the modulated light signal (i.e., $B = 0$ in (9.24)) for both numerical and analytical approaches obtained. Figures 9.26 and 9.27 compare the maximum (worst) phase errors obtained from the above two approaches for the various modulation frequencies $f$ and periodic jitter amplitudes $A_{PJ}$ when the frequency of the periodic jitter is a factor and a non-factor of the modulation frequency, respectively. Four phase error criteria $\phi_{err}$ with first three integration periods are considered in each figure, refer Table 9.2.

It can be seen, 220000 intervals and $2^{19}$ panels used for the trapezoidal and Romberg integration methods are not sufficient to identify the lowest phase error criterion, 0.1 mrad (among the chosen) with the higher integration period, $T = 1$ ms, respectively, when $f_{PJ} = 4.9261$ MHz (see left sub-figure in Fig. 9.26a and b). Other than that, both numerical methods are given similar results with the analytical approach. But, when $T = 0.01$ ms, the lowest phase error curve in the analytical approach is not matched with the corresponding results from the two numerical methods. We believe, it may be due to the unsufficient amount of the orders $m$ and $n$ of the Bessel functions in (9.51), see each right sub-figure in Fig. 9.26.

However, when the frequency of the periodic jitter is a factor of the modulation frequency, $f_{PJ} = 5$ MHz, then the corresponding results from the two approaches are identical, see Fig. 9.27. Here, for each $T$, the same number of intervals $Q = 3000$ in (9.28) and the same number of panels $2^{M-1} = 2^{17} = 131{,}072$ panels in (9.32) are well sufficient, while they were different and relatively large in trapezoidal integration when the $f_{PJ} = 4.9261$ MHz. However, here, for the Romberg integration, one-fourth of the number of panels when $f_{PJ} = 4.9261$ MHz was well sufficient (i.e., $2^{17}$:$2^{19}$) for each $T$. In each case, both numerical approaches are given similar results for different $T$ as well as for all four phase errors $\phi_{err}$. However, they were different for $T = 0.1$ ms than other two periods when $f_{PJ} = 4.9261$ MHz, see each sub-figure in Fig. 9.26(middle). Thus, the correctness of the analytical model developed can be verified with the evidence of the results from the numerical approach.

## 9.7 Execution Time for the Simulation

Tables 9.7 and 9.8 show the approximate execution time for worst (maximum) and mean phase errors in each simulation setup, respectively. Note that these amounts are rounded up to the nearest two decimal places. The corresponding results without

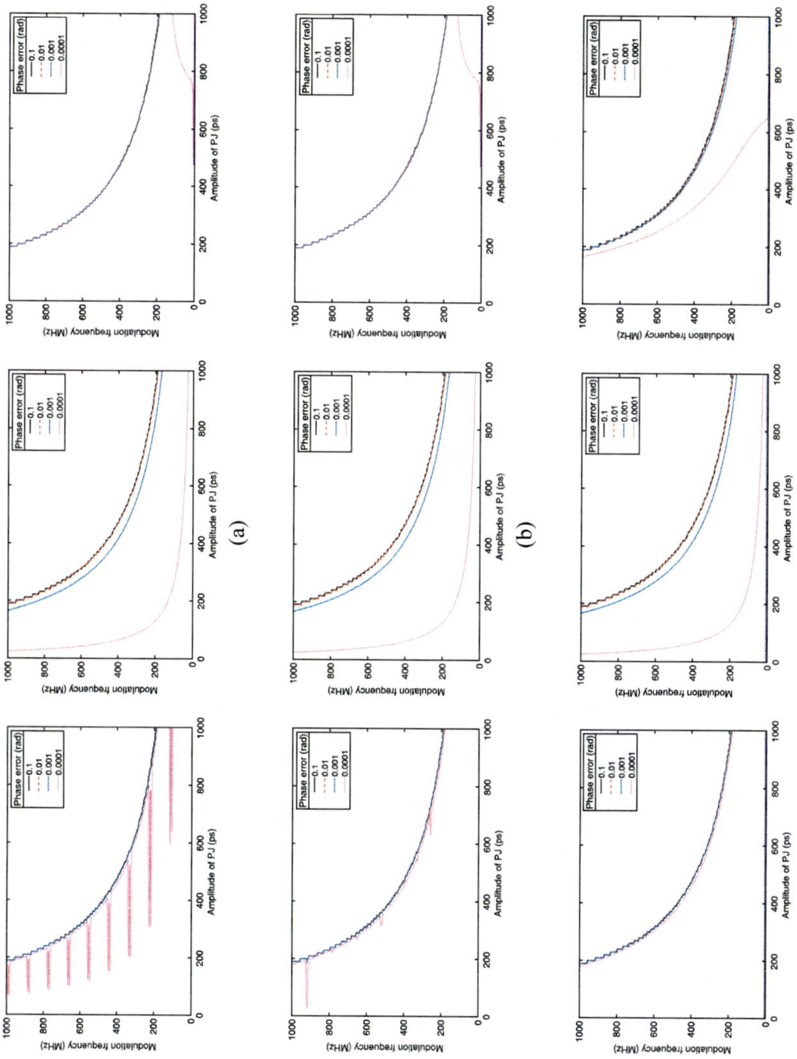


**Fig. 9.26** Four phase error due to the periodic jitter at the frequency $f_{PJ} = 4.9261$ MHz with various modulation frequencies against various jitter amplitudes for the ToF phase shift $\phi = \pi/6$ with $B = 0$ when (left): $T = 1$ ms, (middle): $T = 0.1$ ms, and (right): $T = 0.01$ ms by using **(a)** the trapezoidal integration in (9.28) with the intervals: 220000 for (left), 220000 for (middle) and 21000 for (right); **(b)** Romberg integration in (9.32) with $2^{19} = 524288$ panels for each (left), (middle), and (right); **(c)** the analytical model using (9.51) with order $m, n = -40{:}40$ for each (left), (middle), and (right)

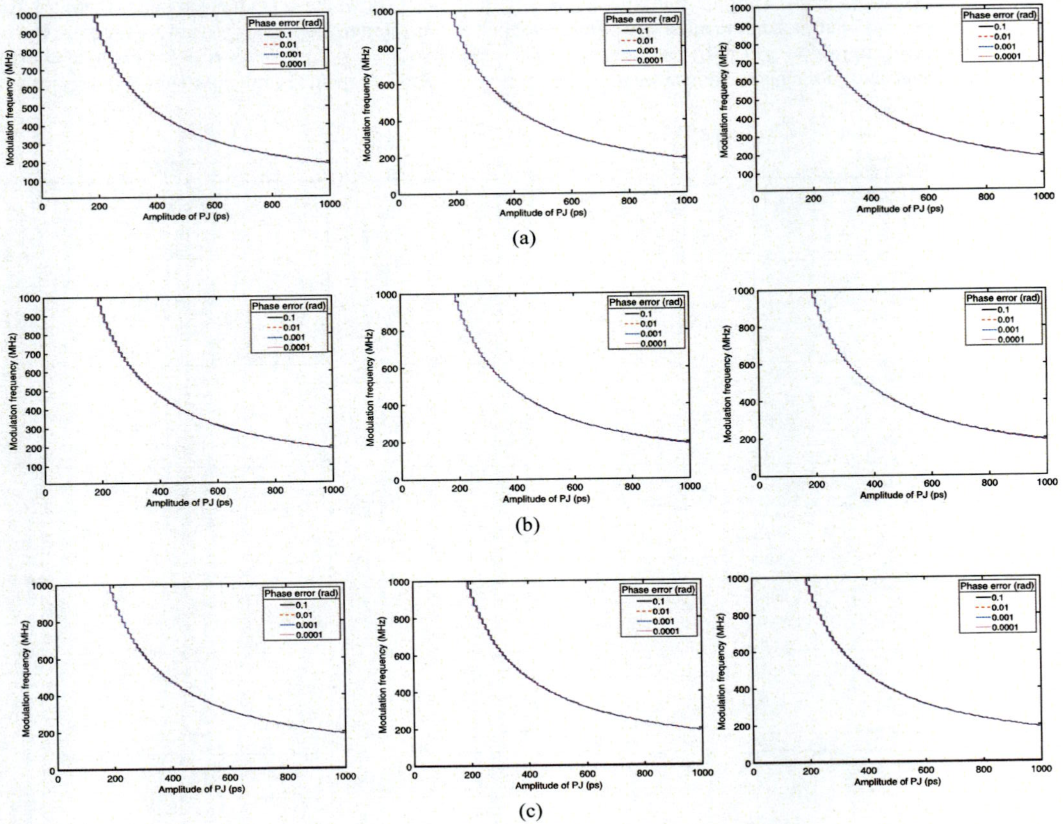


**Fig. 9.27** Four phase error due to the periodic jitter at the frequency $f_{PJ} = 5$ MHz with various modulation frequencies against various jitter amplitudes for the ToF phase shift $\phi = \pi/6$ with $B = 0$ when (left): $T = 1$ ms, (middle): $T = 0.1$ ms, and (right): $T = 0.01$ ms by using (**a**) the trapezoidal integration in (9.28) with 3000 intervals for each (left), (middle), and (right); (**b**) Romberg integration in (9.32) with $2^{17} = 131072$ panels for each (left), (middle), and (right); (**c**) the analytical model using (9.51) with order $m, n = -40{:}40$ for each (left), (middle), and (right)

**Table 9.7** Approximate execution time for the simulations of the maximum phase error computations

| $f_{PJ}$ (MHz) | Approach with the parameter used | | | $T$ (ms) | Execution time (in minutes) for worst (maximum) phase errors when the background illumination $B = 0$ | $B = 1$ | $B = 10$ | $B = 50$ | $B = 100$ |
|---|---|---|---|---|---|---|---|---|---|
| 4.9261 | Analytical model in (9.51), with orders $m, n$ | $-40 \le \{m, n\} \le 40$ 6561 × 2 terms | | 1.0 | 818.89 | 856.22 | 835.85 | 866.16 | 832.57 |
| | | | | 0.1 | 829.81 | 848.83 | 905.20 | 847.97 | 834.66 |
| | | | | 0.01 | 839.53 | 829.98 | 849.30 | 818.84 | 840.03 |
| | | | | 0.001 | 843.33 | 840.49 | 865.89 | 844.13 | 824.09 |
| | Trapezoidal model in (9.28), with intervals $Q$ | 220000 | four times | 1.0 | 271.38 | | | | |
| | | 220000 | | 0.1 | 369.08 | | | | |
| | | 21000 | | 0.01 | 84.59 | | | | |
| | Romberg model in (9.32), with panels $2^{M-1}$ | $2^{19} = 524288$ | | 1.0 | 2320.67 | | | | |
| | | | | 0.1 | 2293.18 | | | | |
| | | | | 0.01 | 2279.27 | | | | |
| 5 | Analytical model in (9.51), with orders $m, n$ | $-40 \le \{m, n\} \le 40$ 6561 × 2 terms | | 1.0 | 858.47 | 833.37 | 904.20 | 843.80 | 838.89 |
| | | | | 0.1 | 834.29 | 832.91 | 935.69 | 862.27 | 838.24 |
| | | | | 0.01 | 873.11 | 825.65 | 854.28 | 842.55 | 849.87 |
| | | | | 0.001 | 830.08 | 852.43 | 952.58 | 827.99 | 846.92 |
| | Trapezoidal model in (9.28), with intervals $Q$ | 3000 | four times | 1.0 | 110.18 | | | | |
| | | | | 0.1 | 108.28 | | | | |
| | | | | 0.01 | 97.57 | | | | |
| | Romberg model in (9.32), with panels $2^{M-1}$ | $2^{17} = 131072$ | | 1.0 | 432.48 | | | | |
| | | | | 0.1 | 418.59 | | | | |
| | | | | 0.01 | 401.36 | | | | |

**Table 9.8** Approximate execution time for the simulations of the mean phase error computations

| $f_{PJ}$ (MHz) | Approach with the parameter used | $T$ (ms) | Execution time (in minutes) for mean phase error computations when the background illumination | | | | |
|---|---|---|---|---|---|---|---|
| | | | $B = 0$ | $B = 1$ | $B = 10$ | $B = 50$ | $B = 100$ |
| 4.9261 | Analytical model in (9.51) with orders $m, n$ | 1.0 | 848.72 | 829.67 | 845.89 | 848.42 | 838.82 |
| | | 0.1 | 835.58 | 832.03 | 853.51 | 830.27 | 826.03 |
| | | 0.01 | 836.61 | 825.41 | 828.73 | 868.75 | 829.85 |
| | | 0.001 | 841.99 | 826.35 | 860.05 | 839.47 | 848.55 |
| 5 | $-40 \leq \{m, n\} \leq 40$ $6561 \times 2$ terms | 1.0 | 840.72 | 850.81 | 843.57 | 859.74 | 824.98 |
| | | 0.1 | 826.53 | 845.82 | 855.51 | 832.49 | 864.36 |
| | | 0.01 | 888.16 | 914.19 | 828.30 | 864.53 | 864.88 |
| | | 0.001 | 881.57 | 846.28 | 837.63 | 842.56 | 838.28 |

the background illumination from the two numerical methods are also embedded in Table 9.7. For each numerical method, the corresponding model is performed four times with their parameters (since four phase step computation), while it is a single step for the analytical model.

As usual, for a greater number of terms, it takes more execution time, i.e., for $2^{19}$ terms (with four times) in Romberg model, it has taken an average of ≈2300 min, while for 3000 terms (with four times) in trapezoidal model has taken an average of 100 min. However, the execution time for the analytical approach is approximately average 850 min, and this amount is similar for both results: the worst and mean phase error computations, see Tables 9.7 and 9.8. Note that these average times may slightly change due to the execution of other concurrent processes in the computer.

All simulations are implemented with MATLAB R2018b (The MathWorks Inc.) and are executed with high performance computing (HPC) cluster supercomputers (Cray®CS400™) provided by New Zealand eScience Infrastructure (NeSI) [19]. The used configuration of the cluster is as follows [20]: on a Linux server with the operating system CentOS 7.4 and 72 logical cores in 2×Broadwell (E5-2695v4, 2.1 GHz, dual-socket 18 cores per socket) nodes; the memory capacity per compute node is 128.0 GB and the disk space allocation of 100.0 GB. All 36 cores and 40 GB of memory space are allocated for each simulation.

## 9.8 Chapter Remarks

The range sensor manufacturing industries are attempting to increase the performance of the cameras, and in the near future one expects to see ToF cameras with high modulation frequencies, so they are introducing relevant electronic components to build these cameras. As a result of that, it may effect time deviation of the signal generated inside, which causes an increase in the jitter in the signals of the camera. At the same time, they are focusing on the accuracy of the camera as well since the accuracy and precision of ToF range cameras are important for many applications. Then, we cannot ignore the effect of the jitter on the range measurement. In this chapter, we proposed an analytical model to clarify the influence of the periodic jitter by using the Fourier analysis on the correlation function of the ToF camera.

Even though we presented the influence of the periodic jitter on four phase error criteria (0.1, 0.01, 0.001, and 0.0001 rad), our model is able to identify the phase error in the range of 0 to $2\pi$ rad. Another advantage of our model is the capability of the usage in heterodyne operation in ToF cameras. We chose the range of $\beta = 0$ to $2\pi$, which is the product of angular modulation frequency (upto $f_{\mathrm{PJ}} = 1000\,\mathrm{MHz}$) and the periodic jitter amplitude (upto $A_{\mathrm{PJ}} = 1000\,\mathrm{ps}$) of the camera.

We proved and found evidence that the phase difference of the periodic jitter between the emitted light and shutter signals of the camera, the modulation frequency, the integration period, the amplitude of the periodic jitter, and the relationship of the periodic jitter frequency to the modulation frequency are the parameters that influence the periodic jitter on range measurement. Also, we found that the background illumination of the system is not affected on the phase error due to the periodic jitter, hence the range measurement. Besides, whether the frequency of the periodic jitter is a factor of the modulation frequency or not, it is caused by different behaviors in the system. However, this influence can be compensated by adjusting the integration period of the camera, since this phenomenon also depends on the integration period. When the integration period is $T = 0.001$ ms (the shortest period among the chosen), then the periodic jitter influences the range measurement for all chosen four phase error criteria for certain amplitudes of periodic jitter $A_{\mathrm{PJ}}$ and modulation frequencies $f$. When the periodic jitter is a non-factor of modulation frequency $f_{\mathrm{PJ}} = 4.9261$ MHz, they are in the range of $[A_{\mathrm{PJ}}, f_{\mathrm{PJ}}]$ = [(200 ps, 1000 MHz)–(1000 ps, 200 MHz)] for the phase errors $\phi_{\mathrm{err}} = 0.01$ and 0.001 rad, while for $\phi_{\mathrm{err}} = 0.1$ rad, they are in the range of $[A_{\mathrm{PJ}}, f_{\mathrm{PJ}}]$ = [450 ps, 1000 MHz] to [1000 ps, 450 MHz]. With the increase of the integration period, the corresponding range of the $[A_{\mathrm{PJ}}, f_{\mathrm{PJ}}]$ due to the PJ is systematically increased. This similar behavior was presented when the $f_{\mathrm{PJ}} = 5$ MHz, but in slightly different amounts. However, in both cases, the starting influence range of the corresponding $A_{\mathrm{PJ}}$ and $f_{\mathrm{PJ}}$ is similar (i.e., (200 ps, 1000 MHz)–(1000 ps, 200 MHz)) for the phase errors $\phi_{\mathrm{err}} = 0.01$ and 0.001 rad.

At present, AMCW ToF cameras use relatively low modulation frequencies and low frame rates. Therefore, the range measurement is not affected by the PJ from current AMCW ToF cameras such as MESA Imaging SwissRanger 4000 (with 30 MHz) and SoftKinetic DepthSense 325 (with 50 MHz), which we used in the experiments in Chaps. 7 and 8. In future, the depth sensor manufacturers increase the resolution range of ToF cameras by using high modulation frequencies and very high frame rates that are the factors that affect integration periods of the cameras. Thus, in order to minimize the influence of the periodic jitter on range measurement, we are recommending when the integration period $T = 0.01$ ms and the modulation frequency of the camera $f = 200$ MHz, then the amplitude of the PJ should be $A_{\mathrm{PJ}} < 700$ ps, while when $f = 1000$ MHz, it should be $A_{\mathrm{PJ}} < 180$ ps. Also, in order to prevent the phase errors of $\phi_{\mathrm{err}} = 0.001$ and 0.01 rad, the parameters should be satisfied when $A_{\mathrm{PJ}} < 400$ ps for $f = 400$ MHz. In addition, these recommended values of $A_{\mathrm{PJ}}$ and $f$ should be further reduced when the integration period of the camera is reduced. Thus, the PJ has significantly affected range measurements apart from the above recommended amounts of the three parameters: the integration period, jitter amplitude, and modulation frequency.

Finally, we clearly found that the periodic jitter in the light signals influences the range measurements in AMCW ToF ranging cameras. By referring to Chaps. 7 and 8, we concluded by experimenting that the random jitter is relatively larger than the periodic jitter in the illumination signals of two commercial AMCW ToF cameras. So, it is important and more interesting to check whether the random jitter has an influence on range measurements in AMCW ToF ranging cameras. This is the focus of the next chapter.

## References

1. Cyrus S Bamji, Swati Mehta, Barry Thompson, Tamer Elkhatib, Stefan Wurster, Onur Akkaya, Andrew Payne, John Godbaz, Mike Fenton, Vijay Rajasekaran, et al. IMpixel 65nm BSI 320MHz demodulated TOF Image sensor with $3\mu$m global shutter pixels and analog binning. In *2018 IEEE International Solid-State Circuits Conference-(ISSCC)*, pages 94–96. IEEE, 2018.
2. *Intel®Stratix®10 Device Datasheet*. Intel Cooperation, 2020. Accessed June 2021.
3. Paul Adrien Maurice Dirac. *The principles of quantum mechanics*. Number 27. Oxford university press, 1981.
4. V Balakrishnan. All about the Dirac delta function (?). *Resonance*, 8(8):48–58, 2003.
5. Milton Abramowitz and Irene A Stegun. *Handbook of mathematical functions: with formulas, graphs, and mathematical tables*, volume 55. Courier Corporation, 1965.
6. Simon S Haykin and Michael Moher. *An introduction to analog and digital communications*. John Wiley & Sons, second edition, 2007. ISBN 9780471432227.
7. Alan V. Oppenheim, Ronald W. Schafer, and John R. Buck. *Discrete-time Signal Processing (2nd Ed.)*. Prentice-Hall, Inc., Upper Saddle River, NJ, USA, 1999. ISBN 0-13-754920-2.
8. Mike Li. *Jitter, Noise, and Signal Integrity at High-speed*. Prentice Hall Press, Upper Saddle River, NJ, USA, first edition, 2007. ISBN 9780132429610.
9. Alejandro L Garcia. *Numerical methods for physics*. Prentice Hall Englewood Cliffs, NJ, 2 edition, 2000.
10. Steven C Chapra and Raymond P Canale. *Numerical Methods for Engineers*, volume 7. McGraw-Hill New York, 2015.
11. Germund Dahlquist and Ake Bjorck. *Numerical Methods in Scientific Computing: Volume 1*, volume 103. Society for Industrial and Applied Mathematics, 2008.
12. R. Lange and P. Seitz. Solid-state time-of-flight range camera. *IEEE Journal of Quantum Electronics*, 37(3):390–397, Mar 2001. ISSN 0018-9197. https://doi.org/10.1109/3.910448.
13. Radu Horaud, Miles Hansard, Georgios Evangelidis, and Clément Ménier. An overview of depth cameras and range scanners based on time-of-flight technologies. *Machine vision and applications*, 27(7):1005–1020, 2016.
14. AA Dorrington, JP Godbaz, MJ Cree, AD Payne, and LV Streeter. Separating true range measurements from multi-path and scattering interference in commercial range cameras. In *IS&T/SPIE Electronic Imaging, Proc. SPIE*, volume 7864, pages 1–10, 2011.
15. T. Kahlmann, F. Remondino, and H. Ingensand. Calibration for increased accuracy of the range imaging camera SwissRanger$^{TM}$. In *Image Engineering and Vision Metrology*, volume 36 of *3*, pages 136–141, 2006.
16. Marvin Lindner and Andreas Kolb. Lateral and depth calibration of PMD-distance sensors. In *Proceedings of the Second International Conference on Advances in Visual Computing - Volume Part II*, ISVC'06, pages 524–533, Berlin, Heidelberg, 2006. Springer-Verlag. https://doi.org/10.1007/11919629_53.
17. Lee Streeter and Adrian A Dorrington. Simple harmonic error cancellation in time of flight range imaging. *Optics Letters*, 40(22):5391–5394, 2015. https://doi.org/10.1364/OL.40.005391.
18. *SR4000 Data Sheet*. MESA Imaging AG, February 2011. URL http://www.adept.net.au/cameras/Mesa/pdf/SR4000.pdf.
19. New Zealand eScience Infrastructure-About Us. https://www.nesi.org.nz/about-us, 2021a. [Accessed August 2021].
20. The NeSI High Performance Computers-Mahuika. https://support.nesi.org.nz/hc/en-gb/articles/360000163575, 2021b. [Accessed August 2021].

# Chapter 10
# Influence of Random Jitter

**Abstract** This chapter describes a benchmark analytical model based on stochastic calculus to investigate the behavior of random jitter on range measurements. The model is developed with the help of fundamental statistical concepts and Fourier analysis. It is assumed that the random jitter follows the Gaussian distribution. The model applies for both heterodyne and homodyne operations in the AMCW ToF range imaging cameras. First, a brief introduction of the standard mathematical expressions with the fundamental concepts in statistics for the main theorems is described. Next, the correlation function with random jitter in both the illumination and shutter signals is considered and a stochastic model is developed. The Monte Carlo simulation is used to analyze the model since this is the best technique to understand the impact of possibility and significant uncertainty in the prediction quantitative model due to the intervention of random variables. The simulation is performed by two non-parametric estimations for the simulated data. The various parameters such as modulation frequency $f$ and amount of random jitter $\sigma_{\mathrm{RJ}}$ with various integration periods are evaluated throughout the simulations. Finally, very interesting results for what corresponding amounts of $\sigma_{\mathrm{RJ}}$ and $f$ that influence range measurements are pointed out. In addition, the uncertainty analysis of the results and execution time for the simulation are examined.

**Keywords** Random jitter · Gaussian distribution · Random variable · Random process · Central limit theorem · Monte Carlo method · Histogram · Kernel density estimation · Probability density function · Law of the unconscious statistician (LOTUS) theorem · Uncertainty (standard error) · Interquartile range · Angular frequency conjugate · Fundamental frequency · Phase and range errors

## 10.1 Background

As described in Chaps. 7 and 8, random jitter was shown to be relatively larger than the periodic jitter in commercial AMCW ToF range imaging cameras. The random effects have an influence on every stage of the correlation process of the camera and finally on the range measurement as well. Random jitter is often modeled as



G. Anthonys, *Timing Jitter in Time-of-Flight Range Imaging Cameras*,
https://doi.org/10.1007/978-3-030-94159-8_10

a Gaussian distribution [1], and we need the help of some statistical techniques to investigate the influence of the random jitter in AMCW ToF range imaging systems.

The rest of this section is focused on the standard statistical expressions used in this chapter, namely the means and variances of a continuous random variable and a linear combination of random variables, the Gaussian probability distribution, the central limit theorem, Monte Carlo method, and two types of density estimates for non-parametric variables: the histogram and kernel density estimations. The definitions and some theorems of Fourier transforms described in Sect. 9.2.2 are also employed.

### 10.1.1 Mean and Variance of a Continuous Random Variable

Let $X$ be a continuous random variable, and the expected value (mean) of $X$ be given by Blitzstein and Hwang [2]

$$\mathbb{E}[X] \triangleq \int_{-\infty}^{\infty} x f_X(x)\, dx, \tag{10.1a}$$

where $f_X(x)$ is the probability density function (PDF) of $X$. Then, the variance of the variable $X$ is given by (see Appendix B.2.3.1 for proof)

$$\mathbb{V}[X] = \mathbb{E}\left[(X - \mathbb{E}[X])^2\right] = \mathbb{E}[X^2] - (\mathbb{E}[X])^2, \tag{10.1b}$$

where $\mathbb{E}[X^2]$ is the second moment of $X$ (or expectation of $X^2$), which can be obtained by using the law of the unconscious statistician (LOTUS) theorem (see Appendix B.1.1). The mean and variance of a real constant variable $\alpha$ is [2]

$$\mathbb{E}[\alpha] = \alpha, \tag{10.2a}$$

$$\mathbb{V}[\alpha] = 0. \tag{10.2b}$$

### 10.1.2 Mean and Variance of a Linear Combination of Random Variables

Let $Y = \sum_{k=1}^{K} \alpha_k X_k$ be the linear combination of $K \in \mathbb{Z}^+$ independent random variables $X_k$ with $\alpha_k$ the real constants. Then, the mean $\mu_Y$ and variance $\sigma_Y^2$ of the linear combination $Y$ are given by Rohatgi and Saleh [3]

$$\mu_Y = \sum_{k=1}^{K} \alpha_k \mu_k, \tag{10.3a}$$

$$\sigma_Y^2 = \sum_{k=1}^{K} \alpha_k^2 \sigma_k^2, \tag{10.3b}$$

where $\mu_k$ and $\sigma_k^2$ are the mean and variance of the $k$th random variable $X_k$ (see Appendix B.2.3.2 for proof).

### *10.1.3 Gaussian (Normal) Probability Distribution*

Random jitter that is not bounded is usefully modeled by a Gaussian (normal) probability distribution. The normal distribution is a probability distribution that associates the random variable $X$ with a cumulative probability (see Appendix B.1.2) and is defined by

$$f_X(x) = \frac{1}{\sigma\sqrt{2\pi}} \exp\left\{ -\frac{1}{2}\left(\frac{x-\mu}{\sigma}\right)^2 \right\}, \tag{10.4}$$

where $x \in X$ with $\mu$ and $\sigma^2$ the mean and variance of the variable $X$, respectively (see Appendix B.2.3.3 for proof). Figure 10.1 shows a normal distribution with the shaded area in the interval $\mu \pm \sigma$ that contains 68.27% of the distribution (see Appendix B.2.2).

In other words, this figure indicates there is approximately a 68.27% possibility that any particular $x$ falls within one standard deviation of the mean. In addition, the

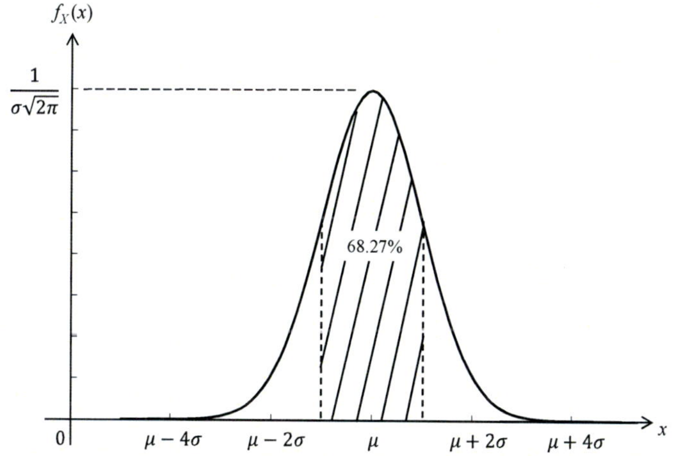


**Fig. 10.1** The Normal distribution with the shaded area of 68.27% from its total, in the interval $\mu \pm \sigma$

possibilities that $x$ falls within two and three standard deviations of the mean are approximately 95.45% and 99.73%, respectively (see Appendix B.2.2).

### 10.1.4 Central Limit Theorem

The central limit theorem states that when there are sufficiently large numbers of random samples from the population with mean $\mu$ and standard deviation $\sigma$, then the distribution of the sample means will be approximately normally (Gaussian) distributed. Therefore, we can use the normal probability model to quantify uncertainty when making inferences about a population mean based on the sample mean $\bar{x}$. Thus, the uncertainty or standard error (SE) of the sample means is defined by Taylor [4]

$$\mathrm{SE}(\bar{x}) = \pm \frac{s}{\sqrt{K}}, \tag{10.5}$$

where $s$ and $K$ are the standard deviation of the samples and the number of samples, respectively.

However, the mean of any finite set of measurements is not to be exactly equal to the actual value of the quantity since the random errors are not likely to perfectly cancel when the number of measurements is relatively small or the distribution is not normal. In that case, the Student's $t$-factor is introduced to compensate for the uncertainty in $s$ and, thus, (10.5) is modified to

$$\mathrm{SE}(\bar{x}) = \pm t_{K-1} \frac{s}{\sqrt{K}}, \tag{10.6}$$

where $t_{K-1}$ is a constant called the Student's $t$-factor for given confidence interval (e.g., 5%, 2%, 1%, etc.) with $K-1$ *degrees of freedom*, which can be obtained from standard $t$ distribution [5].

### 10.1.5 Monte Carlo Method

The Monte Carlo method is a technique used to understand the impact of uncertainty in prediction and forecasting models. These are used to model the probability of different outcomes in a process that cannot easily be predicted due to the intervention of random variables [6]. Monte Carlo simulation (MCS) performs random sampling and conducts a large number of experiments on the computer, then the statistical characteristics of the experiments are observed, and conclusions on these results are drawn based on the statistical process. Thus, when faced with significant uncertainty in the process of making an estimation, rather than just replacing the uncertain variable with a single average number, the Monte Carlo simulation might prove to be a better solution [7].

### 10.1.6 Non-parametric Density Estimates

In order to estimate the density function from the observed data, there are two approaches: parametric and non-parametric estimations. In the parametric approach, the data are drawn from a known distribution while for the other, it is assumed that the distribution has an unknown probability density, then the data are used to estimate its density function. The width of one standard deviation of the probability density function is the corresponding measurement for the parameter. In this chapter, we are focusing on two types of non-parametric density estimations: histogram and kernel density [8]. All estimates of the density function satisfying relatively strong conditions are shown to be biased [9].

Let $\{x_1, \ldots, x_n\}$ be univariate independent and identically distributed $n$ samples drawn from some distribution with an unknown density function $f(x)$. Let $\widehat{f}(x)$ be an estimate of $f(x)$. Now, let us see how the density function can be estimated from these two methods.

#### 10.1.6.1 Histogram Estimation

Histogram estimation is the most widely used density estimator for the distribution of numerical data [8]. To construct the histogram, both an origin $x_0$ and width of the bin $h$ are chosen. The bins of the histogram are defined as the intervals $[x_0 + nh, x_0 + (n+1)h)]$ for $n \in \mathbb{Z}^*$. The histogram is then defined by

$$\widehat{f}(x) = \frac{\text{number of } x_j \text{ in the same bin as } x}{nh}, \quad j = 1, \ldots, n \in \mathbb{Z}^+. \tag{10.7}$$

In order to obtain an unbiased estimation of the PDF, the optimal width of the bin can be found by the Freedman–Diaconis rule, given by Freedman and Diaconis [10]

$$h_{opt}^{hist} = 2\,\frac{\text{IQR}}{\sqrt[3]{n}}, \tag{10.8}$$

where IQR is the interquartile range of the data, which is calculated by the difference between the third and first interquartile $(Q_3 - Q_1)$.

The histogram is an extremely useful class of density estimate, particularly in the univariate case for the presentation and exploration of data. However, the histogram has some drawbacks. As it is not continuous, it is very difficult to find the derivatives. Also, the histogram is not smooth, as the result interpretation depends on the width of the bins and end points of the bins (each of the bin's start). The representation of bivariate and trivariate data by histogram is difficult [8].

#### 10.1.6.2 Kernel Density Estimation

Kernel functions are typically smooth functions with a single mode at $x = 0$. In addition, to remove the dependence on the end points of the bins, kernel estimators center a kernel function at each data point. Here, it uses a simple technique: to make a prediction for any point $x$, just sum, for all points $x_j$ in the sample, all the kernel functions of the difference between $x_j$ and $x$ are used. Therefore, the shape of the function $f(x)$ can be formulated by its kernel density estimator, defined by Silverman [8]

$$\widehat{f}(x) = \frac{1}{nh}\sum_{j=1}^{n}\mathcal{K}\left(\frac{x - x_j}{h}\right), \qquad n \in \mathbb{Z}^+, \tag{10.9}$$

where $h > 0$ is the smoothing parameter called the *bandwidth* and $\mathcal{K}(.)$ is the *kernel* function, which is non-negative and satisfies the condition:

$$\int_{-\infty}^{\infty} \mathcal{K}(x)\, dx = 1. \tag{10.10}$$

Usually, $\mathcal{K}(.)$ will be a symmetric PDF. There are a range of commonly used kernel functions including Gaussian, uniform, triangular, biweight, triweight, and Epanechnikov. The Gaussian kernel function is focused. The optimal bandwidth for this function according to Silverman's *rule of thumb* [8] is

$$h_{opt}^{kde} = \frac{1.06}{\sqrt[5]{n}} \times \min\left(\sigma, \frac{\text{IQR}}{1.34}\right), \tag{10.11}$$

where $\sigma$ is the sample standard deviation and constant 1.34 is the interquartile range of the standard normal distribution. That is, for the Gaussian data samples, IQR $\approx 1.34\,\sigma$ [11].

## 10.2 Random Jitter in the Correlation Model

Random jitter is non-deterministic jitter due to random processes and it often follows a Gaussian distribution $\mathcal{N}$. Therefore, random jitter can be characterized with its root-mean-square (RMS) $\sigma$ and, thus, [1]

$$\Delta t = \varepsilon(t) \sim \mathcal{N}(\mu, \sigma^2), \tag{10.12}$$

where $\varepsilon(t)$ is a random process, which is a function of time and has a Gaussian distribution $\mathcal{N}$ with mean $\mu$ and variance $\sigma^2$. For the heterodyne operation in the camera, we use separate Gaussian random processes for light and shutter signals, respectively, namely

$$l(t,\phi) = \cos(2\pi f_l(t+\varepsilon_l(t)) - \phi), \tag{10.13}$$

and

$$s(t,\theta) = \cos(2\pi f_s(t+\varepsilon_s(t)) + \theta), \tag{10.14}$$

where $f_l$ and $f_s$ are the modulation frequencies of the light and shutter signals, respectively. $\phi$ is the ToF phase shift and $\theta$ is the phase step, which is controlled by the camera. The parameters $\varepsilon_l(t) \sim \mathcal{N}(\mu_l, \sigma_l^2)$ and $\varepsilon_s(t) \sim \mathcal{N}(\mu_s, \sigma_s^2)$ are Gaussian random processes for the random jitter in the light and shutter signals, respectively. Note that both signals are assumed to have a unity amplitude with zero offset coefficient due to the background illumination.

Since the camera integrates a frame capture over an integration period of $T$ for a specific phase step $\theta$, the correlation function $I(\theta)$ with the random jitter is given by

$$I(\theta) = \int_{-T/2}^{T/2} s(t,\theta)\, l(t,\phi)\, dt. \tag{10.15}$$

As stated in Chap. 9, we take the phase step $\theta$ to be continuous, so by taking the Fourier transform of (10.15) with respect to $\theta$, namely let

$$\widehat{I}(\vartheta) = \mathscr{F}[I(\theta)]_{\theta\to\vartheta} = \int_{-T/2}^{T/2} \mathscr{F}[s(t,\theta)]\, l(t,\phi)\, dt, \tag{10.16}$$

where $\vartheta$ is the angular frequency conjugate to $\theta$ in units of per radian. The Fourier transform of the shutter signal can be obtained by application of the Fourier shift theorem (9.14) and Fourier transform of cosine (9.20) as

$$\begin{aligned}
\mathscr{F}[s(t,\theta)] &= \mathscr{F}\left[\cos\left(2\pi f_s(t+\varepsilon_s(t)) + \theta\right)\right]_{\theta\to\vartheta} \\
&= \mathscr{F}\left[\cos\left(\theta + 2\pi f_s(t+\varepsilon_s(t))\right)\right]_{\theta\to\vartheta} \\
&= e^{i2\pi(2\pi f_s(t+\varepsilon_s(t)))\vartheta}\, \mathscr{F}[\cos(\theta)] \\
&= \frac{1}{2} e^{i(2\pi)^2 f_s(t+\varepsilon_s(t))\vartheta} \left[\delta\left(\vartheta - \frac{1}{2\pi}\right) + \delta\left(\vartheta + \frac{1}{2\pi}\right)\right].
\end{aligned} \tag{10.17}$$

Let $\Theta = 2\pi\vartheta$ be the "frequency" conjugate to $\theta$, and with the scaling property of delta function (9.3), then (10.17) becomes

$$\mathscr{F}[s(t,\theta)] = \pi e^{i2\pi f_s(t+\varepsilon_s(t))\,\Theta}\left[\delta(\Theta - 1) + \delta(\Theta + 1)\right]. \tag{10.18}$$

Thus (10.16) becomes

$$\widehat{I}(\Theta) = \pi\left[\delta(\Theta-1)+\delta(\Theta+1)\right]\int_{-T/2}^{T/2} e^{i2\pi f_s(t+\varepsilon_s(t))\,\Theta} \times \cos\left(2\pi f_l(t+\varepsilon_l(t))-\phi\right)dt. \tag{10.19}$$

Considering the fundamental frequency ($\Theta = 1$) only

$$\widehat{I}(1) = \pi\int_{-T/2}^{T/2} e^{i2\pi f_s(t+\varepsilon_s(t))}\cos\left(2\pi f_l(t+\varepsilon_l(t))-\phi\right)dt. \tag{10.20}$$

Then, the phase error $\phi_{\text{err}}$ of the correlation model obtained in (10.20) can be computed by

$$\phi_{\text{err}} = \left|\phi - \angle\left\{\pi\int_{-T/2}^{T/2} e^{i2\pi f_s(t+\varepsilon_s(t))}\cos\left(2\pi f_l(t+\varepsilon_l(t))-\phi\right)dt\right\}\right|, \tag{10.21}$$

where $\angle$ is the angle of the resulted function.

### 10.2.1 An Estimator for the Phase Error Due to the Random Jitter

Now, by using the non-parametric estimation, a corresponding estimator (mean or variance) for the parameter $\phi_{\text{err}}$ can be obtained. In here, the variance $\sigma^2$ (or standard deviation $\sigma$) is more important since the mean $\mu$ of the parameter can be controlled with an offset adjustment in the system. Thus, the standard deviation of the phase error $\sigma_{\phi_{\text{err}}}$, which is the characteristic parameter (since by setting $\mu = 0$) for the estimation of (10.21) that can be obtained by using the histogram (or KDE, refer Sect. 10.1.6), is given by Bennett et al. [5]

$$\sigma_{\phi_{\text{err}}} = \frac{\sqrt{\sum\limits_{j=1}^{M} F_j\left(x_j-\bar{x}\right)^2}}{\sum\limits_{j=1}^{M} F_j - 1}, \tag{10.22}$$

where $F_j$ is the corresponding frequency count of the sample data element $x_j$ and $\bar{x}$ is the sample mean of data elements obtained by

$$\bar{x} = \frac{\sqrt{\sum\limits_{j=1}^{M} F_j x_j}}{\sum\limits_{j=1}^{M} F_j}. \tag{10.23}$$

In order to obtain the typical phase error (mean of the standard deviation of the phase error $\overline{\sigma_{\phi_{\text{err}}}}$), the computation of (10.22) should be repeated many times (say, set $k = 1, 2, \ldots, K$) and then needs to take the mean among them as

$$\overline{\sigma_{\phi_{\text{err}}}} = \sqrt{\sum_{k=1}^{K} \sigma_{\phi_{\text{err}_k}}^2 \Big/ K}, \tag{10.24}$$

where $\sigma_{\phi_{\text{err}_k}}^2$ is the variance of the phase error for the $k$th set. Then, as usual the uncertainty occurred in the result. In order to obtain the uncertainty of the obtained result $\overline{\sigma_{\phi_{\text{err}}}}$ in (10.24), we use (10.5) since it is assumed that the error in the correlation function due to the random jitter is following a Gaussian distribution. Thus, the standard error for $\overline{\sigma_{\phi_{\text{err}}}}$ is given by

$$\text{SE}(\sigma_{\phi_{\text{err}}}) = \pm \frac{\text{std}(\sigma_{\phi_{\text{err}}})}{\sqrt{K}}, \tag{10.25}$$

where std(.) is the standard deviation of $\sigma_{\phi_{\text{err}}}$ among the $K$ sets, which can be calculated as

$$\text{std}(\sigma_{\phi_{\text{err}}}) = \sqrt{\frac{\sum_{k=1}^{K} \left(\sigma_{\phi_{\text{err}_k}} - \overline{\sigma_{\phi_{\text{err}}}}\right)^2}{K-1}}, \tag{10.26}$$

where $\sigma_{\phi_{\text{err}_k}}$ is the standard deviation of the phase error for the $k$th set.

Hence the measurement of the typical phase error $\phi_{\text{err}}$ (mean of the standard deviation of the phase error among the $K$ sets) with the standard error (uncertainty) is found by

$$\sigma_{\phi_{\text{err}}} = \overline{\sigma_{\phi_{\text{err}}}} \pm \frac{\text{std}(\sigma_{\phi_{\text{err}}})}{\sqrt{K}}. \tag{10.27}$$

Now, substituting (10.24) and (10.26) into (10.27), we have

$$\sigma_{\phi_{\text{err}}} = \sqrt{\frac{\sum_{k=1}^{K} \sigma_{\phi_{\text{err}_k}}^2}{K}} \pm \sqrt{\frac{\sum_{k=1}^{K} \left(\sigma_{\phi_{\text{err}_k}} - \sqrt{\sum_{k=1}^{K} \sigma_{\phi_{\text{err}_k}}^2 \Big/ K}\right)^2}{K\,(K-1)}}. \tag{10.28}$$

This is the corresponding estimation for the phase error with the uncertainty due to the presence of random jitter in the correlation function. The random jitter is assumed to be following a Gaussian distribution. The zero means ($\mu_l = \mu_s = 0$) is chosen for the distribution since it can be controlled by an offset adjustment. For the RMS values of the RJ, the set of 21 pairs $(\sigma_l, \sigma_s)$ is used. In order to generalize the result, the amounts for these RMS pairs are chosen that are slightly different in the light and shutter signal such as [(0, 0), (50, 51), ... (950, 951), (1000, 1001)] ps.

Setup with a longer integration period takes a relatively large execution time. Four integration periods that span [1 μs–1 ms] such as $T = \{1, 0.1, 0.01, 0.001\}$ ms are performed during the simulation. Most of the current AMCW ToF cameras operate with 1 ms (or around) integration period. For the purpose of analysis, the two ToF phase shifts ($\phi = \pi/6$ and $\pi/3$) are considered for some of the simulation setups. However, the number of samples during the integration period is more important to identify in the evaluation of the model. In order to identify the sufficient number of samples (steps in the integral) for the simulation, the following computation is carried out.

### 10.2.2 Calculation of Sufficient Number of Samples

Let us consider the modulation frequency $f$ and integration period $T$ for any setup. Then, the number of cycles in the integration period is given by

$$N_{\text{cyl},T} = fT \text{ cycles.} \tag{10.29}$$

Now, the number of samples per cycle $N$ should satisfy the Nyquist criteria (i.e., at least a double of $N_{\text{cyl},T}$) and, thus,

$$N = \frac{N_{int}}{N_{\text{cyl},T}} = \frac{N_{int}}{fT}, \tag{10.30}$$

where $N_{int} \gg 2N_{\text{cyl},T}$ is a sufficient number of samples chosen in the integration period for the simulation. This $N_{int}$ is the corresponding number of steps for the integration period for the model.

### 10.2.3 Simulation Setup

The model (10.21) is evaluated for 14 modulation frequencies in the range of $f = 10$–$1000$ MHz with four integration periods chosen throughout the simulations. In order to maintain the fixed number of samples per cycle ($N = 100$ each), the number of samples in the integration period $N_{int}$ is chosen accordingly. This 100 is a sufficient number of samples per cycle with the chosen integration period for each modulation frequency. Trapezoidal integration is used throughout the simulation since it is a straightforward method. From Chap. 9, we found that trapezoidal integration (and the Romberg integration) can be used for the calculations of complex integrals. However, by comparing the execution time among these two numerical methods (see Table 9.7), we chose trapezoidal integration for our purpose (less execution time).

**Table 10.1** Parameters of the simulation setup for the analysis of random jitter in the correlation model

Time of flight system

Integration period (in millisecond), $T = 1, 0.1, 0.01$ and $0.001$ ms

| Frequency $f$(MHz) | Samples per cycle, $N$ | Number of samples ($N_{int}$) in $T$ (by using $N_{int} = NfT$), | $T = 1$ ms | $T = 0.1$ ms | $T = 0.01$ ms | $T = 0.001$ ms |
|---|---|---|---|---|---|---|
| 10 | 100 | $10^6\times$ | 1 | 0.1 | 0.01 | 0.001 |
| 30 | | | 3 | 0.3 | 0.03 | 0.003 |
| 50 | | | 5 | 0.5 | 0.05 | 0.005 |
| 80 | | | 8 | 0.8 | 0.08 | 0.008 |
| 100 | | | 10 | 1 | 0.1 | 0.01 |
| 200 | | | 20 | 2 | 0.2 | 0.02 |
| 300 | | | 30 | 3 | 0.3 | 0.03 |
| 400 | | | 40 | 4 | 0.4 | 0.04 |
| 500 | | | 50 | 5 | 0.5 | 0.05 |
| 600 | | | 60 | 6 | 0.6 | 0.06 |
| 700 | | | 70 | 7 | 0.7 | 0.07 |
| 800 | | | 80 | 8 | 0.8 | 0.08 |
| 900 | | | 90 | 9 | 0.9 | 0.09 |
| 1000 | | | 100 | 10 | 1 | 0.1 |

Random jitter with Gaussian distribution $\mathcal{N}(\mu, \sigma^2)$

Means (in picosecond), $\mu_l = \mu_s = 0$ ps

Standard deviations (in picosecond), 21 RMS pairs $(\sigma_l, \sigma_s)$ as [(0, 0); (50, 51); (100, 101); ... (950, 951); (1000, 1001)] ps

Other parameters—Number of:

Independent evaluations for the MCS of the model, nEval = 500

Repetitions in order to analyze the uncertainty of the model, nSet = 50

For the purpose of analysis:

Choose two typical phase error criteria $\overline{\sigma_{\phi_{\text{err}}}} = 0.1$ and 1 milliradian (mrad)

The MCS is performed for 500 independent evaluations of the model for each setup. Then, the standard deviation of the corresponding histogram (or KDE) is used for analyzing the result of the phase error obtained ($\sigma_{\phi_{\text{err}}}$). In order to analyze the uncertainty of the resultant phase error, fifty repetitions are performed for each simulation setup. All corresponding parameters of the simulation setup are tabulated in Table 10.1, while the simulation setup is visualized in Fig. 10.2.

## 10.3 Results and Discussion

For the purpose of better understanding, the results obtained from this investigation are divided into three sections: the calculations of the phase error and range error, and the uncertainty analysis of the results.

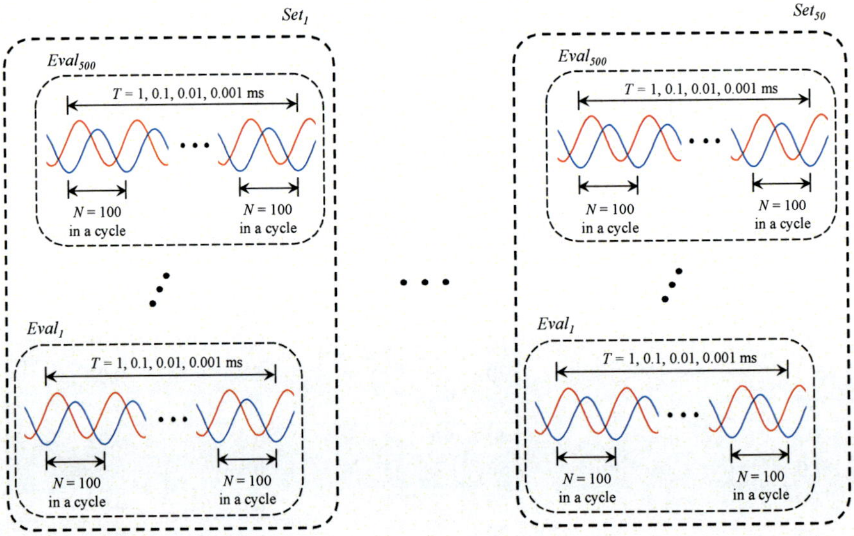


**Fig. 10.2** Illustration of the simulation setup for the analysis of random jitter in the correlation model

### *10.3.1 Phase Error Due to the Random Jitter*

By using the Monte Carlo simulation, the corresponding set of data is generated and then these data are simulated by using two non-parametric estimations stated in Sect. 10.1.6 (see Sect. C.4 in Appendix C for Matlab script). The corresponding probability density functions by the two estimations for two ToF phase shifts $\phi = \pi/6$ and $\pi/3$ due to the random jitter of a pair of RMS $(\sigma_l, \sigma_s) = (50, 51)$ ps with a single set and $T = 1$ ms for the first four modulation frequencies are shown in Fig. 10.3. In addition, the relevant standard deviation of the phase error (i.e., $\sigma_{\phi_{\text{err}_k}}$ with $k = 1$) for each case is also pointed in the caption of the figure. By comparison of the standard deviations, it is observed that the phase errors at high modulation frequencies are larger than those at low modulation frequencies.

Similarly, one can obtain these estimations for other modulation frequencies as well. Table 10.2 tabulates the corresponding standard deviation of the phase error for all 14 modulation frequencies and two ToF phase shifts by performing 500 independent evaluations of MCS each with a pair of RMS $(\sigma_l, \sigma_s) = (50, 51)$ ps for a single set when $T = 1$ ms.

The standard deviation of the phase error increases when the modulation frequency increases. In addition, for the same modulation frequency with a single set, the phase errors are not reasonably changed for the two ToF phase shifts chosen and there is no pattern among them. Note that the results for each setup are generated with a single set. In each setup, an independent MCS is performed for each phase shift.

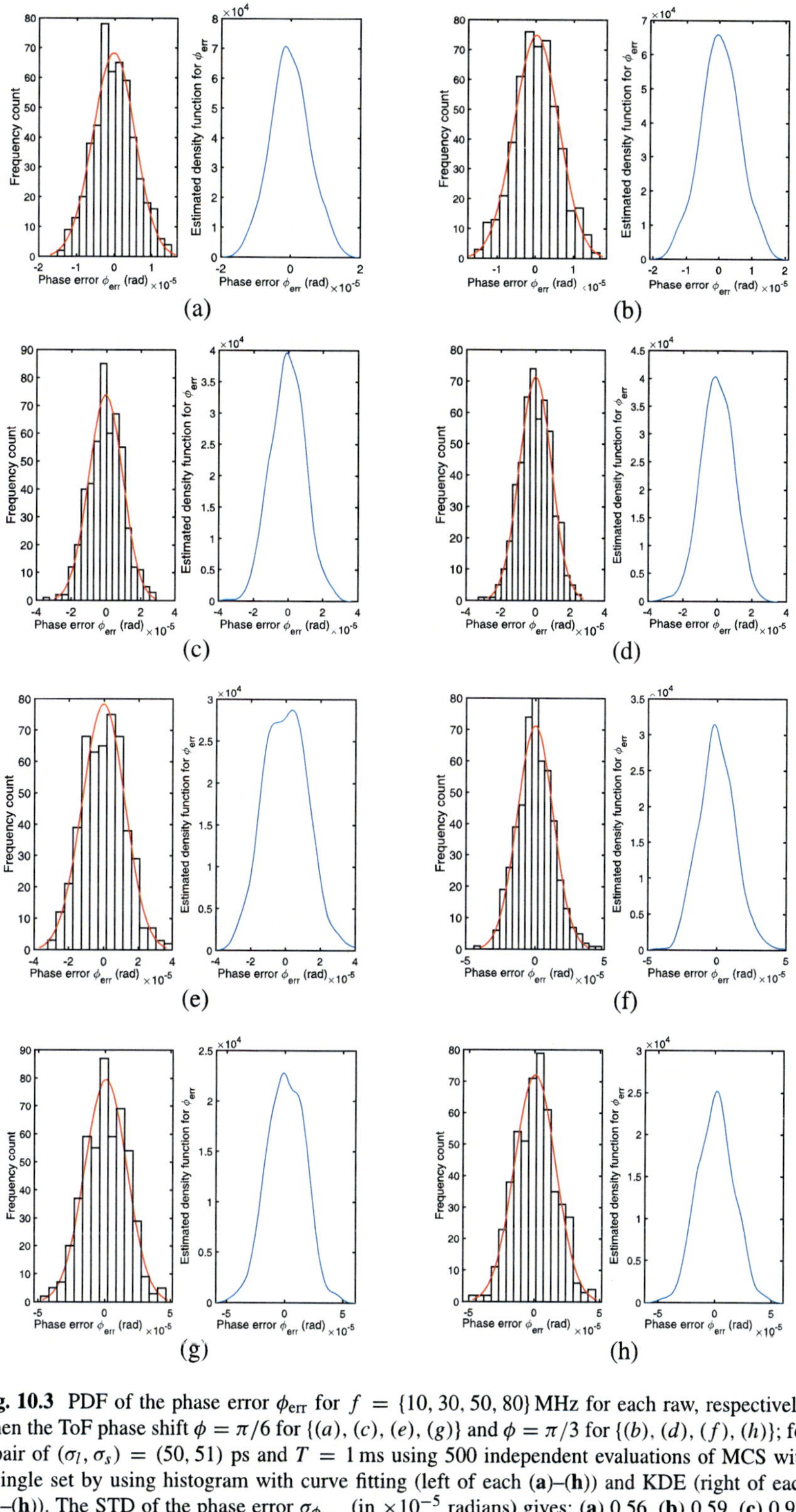


**Fig. 10.3** PDF of the phase error $\phi_{\text{err}}$ for $f = \{10, 30, 50, 80\}$ MHz for each raw, respectively, when the ToF phase shift $\phi = \pi/6$ for $\{(a), (c), (e), (g)\}$ and $\phi = \pi/3$ for $\{(b), (d), (f), (h)\}$; for a pair of $(\sigma_l, \sigma_s) = (50, 51)$ ps and $T = 1$ ms using 500 independent evaluations of MCS with a single set by using histogram with curve fitting (left of each **(a)**–**(h)**) and KDE (right of each **(a)**–**(h)**). The STD of the phase error $\sigma_{\phi_{\text{err}_1}}$ (in $\times 10^{-5}$ radians) gives: **(a)** 0.56, **(b)** 0.59, **(c)** 0.97, **(d)** 0.92, **(e)** 1.22, **(f)** 1.29, **(g)** 1.58, and **(h)** 1.57

**Table 10.2** Standard deviation of phase error for each modulation frequency with an RMS pair of $(\sigma_l, \sigma_s) = (50, 51)$ ps and 500 independent MCS for a single set for the two ToF phase shifts when $T = 1$ ms

| $f$ (MHz) | Parameters used | Phase error $\sigma_{\phi_{\text{err}_1}}$ (in $\times 10^{-5}$ rad) for $\phi = \pi/6$ | $\phi = \pi/3$ |
|---|---|---|---|
| 10 | | 0.56 | 0.59 |
| 30 | | 0.97 | 0.92 |
| 50 | | 1.22 | 1.29 |
| 80 | $T = 1$ ms | 1.58 | 1.57 |
| 100 | $(\sigma_l, \sigma_s) = (50, 51)$ ps | 1.72 | 1.72 |
| 200 | | 2.43 | 2.43 |
| 300 | | 3.18 | 3.09 |
| 400 | Independent MCS | 3.60 | 3.40 |
| 500 | with | 3.97 | 3.73 |
| 600 | nEval = 500, | 4.20 | 4.56 |
| 700 | nSet = 1 | 4.58 | 4.47 |
| 800 | | 5.10 | 5.12 |
| 900 | | 5.39 | 5.34 |
| 1000 | | 5.56 | 5.12 |

Now, fifty repetitions are performed to estimate the uncertainty of the phase error for all 21 RMS pairs $(\sigma_l, \sigma_s)$. First, let us consider the first four modulation frequencies for the ToF phase shift $\pi/6$ rad. Figure 10.4 shows the typical phase error that is computed by the mean of the standard deviation against the RMS of the injected random jitter for $f = 10, 30, 50$, and 80 MHz with 500 independent MCS for each evaluation in 1 ms integration period. According to Table 10.2, within the same frequency, the corresponding results for any ToF phase shifts are approximately similar.

For each of these four frequencies, an approximately linear relationship between the RMS of RJ and phase error has occurred. For $f = 80$ MHz, a slightly curving upward (see Fig. 10.4d) can be seen by the naked eye. For the maximum RMS of RJ (i.e., at $\sigma_{\text{RJ}} = 1000$ ps), the first four modulation frequencies $f = \{10, 30, 50, 80\}$ MHz are given as $\overline{\sigma_{\phi_{\text{err}}}} = \{109.05, 190.18, 249.41, 328.56\}$ μrad of the (typical) standard deviation of phase errors, respectively (refer Fig. 10.4). These are relatively very low values that influence range measurements. Thus, for those four modulation frequencies, the RJ is unable to impact range measurements, even its RMS is a relatively large value of 1000 ps each. Now, it is interesting to see whether this behavior is the same for the rest of the modulation frequencies in Table 10.1. Let us consider the behavior of the RJ when $f$ is increasing.

Figure 10.5 compares (typical phase error) the mean of the standard deviation of the phase error $\overline{\sigma_{\phi_{\text{err}}}}$ against the RMS of random jitter $\sigma_{\text{RJ}}$ for all modulation frequencies chosen (see Table 10.1) for the ToF phase shift $\phi = \pi/6$. Note that in order to compare the results, the subplots shown in Fig. 10.4 are also included in the figure. In practice, we can measure the phase errors to about 1000th or 10,000th of a

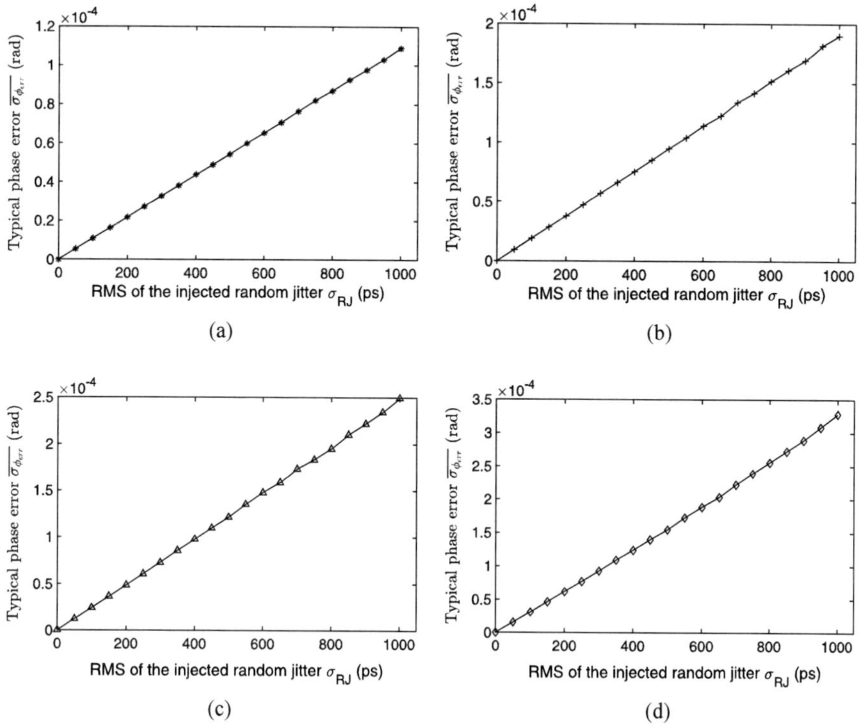


**Fig. 10.4** Typical phase error $\overline{\sigma_{\phi_{\text{err}}}}$ against the RMS of random jitter when the modulation frequency $f$ (MHz): (**a**) 10, (**b**) 30, (**c**) 50, and (**d**) 80, for the ToF phase shift $\phi = \pi/6$ by using 500 independent MCS with 50 repetitions and $T = 1$ ms each

cycle (i.e., $2\pi/10^3 \approx 6.283$ mrad and $2\pi/10^4 \approx= 0.6283$ mrad) in the system. The corresponding levels of the phase errors $\overline{\sigma_{\phi_{\text{err}}}} = 10^{-3}$ and $10^{-4}$ rad criteria for all RMS values are also marked (see lines ··· and - - - -) in the figure. The analysis of the intersection points between each curve with $2\pi/10^3$ rad and $2\pi/10^4$ rad criteria lines is interesting.

It is clear that curves are behaving in non-linear shapes when $f$ increases. Also, for each modulation frequency $f$, the corresponding $\overline{\sigma_{\phi_{\text{err}}}}$ increases when $\sigma_{\text{RJ}}$ increases. Among the two cameras experimented with in Chap. 7, it was found the maximum random jitter with RMS of $\sigma_{\text{RJ}} = (159.6 \pm 0.1)$ ps in MESA SR4000. For the purpose of analysis, let us first consider the RMS of $\sigma_{\text{RJ}} = 200$ ps for each modulation frequency in the figure. Then, the typical phase error of $\overline{\sigma_{\phi_{\text{err}}}} = 0.1$ mrad criterion is starting to occur at least from $f = 200$ MHz. The phase error $\overline{\sigma_{\phi_{\text{err}}}} = 1$ mrad criterion is occurring at $\sigma_{\text{RJ}} \approx 950$ ps for the signal with frequency $f = 200$ MHz. In addition, only for the modulation frequency of 10 MHz, the phase error $\overline{\sigma_{\phi_{\text{err}}}} = 0.1$ mrad criterion occurred at $\sigma_{\text{RJ}} = 280$ ps, while for all other frequencies, the same criterion occurred at $\sigma_{\text{RJ}} < 200$ ps when $T = 1$ ms.

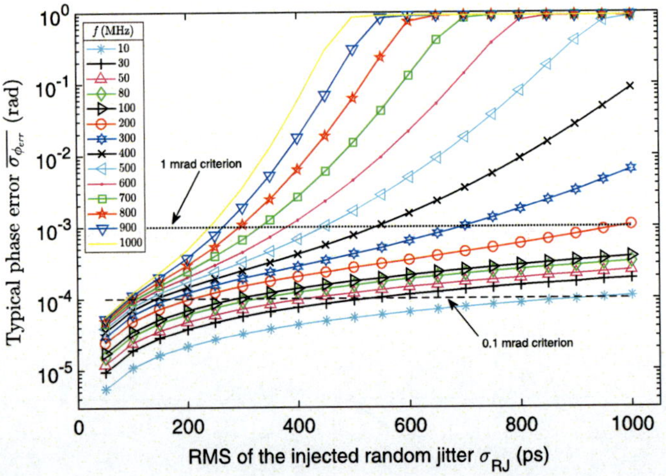


**Fig. 10.5** Typical phase error $\overline{\sigma_{\phi_{\text{err}}}}$ against the RMS of random jitter for modulation frequencies in Table 10.1 for a ToF phase shift $\phi = \pi/6$ by using 500 independent MC simulation with $T = 1$ ms and 50 repetitions. The lines $\cdots$ and - - - indicate the phase error criterion $\overline{\sigma_{\phi_{\text{err}}}} = 10^{-3}$ and $10^{-4}$ rad, respectively. The uncertainties of the plotted points are smaller than the width of a line

In contrast, when the modulation frequency is greater than 100 MHz, there is a rising trend of the typical phase error at the RMS $\sigma_{\text{RJ}}$ = 300 ps. An example, for $f = 400$ MHz, the sharp rising of 0.1 mrad phase error occurred at the RMS $\sigma_{\text{RJ}}$ = 300 ps. For the last four modulation frequencies, the phase error $\overline{\sigma_{\phi_{\text{err}}}} = 1$ mrad criterion is starting to occur from the RMS $\sigma_{\text{RJ}} \approx \{550, 375, 300, 240\}$ ps, respectively. Furthermore, when $f = 600$ MHz, the phase error 0.1 mrad criterion is dramatically increased up to 1 rad at the RMS $\sigma_{\text{RJ}}$ = 850 ps and then shows stable behavior for the rest of the RMS values. This similar behavior occurred at $\sigma_{\text{RJ}}$ = 650 and 550 ps for the modulation frequencies $f = 800$ and 1000 MHz, respectively.

On the other hand, for 1 mrad criterion of the phase error, the RMS amounts of the random jitter $\sigma_{\text{RJ}} \approx \{690, 540, 450, 375, 325, 290, 270, 240\}$ ps occurred when $f = \{300, 400, \ldots, 1000\}$ MHz, respectively. So far, we have analyzed the results generated with the higher integration period chosen (i.e., when $T = 1$ ms). Similarly, we can generate the typical phase errors for other three integration periods, $T = 0.1, 0.01$, and 0.001 ms as shown in Figs. 10.6, 10.7, and 10.8, respectively.

By comparing the results with four integration periods for each corresponding frequency, the phase errors for the same modulation frequency are increased when $T$ is decreased, i.e., the phase error is large for the short integration period. This can be clearly seen when those results are reformed in modulation frequency-wise and integration period-wise as shown in Fig. 10.9. This figure is drawn by using the results in Figs. 10.5, 10.6, 10.7, and 10.8 for each corresponding frequency. When the modulation frequency is increased, the gaps between the phase error for the chosen integration periods are dramatically decreased. Those phase error

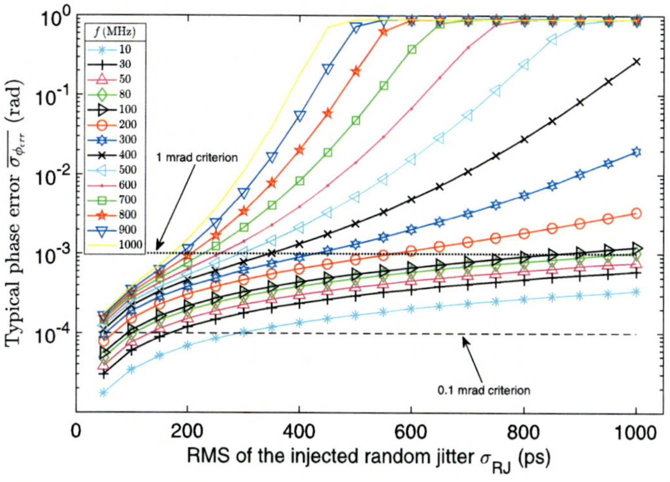


**Fig. 10.6** Typical phase error $\overline{\sigma_{\phi_{\mathrm{err}}}}$ against the RMS of random jitter for modulation frequencies in Table 10.1 for a ToF phase shift $\phi = \pi/6$ by using 500 independent MC simulation with $T =$ 0.1 ms and 50 repetitions. The lines $\cdots$ and - - - indicate the phase error criterion $\overline{\sigma_{\phi_{\mathrm{err}}}} = 10^{-3}$ and $10^{-4}$ rad, respectively. The uncertainties of the plotted points are smaller than the width of a line

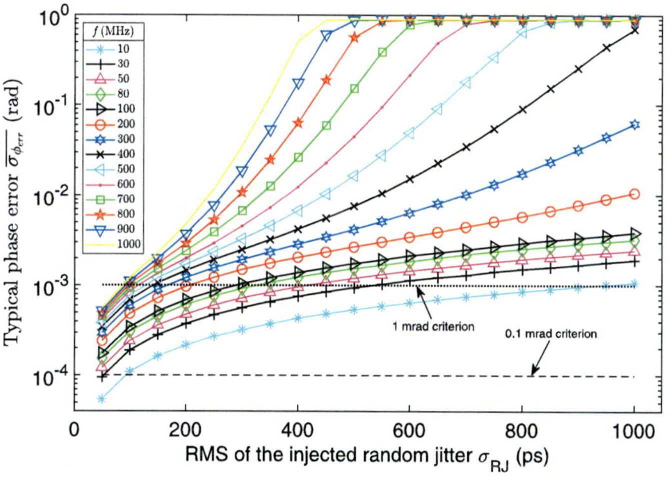


**Fig. 10.7** Typical phase error $\overline{\sigma_{\phi_{\mathrm{err}}}}$ against the RMS of random jitter for modulation frequencies in Table 10.1 for a ToF phase shift $\phi = \pi/6$ by using 500 independent MC simulation with $T =$ 0.01 ms and 50 repetitions. The lines $\cdots$ and - - - indicate the phase error criterion $\overline{\sigma_{\phi_{\mathrm{err}}}} = 10^{-3}$ and $10^{-4}$ rad, respectively. The uncertainties of the plotted points are smaller than the width of a line

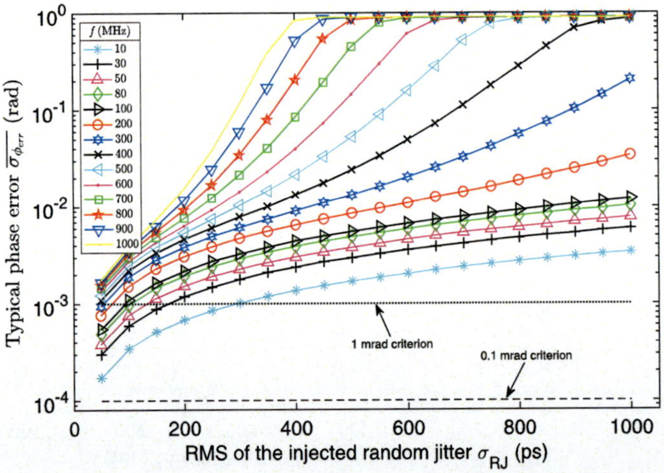


**Fig. 10.8** Typical phase error $\overline{\sigma_{\phi_{\text{err}}}}$ against the RMS of random jitter for modulation frequencies in Table 10.1 for a ToF phase shift $\phi = \pi/6$ by using 500 independent MC simulation with $T = 0.001$ ms and 50 repetitions. The lines $\cdots$ and - - - indicate the phase error criterion $\overline{\sigma_{\phi_{\text{err}}}} = 10^{-3}$ and $10^{-4}$ rad, respectively. The uncertainties of the plotted points are smaller than the width of a line

gaps are approximately similar up to the modulation frequency 200 MHz while they are starting to decrease from the frequency 300 MHz onward. However, for the frequency greater than 600 MHz, the phase error gaps between the integration periods are relatively low; that is, approximately similar behavior is presented for high modulation frequencies.

Using the corresponding data in Figs. 10.5, 10.6, 10.7, and 10.8, the behavior of the random jitter on the measurement of phase against the modulation frequency for chosen integration periods is shown in Fig. 10.10. The upper side of the each fitted curve is the corresponding affecting area on range measurements for the integration period chosen due to the presence of random jitter. By comparison, it is noted that the curves for $T = 1$ ms and 0.1 ms on the 0.1 mrad criterion (see Fig. 10.10a) are approximately similar for $T = 0.01$ ms and 0.001 ms on the 1 mrad criterion (see Fig. 10.10b), respectively.

Now, consider the results on the 0.1 mrad criterion with longer integration period $T = 1$ ms, it is clear that at $f = \{30, 50, 80\}$ MHz, the error can tolerate the RJ with RMS $\sigma_{\text{RJ}} = \{530, 410, 325\}$ ps, while at $f = \{800, 1000\}$ MHz only $\sigma_{\text{RJ}} = \{97, 87\}$ ps can be tolerated, respectively. On the other hand, the phase errors for low modulation frequencies (less than 100 MHz) when $T = 1$ ms are not presented on the 1 mrad criterion; see Fig. 10.10b. Thus, the random jitter in two ToF cameras experimented in Chap. 7 is unable to influence range measurements (since $\sigma_{\text{RJ}} = (159.5 \pm 20.5)$ ps and $(40.37 \pm 0.04)$ ps for $f = 30$ MHz and 50 MHz, respectively, when $T = 1$ ms). For currently available AMCW ToF flight range imaging cameras (for $f < 100$ MHz), the random jitter has not a measurable effect

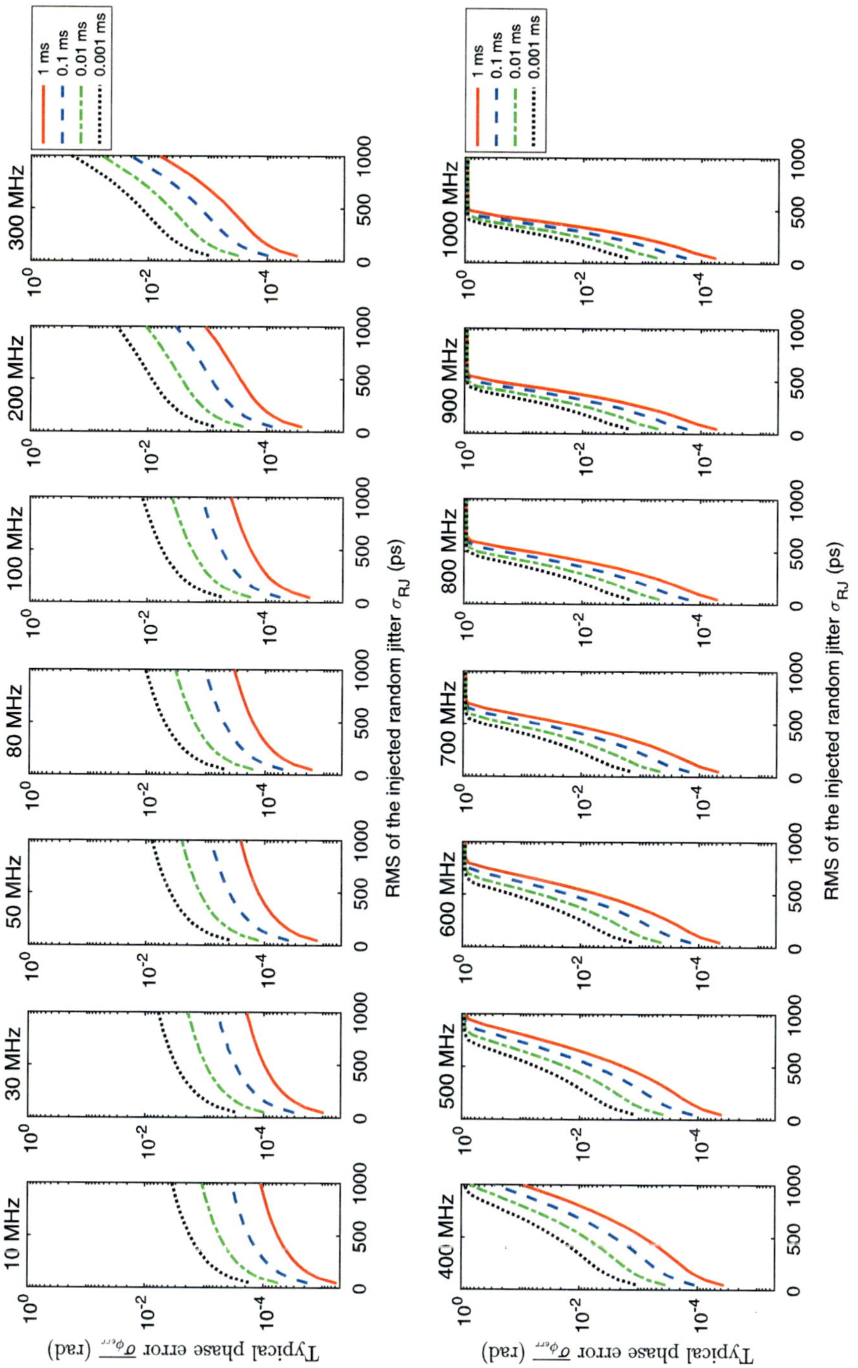


**Fig. 10.9** Typical phase error $\overline{\sigma_{\phi_{\rm err}}}$ against the RMS of random jitter for a ToF phase shift $\phi = \pi/6$ by using 500 independent MC simulation with 50 repetitions for 14 modulation frequencies (10 MHz to 1 GHz) and four integration periods as shown in Table 10.1

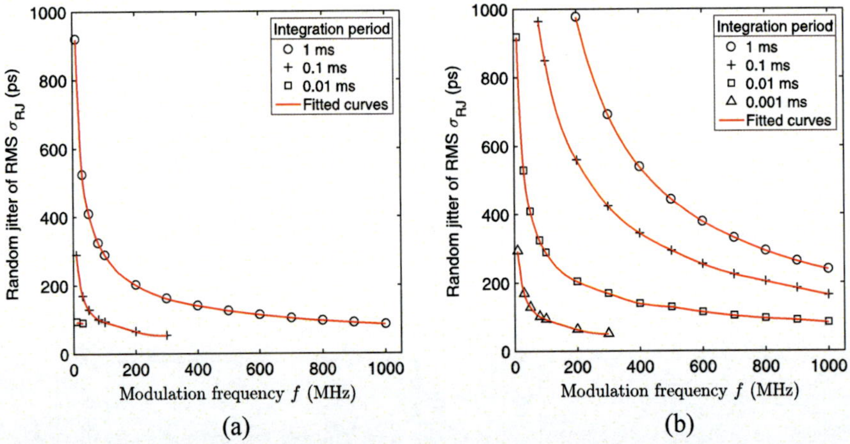


**Fig. 10.10** The domain of the random jitter that affects the measurement of phase obtained by using Figs. 10.5, 10.6, 10.7, and 10.8 with the phase error criterion $\overline{\sigma_{\phi_{\text{err}}}}$ : (**a**) 0.1 mrad and (**b**) 1 mrad, for the chosen four integration periods $T$. Above each curve, the random jitter has a potentially measurable effect on ranging

on range measurement. But increasing the modulation frequency further, the random jitter has a measurable effect on phase, thus in range. As examples, 400 MHz with random jitter of RMS $\sigma_{\text{RJ}} = \{140$ and $540\}$ ps while 600 MHz with $\sigma_{\text{RJ}} = \{115$ and $380\}$ ps have measurable effect on ranging when $T = 1$ ms for the criteria 0.1 mrad and 1 mrad, respectively (see Fig. 10.10). In addition, these RMS amounts are further dropped when $T$ is shorter, which have more effect on range measurements. As examples, the random jitter with RMS $\sigma_{\text{RJ}} = 140$ and 115 ps have effect on ranging when $T = 0.01$ ms for 400 and 600 MHz, respectively, on the 0.1 mrad criterion (see Fig. 10.10b).

### 10.3.2 *Range Error Due to the Random Jitter*

The results obtained in Sect. 10.3.1 are used for the computation here. Now the standard deviation of the range error for each corresponding typical phase error can be computed by using the fundamental relation of the ToF ranging (i.e., (2.12)). An example, from Fig. 10.5, for $T = 1$ ms and $f = 30$ MHz, the standard deviations of phase errors when the random jitter pairs of RMS $(\sigma_l, \sigma_s) = \{(200, 201), (400, 401), (600, 601), (800, 801), (1000, 1001)\}$ ps are given as $\sigma_{\phi_{\text{err}}} \approx \{37.77, 75.55, 113.98, 151.72, 191.03\}$ μrad, respectively, for the ToF phase shift $\phi = \pi/6$. Then, the standard deviations of corresponding range errors (typical) can be computed as $\sigma_{d_{\text{err}}} \approx \{0.03, 0.06, 0.09, 0.12, 0.15\}$ mm, respectively, which are an approximately linear relationship as shown in Fig. 10.4b. These values are extremely insignificant in their influence on range measurements. That is, even in the presence of maximum

RMS $(\sigma_l, \sigma_s) = (1000, 1001)$ ps of random jitter is unable to sufficiently influence range measurements when $f = 30$ MHz (since $\sigma_{d_{\text{err}}} \approx 0.15$ mm). Similarly, with (2.12) and Figs. 10.5, 10.6, 10.7, and 10.8, the calculated typical range errors for each modulation frequency for the selected five RMS pairs of random jitter for the chosen $T = 1, 0.1, 0.01$, and 0.001 ms can be computed as shown in Tables 10.3, 10.4, 10.5, and 10.6, respectively. Note that these values are rounded to two decimal places and the cells inside the thick border are the potentially measurable range error.

Now, let us analyze the results in those tables. First consider the results for the integration period $T = 1$ ms, which is tabulated in Table 10.3. It is straightforward to see, for each modulation frequency, the standard deviation of range errors $\sigma_{d_{\text{err}}}$ is increased when the injected random jitter amounts $\sigma_{\text{RJ}}$ are increased. However, at $f = 400$ MHz, the potential measurable amount $\sigma_{d_{\text{err}}} = 5.24$ mm is starting to occur with the presence of the RJ with $\sigma_{\text{RJ}} = 1$ ns. It is assumed that the measurable range error is at least 2 mm. In addition, it is well clear that the measurable range errors $\sigma_{d_{\text{err}}}$ started to occur for low RMS values when the modulation frequency is increased (see the cells inside the thick border of the table). As examples, when the RMS of the RJ $\sigma_{\text{RJ}} = (600, 601)$ ps, there are measurable amounts of range errors $\sigma_{d_{\text{err}}} = \{4.40, 22.24, 23.93, 21.62\}$ mm for the last four modulation frequencies, respectively.

Also, it is noted that for the last three RMS pairs chosen in the table, the maximum range errors $\sigma_{d_{\text{err}}} = \{23.92, 31.12, 42.11\}$ are appeared in different

**Table 10.3** The typical range error (standard deviation $\sigma_{d_{\text{err}}}$ in mm) when $T = 1$ ms, due to the selected RMS of random jitter $\sigma_{\text{RJ}}$ in the correlation model for all modulation frequencies. The cells inside the thick border are the only potentially measurable $\sigma_{d_{\text{err}}}$ amounts. Assume that more than 2 mm is measurable

| $f$ (MHz) | Range error $\sigma_{d_{\text{err}}}$ (in mm), by using Fig. 10.5 and (2.12) | Injected random jitter of RMS pair $(\sigma_l, \sigma_s)$ in ps | | | | |
|---|---|---|---|---|---|---|
| | | (200, 201) | (400, 401) | (600, 601) | (800, 801) | (1000, 1001) |
| 10 | | 0.05 | 0.10 | 0.16 | 0.21 | 0.26 |
| 30 | | 0.03 | 0.06 | 0.09 | 0.12 | 0.15 |
| 50 | | 0.02 | 0.05 | 0.07 | 0.09 | 0.12 |
| 80 | | 0.02 | 0.04 | 0.06 | 0.08 | 0.10 |
| 100 | | 0.02 | 0.03 | 0.05 | 0.07 | 0.09 |
| 200 | | 0.01 | 0.02 | 0.04 | 0.07 | 0.13 |
| 300 | | 0.01 | 0.02 | 0.05 | 0.14 | 0.51 |
| 400 | | 0.01 | 0.02 | 0.09 | 0.54 | 5.24 |
| 500 | | 0.01 | 0.03 | 0.24 | 3.78 | 42.11 |
| 600 | | 0.01 | 0.05 | 0.86 | 31.12 | 36.00 |
| 700 | | 0.01 | 0.09 | 4.40 | 30.86 | 30.87 |
| 800 | | 0.01 | 0.19 | 22.24 | 26.95 | 26.98 |
| 900 | | 0.01 | 0.47 | 23.92 | 24.08 | 24.10 |
| 1000 | | 0.01 | 1.34 | 21.62 | 21.59 | 21.61 |

**Table 10.4** The typical range error (standard deviation $\sigma_{d_{err}}$ in mm) when $T = 0.1$ ms, due to the selected RMS of random jitter $\sigma_{RJ}$ in the correlation model for all modulation frequencies. The cells inside the thick border are the only potentially measurable $\sigma_{d_{err}}$ amounts. Assume that more than 2 mm is measurable

| $f$ (MHz) | Range error $\sigma_{d_{err}}$ (in mm), by using Fig. 10.6 and (2.12) | Injected random jitter of RMS pair $(\sigma_l, \sigma_s)$ in ps | | | | |
|---|---|---|---|---|---|---|
| | | (200, 201) | (400, 401) | (600, 601) | (800, 801) | (1000, 1001) |
| 10 | | 0.17 | 0.33 | 0.50 | 0.66 | 0.83 |
| 30 | | 0.10 | 0.19 | 0.28 | 0.38 | 0.48 |
| 50 | | 0.07 | 0.15 | 0.22 | 0.30 | 0.37 |
| 80 | | 0.06 | 0.12 | 0.18 | 0.24 | 0.31 |
| 100 | | 0.05 | 0.11 | 0.16 | 0.22 | 0.29 |
| 200 | | 0.04 | 0.08 | 0.13 | 0.22 | 0.40 |
| 300 | | 0.03 | 0.07 | 0.16 | 0.45 | 1.61 |
| 400 | | 0.03 | 0.08 | 0.29 | 1.70 | 16.34 |
| 500 | | 0.03 | 0.10 | 0.75 | 11.90 | 43.33 |
| 600 | | 0.03 | 0.16 | 2.73 | 35.57 | 35.92 |
| 700 | | 0.03 | 0.29 | 13.16 | 31.04 | 30.77 |
| 800 | | 0.03 | 0.61 | 26.52 | 27.14 | 26.91 |
| 900 | | 0.03 | 1.49 | 24.07 | 24.03 | 24.09 |
| 1000 | | 0.04 | 4.28 | 21.62 | 21.73 | 21.65 |

**Table 10.5** The typical range error (standard deviation $\sigma_{d_{err}}$ in mm) when $T = 0.01$ ms, due to the selected RMS of random jitter $\sigma_{RJ}$ in the correlation model for all modulation frequencies. The cells inside the thick border are the only potentially measurable $\sigma_{d_{err}}$ amounts. Assume that more than 2 mm is measurable

| $f$ (MHz) | Range error $\sigma_{d_{err}}$ (in mm), by using Fig. 10.7 and (2.12) | Injected random jitter of RMS pair $(\sigma_l, \sigma_s)$ in ps | | | | |
|---|---|---|---|---|---|---|
| | | (200, 201) | (400, 401) | (600, 601) | (800, 801) | (1000, 1001) |
| 10 | | 0.52 | 1.05 | 1.56 | 2.07 | 2.60 |
| 30 | | 0.30 | 0.60 | 0.91 | 1.21 | 1.52 |
| 50 | | 0.23 | 0.47 | 0.70 | 0.94 | 1.18 |
| 80 | | 0.19 | 0.37 | 0.56 | 0.76 | 0.97 |
| 100 | | 0.17 | 0.33 | 0.51 | 0.70 | 0.92 |
| 200 | | 0.12 | 0.25 | 0.42 | 0.70 | 1.29 |
| 300 | | 0.10 | 0.23 | 0.51 | 1.41 | 5.07 |
| 400 | | 0.09 | 0.25 | 0.91 | 5.45 | 41.92 |
| 500 | | 0.08 | 0.32 | 2.37 | 31.87 | 43.42 |
| 600 | | 0.08 | 0.50 | 8.67 | 35.85 | 35.95 |
| 700 | | 0.08 | 0.91 | 27.14 | 30.92 | 31.10 |
| 800 | | 0.09 | 1.91 | 27.07 | 27.11 | 26.98 |
| 900 | | 0.10 | 4.75 | 24.15 | 24.14 | 23.95 |
| 1000 | | 0.11 | 12.39 | 21.62 | 21.64 | 21.66 |

**Table 10.6** The typical range error (standard deviation $\sigma_{d_{err}}$ in mm) when $T = 1\mu s$, due to the selected RMS of random jitter $\sigma_{RJ}$ in the correlation model for all modulation frequencies. The cells inside the thick border are the only potentially measurable $\sigma_{d_{err}}$ amounts. Assume that more than 2 mm is measurable

| $f$ (MHz) | Range error $\sigma_{d_{err}}$ (in mm), by using Fig. 10.8 and (2.12) | Injected random jitter of RMS pair $(\sigma_l, \sigma_s)$ in ps | | | | |
|---|---|---|---|---|---|---|
| | | (200, 201) | (400, 401) | (600, 601) | (800, 801) | (1000, 1001) |
| 10 | | 1.64 | 3.29 | 4.90 | 6.56 | 8.23 |
| 30 | | 0.95 | 1.89 | 2.87 | 3.80 | 4.77 |
| 50 | | 0.74 | 1.46 | 2.22 | 2.96 | 3.75 |
| 80 | | 0.58 | 1.17 | 1.78 | 2.41 | 3.08 |
| 100 | | 0.52 | 1.06 | 1.60 | 2.19 | 2.90 |
| 200 | | 0.37 | 0.78 | 1.32 | 2.24 | 4.07 |
| 300 | | 0.31 | 0.73 | 1.63 | 4.48 | 16.35 |
| 400 | | 0.28 | 0.79 | 2.92 | 16.72 | 52.69 |
| 500 | | 0.26 | 1.03 | 7.54 | 41.80 | 43.24 |
| 600 | | 0.25 | 1.59 | 24.52 | 36.06 | 36.01 |
| 700 | | 0.26 | 2.88 | 30.49 | 30.97 | 30.93 |
| 800 | | 0.28 | 6.11 | 27.17 | 27.04 | 27.00 |
| 900 | | 0.31 | 13.87 | 24.01 | 24.06 | 24.06 |
| 1000 | | 0.36 | 20.25 | 21.71 | 21.67 | 21.66 |

modulation frequencies $f = \{900, 600, 500\}$, respectively. Then, the range errors are systematically decreasing when $f$ increases. However, the rest of the range errors are keeping approximately the same amounts within the frequency. As examples, the range errors $\sigma_{d_{err}} = \{5.24, 3.78, 4.40\}$ mm are increased to the amounts $\sigma_{d_{err}} = \{42.11, 31.12, 22.24\}$ mm, for the frequencies from $f = \{400, 500, 700\}$ MHz to $f = \{500, 600, 800\}$ MHz, respectively. Also, for the last four frequencies, the corresponding range errors can be rounded to integer values $\sigma_{d_{err}} \approx \{31, 27, 24, 22\}$ mm, respectively. Note that these results are corresponding to the integration period $T = 1$ ms and it is clear that the potentially measurable amounts of the range error are within the region of $\sigma_{d_{err}} \approx [4–42]$ mm with the spanning of the RMS $\sigma_{RJ} = [600 - 1000]$ ps and the frequencies $f = [400 - 1000]$ MHz (consider the results inside the thick border of Table 10.3).

Now, let us move to the results with a short integration period $T = 0.1$ ms. As before, at $f = 400$ MHz, the potential measurable amount $\sigma_{d_{err}} = 16.34$ mm is starting to occur with the presence of the RJ with $\sigma_{RJ} = 1$ ns. Here this is a relatively larger amount than 5.24 mm when $T = 1$ ms; see Table 10.3. However, the potential measurable amount $\sigma_{d_{err}} = 4.28$ mm is starting to occur when $\sigma_{RJ} = 400$ ps for the signal with $f = 1000$ MHz. In addition, all the starting amounts of the measurable range errors are relatively larger than the previous case. But, for the last four frequencies, the corresponding range errors can be rounded to the same integer values that occurred when $T = 1$ ms, which are given as $\sigma_{d_{err}} \approx \{31, 27, 24, 22\}$ mm, respectively. The potential measurable range errors are given $\sigma_{d_{err}} \approx [3–$

43] mm with the spanning of the RMS $\sigma_{RJ} = [400 - 1000]$ ps and the frequencies $f = [400 - 1000]$ MHz.

A similar kind of behavior is present when $T = 0.01$ ms with medium to high frequencies. In addition, at $f = 10$ MHz, the $\sigma_{d_{err}} > 2$ mm range error is available from $\sigma_{RJ} = 800$ ps onward. At $f = 300$ MHz, the potential measurable amount $\sigma_{d_{err}} = 5.07$ mm is starting to occur with the presence of the RJ with $\sigma_{RJ} = 1$ ns. However, at $f = 900$ MHz, the 4.75 mm of range error has occurred when $\sigma_{RJ} = 400$ ps. Here also, the potential measurable range errors are spanned $\sigma_{d_{err}} \approx [2$–$43]$ mm within the range of the RMS $\sigma_{RJ} = [400 - 1000]$ ps and the frequencies $f = [300 - 1000]$ MHz.

Let us consider the results with the shortest integration period $T = 1$ µs, which is given in Table 10.6. It can be seemed a large region of potentially measurable range error amounts that span to all modulation frequencies. Thus, when the integration period becomes shorter (i.e., high frame rates), there are potentially measurable range errors even for the low modulation frequencies with the presence of reasonable amounts of random jitter. In other words, for the signals with 10 MHz (and the last four frequencies in the table), the measurable range error has occurred when the RMS of the random jitter from $\sigma_{RJ} > 200$ ps. The maximum range error is presented for the modulation frequency 40 MHz when $\sigma_{RJ} = 1$ ns. However, the potential measurable range errors are spanned $\sigma_{d_{err}} \approx [2 - 53]$ mm within the range of the RMS $\sigma_{RJ} = [400 - 1000]$ ps among all frequencies considered, i.e., $f = [10 - 1000]$ MHz.

### 10.3.3 Uncertainty Analysis of the Phase Errors

Figures 10.11, 10.12, 10.13, and 10.14 show the uncertainty (standard error) as error bars for the typical phase error of the eight modulation frequencies chosen separately from Figs. 10.5, 10.6, 10.7, and 10.8 for the four integration periods, respectively. The error bars are computed by using (10.25) with the corresponding phase errors obtained. However, with the presence maximum amounts of RJ ($\sigma_{RJ} = 1$ ns), for the frequencies up to 300 MHz, the uncertainty is in the range [0.5–30] µrad, while for the frequencies from 400 MHz to 1000 MHz they are within the range [0.4–3] mrad when $T = 1$ ms (see Fig. 10.11). However, for the signals with high frequencies, the uncertainty amounts are starting to increase from 500 ps of random jitter. Those starting amounts are decreased when the integration period is decreased (compare Figs. 10.11, 10.12, 10.13, and 10.14). The uncertainty values (length of the error bars) are increasing when the modulation frequency increases and the RJ amounts increase. This is verified in Figs. 10.15 and 10.16, which compare the corresponding uncertainty (standard error) for each modulation frequency and each RMS of random jitter for the integration periods, $T = \{1, 0.1, 0.01, 0.001\}$ ms, respectively. In addition, for the purpose of comparison, an indicator for the uncertainty level of 1 $\mu$rad is marked in each figure.

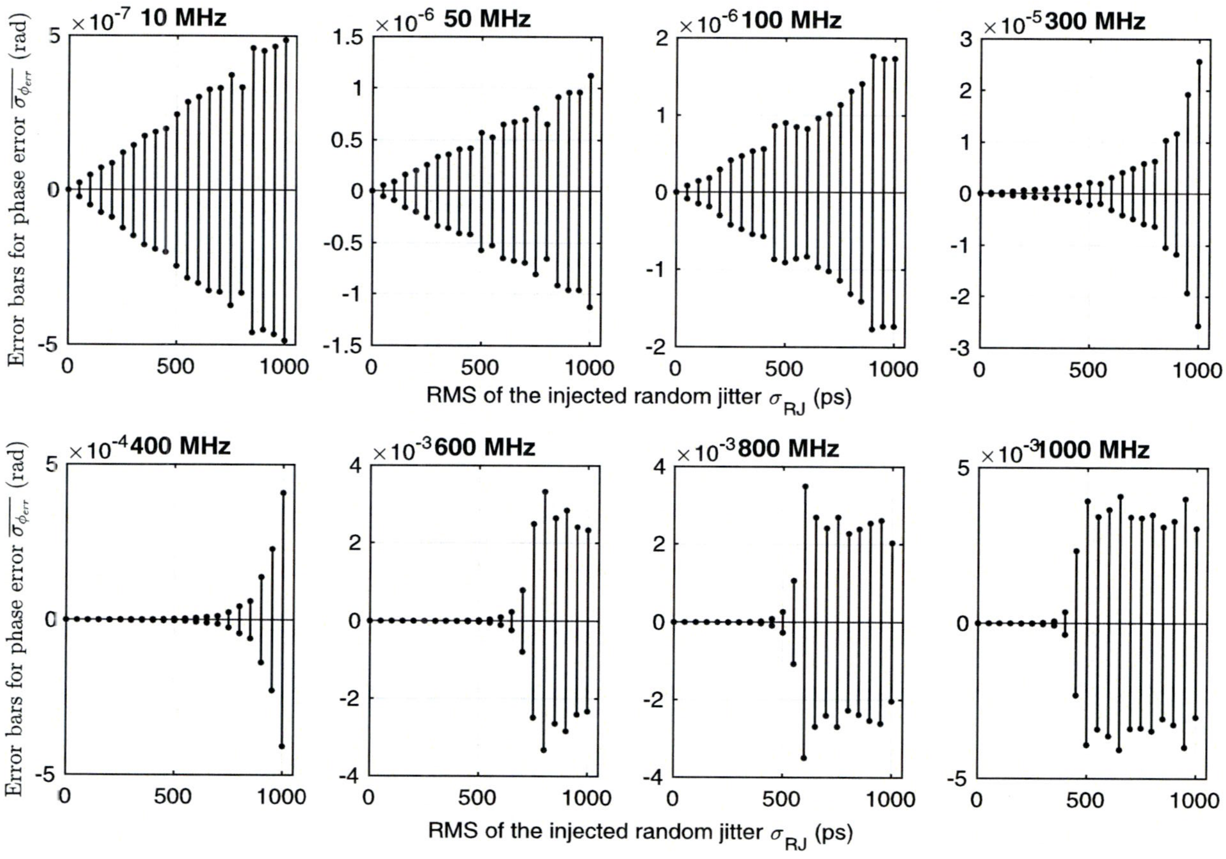


**Fig. 10.11** Uncertainty SE($\sigma_{\phi_{\text{err}}}$) of the typical phase error $\overline{\sigma_{\phi_{\text{err}}}}$ against the RMS of the random jitter by using (10.25) with fifty repetitions for 8 modulation frequencies chosen when $\phi = \pi/6$ and $T = 1$ ms

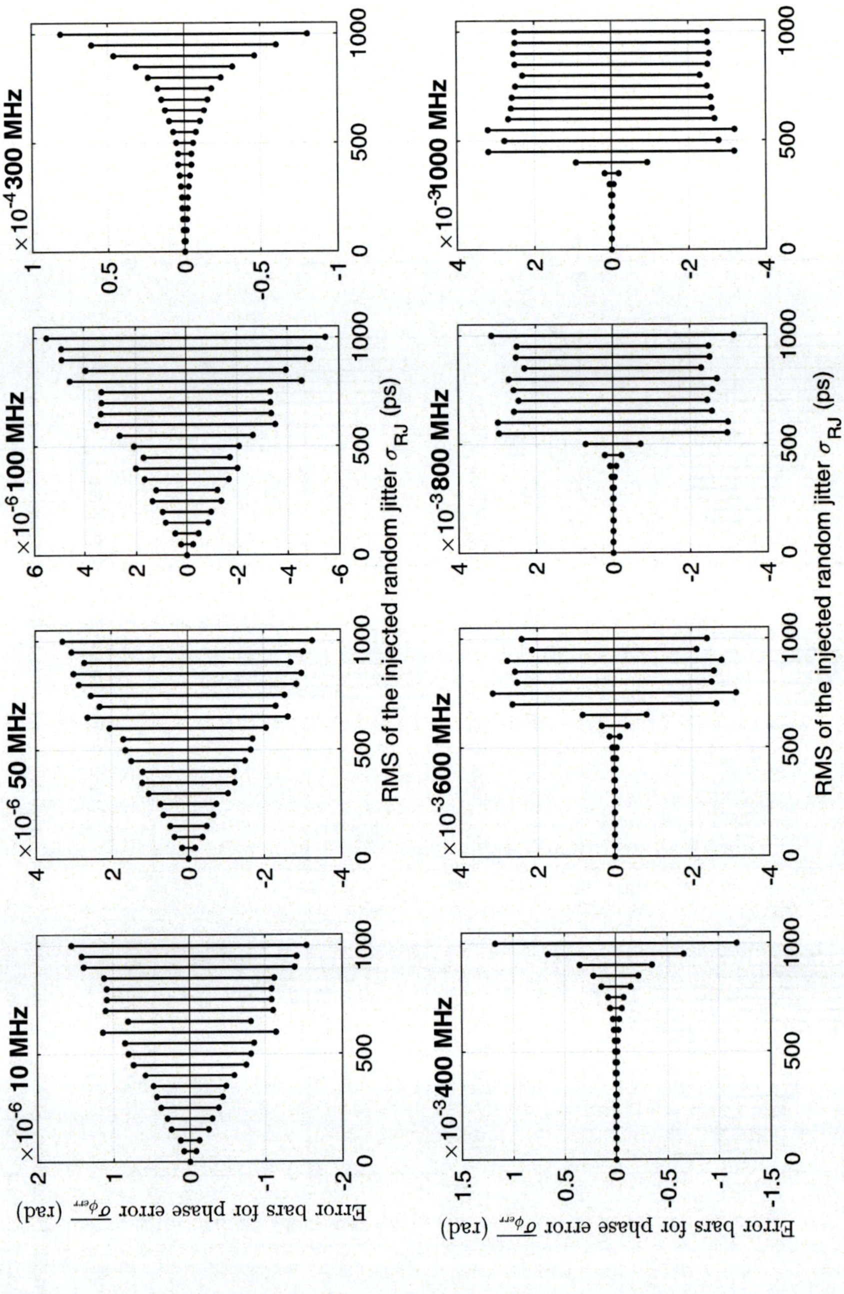


**Fig. 10.12** Uncertainty SE($\sigma_{\phi_{err}}$) of the typical phase error $\overline{\sigma_{\phi_{err}}}$ against the RMS of the random jitter by using (10.25) with fifty repetitions for 8 modulation frequencies chosen when $\phi = \pi/6$ and $T = 0.1$ ms

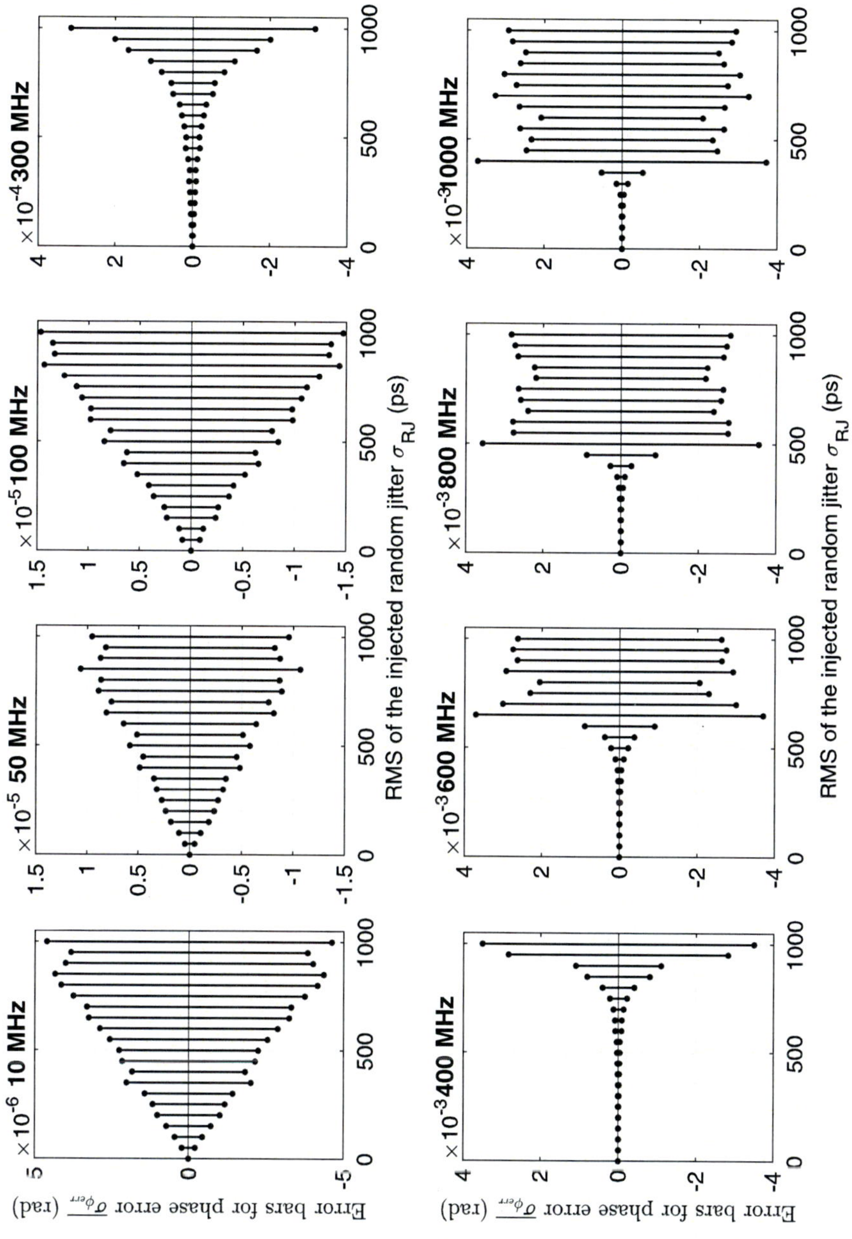


**Fig. 10.13** Uncertainty SE($\sigma_{\phi_{\mathrm{err}}}$) of the typical phase error $\overline{\sigma_{\phi_{\mathrm{err}}}}$ against the RMS of the random jitter by using (10.25) with fifty repetitions for 8 modulation frequencies chosen when $\phi = \pi/6$ and $T = 0.01$ ms

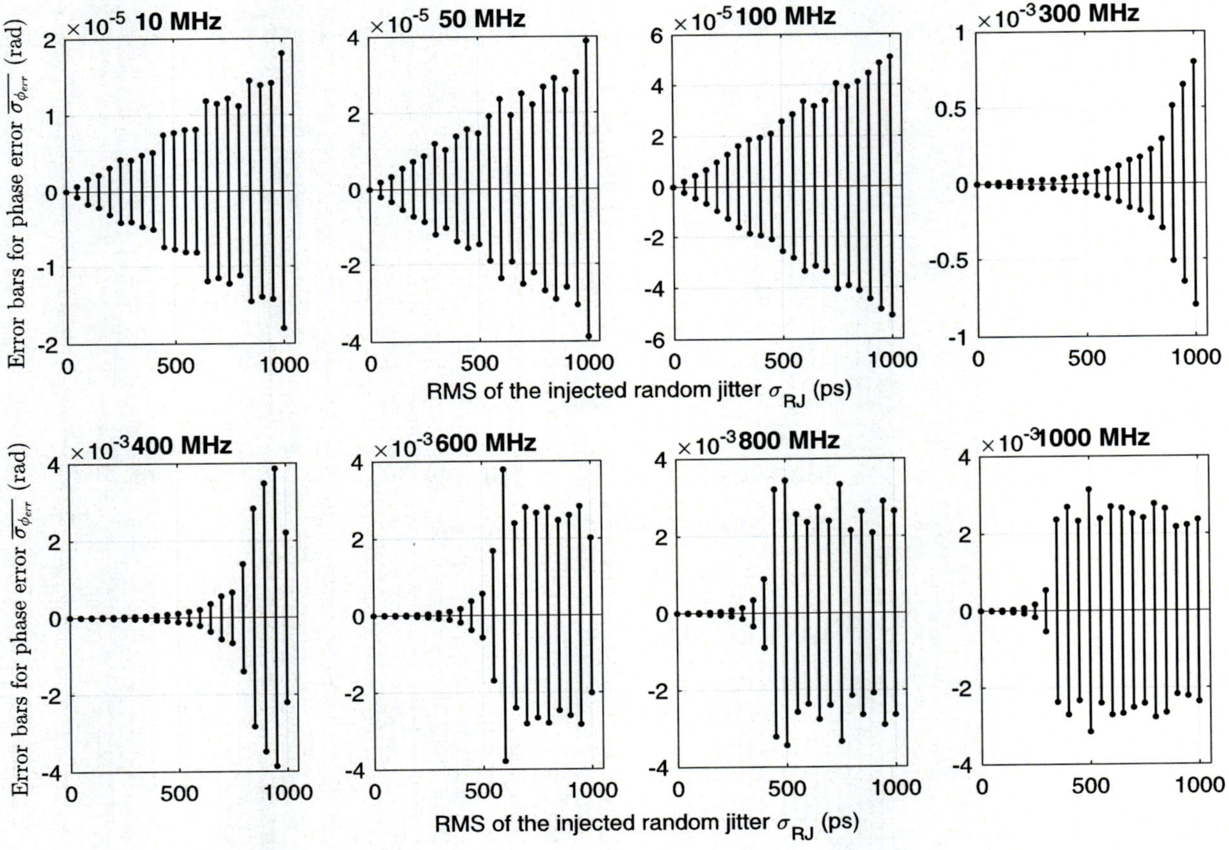


**Fig. 10.14** Uncertainty SE($\sigma_{\phi_{\text{err}}}$) of the typical phase error $\overline{\sigma_{\phi_{\text{err}}}}$ against the RMS of the random jitter by using (10.25) with fifty repetitions for 8 modulation frequencies chosen when $\phi = \pi/6$ and $T = 0.001$ ms

By comparison, when the random jitter is greater than 500 ps in Figs. 10.15 and 10.16, the standard errors are given within the range of [0.2–50] µrad for the first five modulation frequencies, while they are in the range of [0.4–4] mrad for the last five modulation frequencies (i.e., 600–1000 MHz). This milliradian range for the standard error has relatively large deviations with respect to the typical phase errors obtained (compare Figs. 10.5, 10.6, 10.7, 10.8, 10.15, and 10.16, respectively), especially for the high modulation frequencies. That is, fifty repetitions are not significant enough for the uncertainty analysis of the results when $f$ increases. Thus, in order to increase the accuracy of the results, it needs more repetitions than the current value ($\gg 50$). Then, as usual, the execution time for each simulation setup will be increased accordingly. Execution time of the simulations is described in the next section (Sect. 10.4).

It is clear when $T$ decreases, the standard error levels for the phase error are climbing (compare the curves with respect to the 1 µrad level indicator in Figs. 10.15 and 10.16). As example, for the signal with 10 MHz and 50 ps random jitter, the uncertainty level is nearly at $10^{-8}$ rad when $T = 1$ ms, while it is nearly at $10^{-6}$ rad when $T = 0.001$ ms. As before, this behavior can be clearly seen when the corresponding results are redrawn in modulation frequency-wise and integration period-wise as shown in Fig. 10.17. When the modulation frequency has increased, the gaps between the uncertainties of the phase error are also dramatically decreased for the chosen integration periods. As usual, the phase error and its uncertainty (standard error) are given similar behavior. Thus, the standard errors of the phase error gaps are approximately similar up to the modulation frequency 200 MHz, while they are starting to decrease from the frequency 300 MHz onward. However, for the frequency greater than 600 MHz, the standard error of the phase error gaps between the integration periods is relatively less.

## 10.4 Execution Time for the Simulation

Table 10.7 shows the approximate execution time in hours for the main results obtained in this chapter that correspond to Figs. 10.5, 10.6, 10.7, and 10.8. Note that these amounts are rounded up to the nearest two decimal places. In addition, the two forms of graphical representation of this execution time are shown in Fig. 10.18. The decimal scale representation is shown in Fig. 10.18a, while the logarithm scale with base 10 is shown in Fig. 10.18b.

As in the previous chapter, all simulations are implemented with MATLAB R2018b (MathWorks Inc.) and are executed with high-performance computing (HPC) cluster supercomputers [12] (refer Sect. 9.7). For each simulation, all 36 cores are performed. In addition, the allocated memory space for each simulation is varied between 30 GB and 90 GB. When the modulation frequency is increased, the memory space allocation is also increased. For example, for the modulation frequencies less than 100 MHz, 30 GB of memory space is allocated while for the

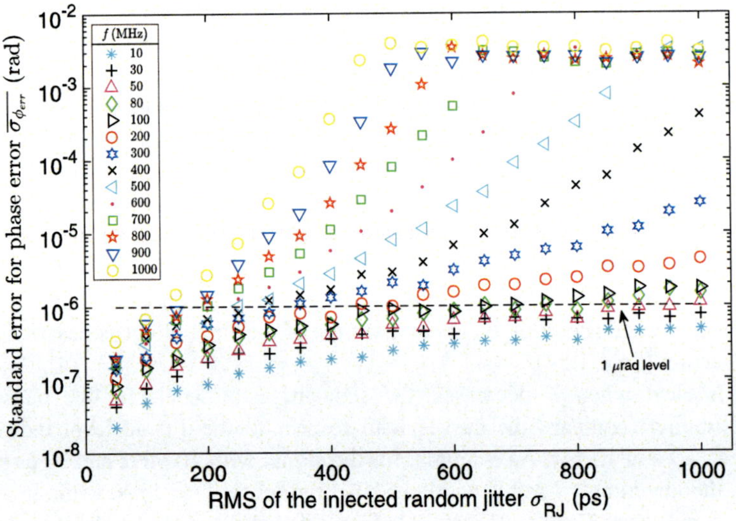


(a)

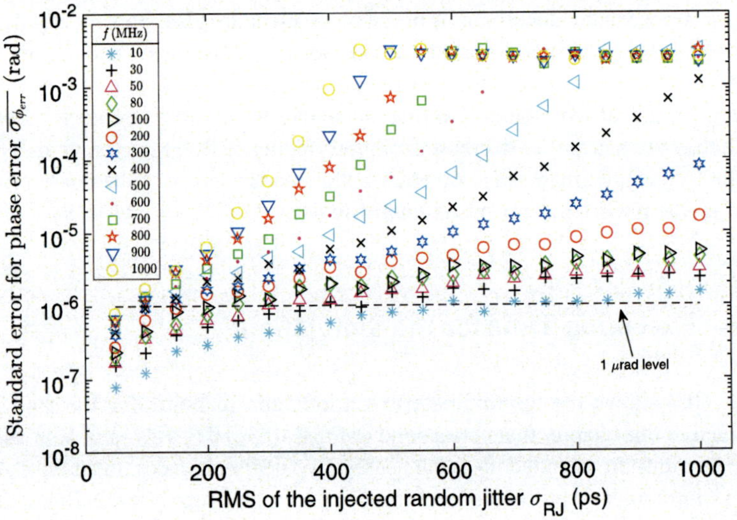


(b)

**Fig. 10.15** Comparison of the uncertainty (standard error) for the typical phase error $\overline{\sigma_{\phi_{err}}}$ against the RMS of the random jitter by using (10.25) with the repetitions nSet = 50 for all modulation frequencies when ToF phase shift $\phi = \pi/6$ with the integration period: **(a)** $T = 1$ ms and **(b)** $T = 0.1$ ms

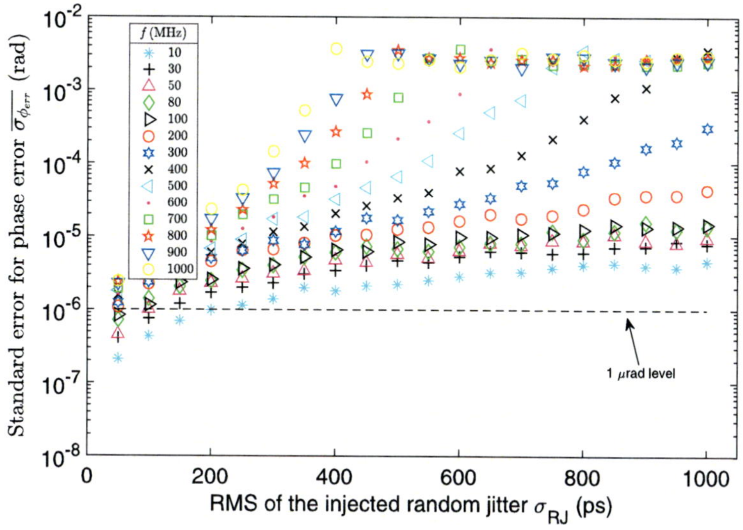


(a)

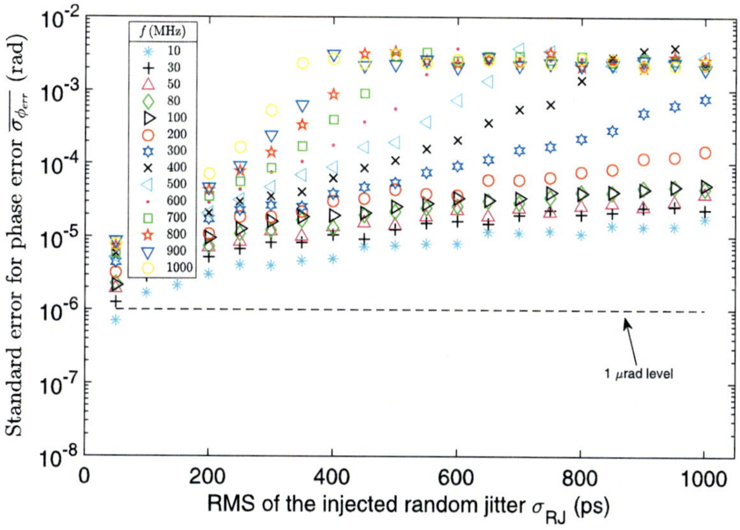


(b)

**Fig. 10.16** Comparison of the uncertainty (standard error) for the typical phase error $\overline{\sigma_{\phi_{\text{err}}}}$ against the RMS of the random jitter by using (10.25) with the repetitions nSet = 50 for all modulation frequencies when ToF phase shift $\phi = \pi/6$ with the integration period: (**a**) $T = 0.01$ ms and (**b**) $T = 0.001$ ms

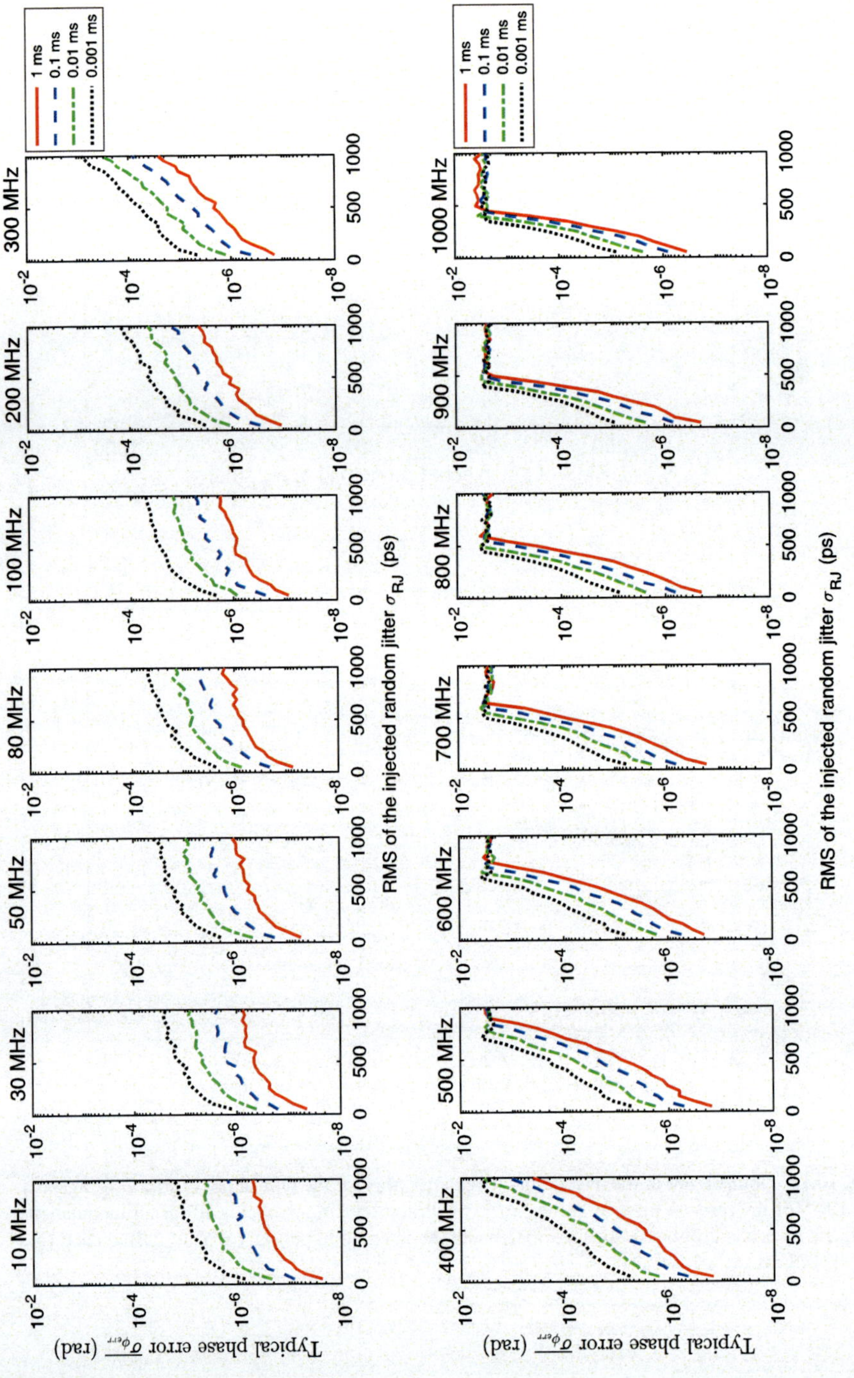


**Fig. 10.17** Uncertainty of the phase error $\overline{\sigma_{\phi_{err}}}$ against the RMS of the RJ for a ToF phase shift $\phi = \pi/6$ by using 500 independent MC simulation with 50 repetitions for 14 modulation frequencies and four integration periods as tabulated in Table 10.1

**Table 10.7** Approximate execution time for simulations on the influence of random jitter on range measurements

| $f$ (MHz) | Parameters used | Execution time (in hours) when $T$ | | | |
|---|---|---|---|---|---|
| | | 1 ms | 0.1 ms | 0.01 ms | 0.001 ms |
| 10 | | 3.51 | 0.33 | 0.07 | 0.05 |
| 30 | ToF phase shift $\phi = \pi/6$, | 9.43 | 0.95 | 0.12 | 0.06 |
| 50 | | 19.71 | 1.54 | 0.17 | 0.06 |
| 80 | $\mu_l = \mu_s = 0\,\text{ps}$, | 26.40 | 2.54 | 0.27 | 0.06 |
| 100 | $(\sigma_l, \sigma_s) = [(0, 0); (50, 51); \ldots$ | 33.56 | 3.12 | 0.33 | 0.07 |
| 200 | $\ldots (950, 951); (1000, 1001)]\,\text{ps}$, | 64.25 | 7.92 | 0.65 | 0.09 |
| 300 | | 95.94 | 9.78 | 0.96 | 0.12 |
| 400 | Samples per cycle $N = 100$, | 139.79 | 13.14 | 1.27 | 0.15 |
| 500 | Number of samples in $T$ is given | 167.75 | 17.01 | 1.57 | 0.18 |
| 600 | as $N_{int} = NfT$; by using (10.30), | 258.34 | 19.32 | 1.91 | 0.20 |
| 700 | | 309.57 | 22.84 | 2.24 | 0.25 |
| 800 | Independent MCS with nEval = 500, | 389.51 | 26.03 | 2.64 | 0.28 |
| 900 | Number of repetitions nSet = 50. | 472.89 | 30.06 | 2.87 | 0.32 |
| 1000 | | 565.11 | 33.89 | 3.10 | 0.34 |

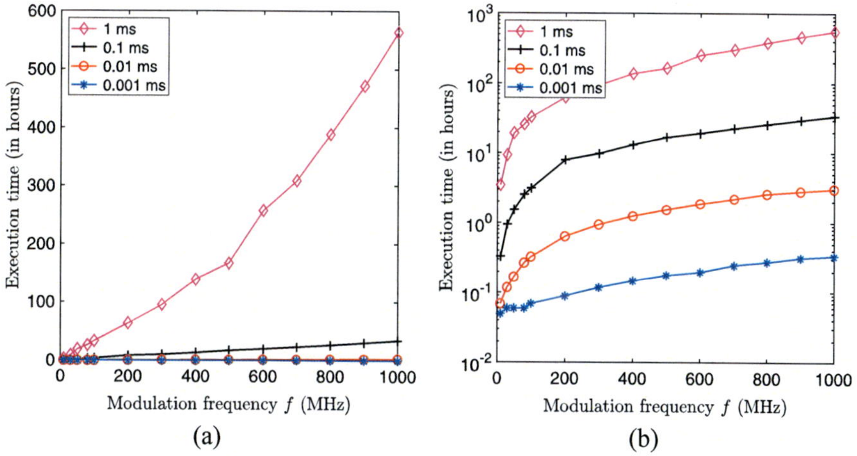


**Fig. 10.18** The graphical representation: **(a)** in decimal scale and **(b)** in logarithm scale with base 10, of the execution time tabulated in Table 10.7

modulation frequencies greater than 800 MHz, 90 GB of memory space is allocated for each simulation setup.

As usual, when the number of samples in the integration period $N_{int}$ increases, the execution time increases. Since a fixed number of samples per cycle ($N = 100$) are used for each simulation, we can obtain the relation $N_{int} \propto fT$ from (10.30). That is, the number of samples in the integration period is proportional to the product of modulation frequency and integration period. Then, for the same modulation frequency, the execution time decreases when the integration period decreases (compare the values from left to right of Table 10.7), while for the same integration period, the execution time increases when the modulation frequency increases (compare the values from top to bottom of Table 10.7). When the integration period is reduced from 1 ms to 0.1 ms (i.e., 10% reduction), the execution time is also decreased by nearly 10% for the modulation frequencies up to 500 MHz. However, this percentage amount has been further reduced approximately from 7.5% to 6% for the frequencies from 600 MHz to 1 GHz, respectively (see Fig. 10.18a). Furthermore, from Fig. 10.18b, the increase of the execution time is reduced when the modulation frequency is increased for each integration period, separately, and they are not in a linear relationship.

## 10.5 Chapter Remarks

The analysis of the influence of random jitter in the correlation function of AMCW ToF cameras was investigated in this chapter. It was assumed the random jitter in the emitted and shutter signals of the correlation function follow two separate Gaussian distributions. The Monte Carlo method would evaluate the correlation function with random jitter, and non-parametric estimations were used to obtain the PDF of the standard deviation of the phase error.

Due to the low modulation frequencies of current AMCW ToF cameras (less than 100 MHz), we found that the random jitter is unable to sufficiently influence the range measurements even in the presence of a relatively large random jitter. As examples, for the modulation frequencies $f = \{30, 50, 100\}$ MHz, the typical phase error $\overline{\sigma_{\phi_{\text{err}}}} = 0.1$ mrad criterion was starting to occur from $\sigma_{\text{RJ}} = \{500, 400, 300\}$ ps of random jitter, respectively, when the integration period $T = 1$ ms, which is the most present ToF cameras operate. Then, the corresponding range errors for this typical phase error 0.1 mrad criterion are given by $\sigma_{d_{\text{err}}} \approx \{0.08, 0.05, 0.02\}$ mm, respectively, which are not sufficient amounts to be detectable in range measurements. Even of RMS $\sigma_{\text{RJ}} = 1$ ns, the typical range errors (standard deviation) for first three modulation frequencies are $\sigma_{d_{\text{err}}} \approx \{0.15, 0.12, 0.09\}$ mm, respectively. If we take $10^{-4}$ rad criterion as when errors could be measured, then at $f = \{30, 50, 100\}$ MHz $\sigma_{\text{RJ}} = \{530, 410, 288\}$ ps can be tolerated but at $f = \{600, 800, 1000\}$ MHz only $\sigma_{\text{RJ}} = \{115, 97, 87\}$ ps can be tolerated, respectively. However, at $f = 200$ MHz, the typical phase error of $\overline{\sigma_{\phi_{\text{err}}}} = 1$ mrad criterion (corresponding range error $\sigma_{d_{\text{err}}} \approx 0.12$ mm) was starting to occur but due to the large RMS $\sigma_{\text{RJ}} \approx 950$ ps.

We found increasing the modulation frequency above approximately 400 MHz with random jitter of 140 ps has a measurable effect on ranging when the criterion of the phase error 0.1 mrad is chosen for the integration period, $T = 1$ ms. This is further reduced when the modulation frequency 600 MHz is given the random jitter of 115 ps. Thus, the potential measurable range errors occurred in the depth measurements. When $T$ is shorter than the current ToF camera's integration period (i.e., less than 1 ms), then this behavior is worst. For example, within the range of the RMS $\sigma_{\mathrm{RJ}} = [400 - 1000]$ ps is occurred among all frequencies considered and the potential measurable range errors are spanned $\sigma_{d_{\mathrm{err}}} \approx [2 - 53]$ mm when $T = 1\,\mu$s.

In addition, by analyzing the uncertainty of the results (the phase error $\overline{\sigma_{\phi_{\mathrm{err}}}}$, hence the range error $\sigma_{d_{\mathrm{err}}}$), we found that fifty repetitions of the simulation set up for higher RMS values especially with higher modulation frequencies are not sufficient. An example, when $T = 1$ ms and the RMS $\sigma_{\mathrm{RJ}} > 800$ ps, the typical phase errors are in the range of (0.6–4) mrad for the modulation frequencies 400 and 600 MHz, which are relatively large values with respect to the corresponding phase error $\overline{\sigma_{\phi_{\mathrm{err}}}}$.

## References


1. Mike Li. *Jitter, Noise, and Signal Integrity at High-speed*. Prentice Hall Press, Upper Saddle River, NJ, USA, first edition, 2007. ISBN 9780132429610.
2. Joseph K Blitzstein and Jessica Hwang. *Introduction to probability*. Chapman and Hall/CRC, 2014.
3. Vijay K Rohatgi and AK Md Ehsanes Saleh. *An introduction to probability and statistics*. John Wiley & Sons, 2015.
4. John R Taylor. *An Introduction to Error Analysis*. University Science Books, Sausalito, California, 1997.
5. Carl Allen Bennett, Norman L Franklin, et al. *Statistical analysis in chemistry and the chemical industry*. Wiley, 1954.
6. Christian Robert and George Casella. *Monte Carlo statistical methods*. Springer Science & Business Media, 2013.
7. Christopher Z Mooney. *Monte Carlo simulation*. Number 116. Sage, 1997.
8. Bernard W Silverman. *Density estimation for statistics and data analysis*. Chapman and Hall/CRC, 1998.
9. Murray Rosenblatt. Remarks on some nonparametric estimates of a density function. *The Annals of Mathematical Statistics*, pages 832–837, 1956.
10. David Freedman and Persi Diaconis. On the histogram as a density estimator: $L_2$ theory. *Probability theory and related fields*, 57(4): 453–476, 1981.
11. Wolfgang Härdle et al. *Smoothing techniques: with implementation in S*. Springer Science & Business Media, 1991.
12. The NeSI High Performance Computers-Mahuika. https://support.nesi.org.nz/hc/en-gb/articles/360000163575, 2021. [Accessed August 2021].

# Chapter 11
# Conclusions and Outlook

Three main sections of the study are covered in Parts II and III of this book. These are the development of a common methodology for jitter extraction and calculation, an alternative cost effective technique for jitter measurement, and the analysis of jitter influence on range measurements in AMCW ToF range imaging cameras. In Part I, a comprehensive literature survey is provided that places the presented study in a context relevant to the task of understanding gaps in the noise and jitter investigation in the AMCW ToF range imaging cameras. Over the last couple of decades, these cameras are commercially available at reasonable prices and have been used in many applications. These cameras have relatively high error sources due to a variety of reasons. The reasonable noise investigations and a couple of studies for the jitter analysis on the AMCW ToF range imaging are found in the literature. These are thoroughly described in Chaps. 2–4, which are the literature review section.

Before this book, there was a huge gap in the analysis of jitter present in the AMCW ToF ranging systems, namely, the jitter extraction, the effect of jitter, what types and what amounts of jitter have more effect on range measurements related to the AMCW ToF ranging. Most of these were largely uninvestigated in the literature. Because of this lack of jitter investigation, the objective of this book became analyzing, experimenting, and designing theoretical models that have not been performed previously. Based on this undertaking, enhancements have been developed and successfully incorporated to significantly improve the quality of the achievements. Thus, this book has contributed to minimizing the abovementioned gaps as much as possible. These are analyzed in the next two parts that correspond to jitter extraction with measurement explored in Chaps. 5–8, and the jitter influence on the AMCW ToF cameras was inspected in Chaps. 9 and 10.

Throughout this book, real and simulated data was used for empirical analysis and verification of algorithms. Thus, the algorithms and the results in this book can be supported to investigate the periodic and random jitter on the illumination signal in most of the AMCW ToF range imaging cameras. The remainder of this

 
G. Anthonys, *Timing Jitter in Time-of-Flight Range Imaging Cameras*,
https://doi.org/10.1007/978-3-030-94159-8_11

last chapter is divided into two sections: a summary of the obtained results with achieved improvements. Next, some interesting extensions for future investigation on the jitter in AMCW ToF ranging are pointed out.

## 11.1 Summary of Findings

The extraction and measurement of jitter in AMCW ToF range imaging cameras were investigated in Part II (Chaps. 5–8). Since it is difficult to access signals inside the camera, the light source of the camera was investigated. An algorithm was proposed for real-time jitter extraction on the emitted light source signal of the AMCW ToF range imaging cameras. The most important advantage of this proposed algorithm is that a reference clock signal is not needed to trigger the time measurement unit and Fourier analysis with some signal processing techniques is used instead. In addition, this algorithm is camera independent and can be used to extract jitter on both digital and analogue signals of any kind of shape (square, sinusoidal, triangular). However, because the modulated light source does not carry a data stream, only RJ and PJ were decomposed from the illumination signal. Other types of jitter are only relevant to communication channels. In order to verify the proposed algorithm, first it was tested on simulated data and the results showed a reasonable accuracy in the jitter measurements. As examples, it was given $(4.5 \pm 0.3)$ ps and $(4.93 \pm 0.04)$ ps when we injected only a RJ of RMS 5 ps and only a PJ of 5 ps, respectively. However, it proved difficult sometimes to separate the PJ accurately when RJ is a large fraction of the total jitter.

Then, the proposed algorithm was experimentally tested with the emitted light signal of two AMCW ToF cameras, namely, MESA Imaging SR4000 and SoftKinetic DS325. For control, a signal with the corresponding modulation frequency of the cameras was generated with a signal generator (HP Agilent 8648B) and measured with the oscilloscope. In order to control the uncertainties, 100 acquisitions for each experimental setup were captured. From the experiments, the periodic jitter was found at two frequencies in both cameras. For the MESA SR4000, the PJ of frequencies of 0.064 MHz and $(5.09 \pm 0.06)$ MHz with the amplitudes of $(71.5 \pm 0.3)$ ps and $(14.0 \pm 1.4)$ ps, respectively. For the DS325, PJ was found at $(0.399 \pm 0.003)$ MHz and $(2.905 \pm 0.009)$ MHz with the amplitudes of $(61.0 \pm 5.2)$ ps and $(36.3 \pm 3.5)$ ps, respectively. Between the two cameras, a relatively larger random jitter was found in the SR4000 with RMS of $(159.6 \pm 0.1)$ ps than for the DS325 with RMS of $(40.37 \pm 0.04)$ ps. We found evidence of flicker noise in the MESA SR4000 camera that was found to behave as $(f^{-0.94 \pm 0.02} \approx 1/f)$. As expected, there was no periodic jitter in the signal generator, and we believe the oscilloscope measurements were very accurate.

Next, the real-time oscilloscope was replaced by a much cheaper measuring instrument, a SDR USB dongle, to measure the jitter on the same two depth cameras. This dongle is based on one of the emerging technologies in telecommunication: software defined radio technology. Interesting results were found in the experiments.

With the comparison of the results from the oscilloscope, we were able to obtain some accurate results with the limitations of the measuring device. They are: (1) the SDR dongle can be used to measure the periodic and random jitter only up to half of the intermediate frequency (IF) obtained from the down shift of the amplified radio frequency (e.g., RF = 30 MHz) with the local oscillator (e.g., LO = 29.6 MHz), which can be substantially less than the Nyquist frequency (e.g., 3.2/2 = 1.6 MHz) of the dongle, (2) when the sampling frequency is less (i.e., if the number of sample points is an insufficient amount in a cycle), then this SDR dongle cannot be used for jitter measurement since the signal is not properly (smoothly) represented. In that situation, it can give incorrect jitter amounts. From the experiments, for SR4000, the periodic jitter at $(62.5 \pm 1.1)$ kHz with the amplitude $(63.5 \pm 1.8)$ ps and random jitter with RMS of $(70.0 \pm 0.2)$ ps are found, which verifies the above limitation (1) where the frequencies are less than half of the IF, that is, $(30 - 29.6)/2\,\text{MHz} = 200\,\text{kHz}$. However, due to the above limitation (2), the experiment on the DS325 camera with this dongle was not carried out. Note that it was needed to use the ratio of the number of samples per cycle between the oscilloscope and SDR dongle, for the calculation of jitter amounts.

The period and random jitter influence on the range measurements is reported in Chaps. 9 and 10, respectively. An analytical model to investigate the influence of the periodic jitter on the range measurements under the heterodyne and homodyne operations in AMCW ToF range imaging cameras was obtained with the help of the Fourier analysis and some standard mathematical functions. The obtained model was tested with simulated data. We found that the phase difference of the periodic jitter between the emitted light and shutter signals of the camera, the integration period and modulation frequency of the camera, the amplitude of the periodic jitter and the relationship of the periodic jitter frequency to the modulation frequency are the factors that influence the periodic jitter on range measurements. However, the influence can be compensated by adjusting the integration period of the camera.

For the chosen phase error criteria $\phi_{\text{err}} = \{0.1, 0.01, 0.001, 0.0001\}$ rad the periodic jitter influences range measurements when the integration period $T = 0.01$ ms for certain amplitudes of the periodic jitter $A_{\text{PJ}}$ and modulation frequencies $f_{\text{PJ}}$. When the periodic jitter is at $f_{\text{PJ}} = 4.9261$ MHz and at $f_{\text{PJ}} = 5$ MHz, there is evidence that the starting influence range of $[A_{\text{PJ}}, f_{\text{PJ}}]$ is proportional to the $T$ with slightly different amounts in each $f_{\text{PJ}}$. Since the two AMCW ToF cameras used have low modulation frequencies (30 MHz for SR4000 and 50 MHz for DS325) with low amplitudes of periodic jitter ($A_{\text{PJ}} = (71.5 \pm 0.3)$ ps and $(14.0 \pm 1.4)$ ps for SR4000, while $A_{\text{PJ}} = (61.0 \pm 5.2)$ and $(36.3 \pm 3.5)$ ps for DS325), we concluded that the periodic jitter in the light sources of each camera do not influence range measurements.

We are recommending when the integration period $T = 0.01$ ms and the modulation frequency of the camera is $f = 200$ MHz, then the amplitude of the PJ should be $A_{\text{PJ}} < 700$ ps, while when $f = 1000$ MHz, it should be $A_{\text{PJ}} < 180$ ps. For the middle values of the parameters, the periodic jitter does not influence when $A_{\text{PJ}} < 400$ ps for $f = 400$ MHz in order to prevent the phase errors of $\phi_{\text{err}} = 0.001$ and 0.01 rad on range measurements. In addition, these recommended values of $A_{\text{PJ}}$

for modulation frequency $f$ should be further reduced when the integration period of the camera is reduced. Thus, the periodic jitter influences current AMCW ToF ranging cameras (since $f < 100\,\text{MHz}$) if the integration period $T \leq 1\,\mu\text{s}$. In future, the depth sensor manufacturers will increase the resolution range of ToF cameras by using high modulation frequencies and very high frame rates that are the factors that affect integration periods of the cameras.

In Chap. 10, the stochastic model for the correlation function of the AMCW range imaging was investigated to identify the influence of the random jitter on range measurements. It was assumed that the random jitter follows the Gaussian distribution. The simulated data was used with the help of 500 independent evaluations of the Monte Carlo method to estimate the result for the model obtained. Two non-parametric estimations were used to identify the PDF for the standard deviation of the phase error $\sigma_{\phi_{\text{err}}}$. Fifty repetitions for each setup with the independent evaluations were performed for various parameters such as modulation frequency $f$, RMS of random jitter $\sigma_{\text{RJ}}$, and four integration periods: $T = \{1, 0.1, 0.01 \text{ and } 0.001\}$ ms. For the random jitter, slightly different RMS amounts $(\sigma_l, \sigma_s)$ for light and shutter signals were used for each setup throughout the simulations. However, for the mean of the random jitter $(\mu_l, \mu_s)$, zero values were chosen since they are unbiased with data; hence, the phase error can be settled by an offset adjustment.

It is found that the random jitter is not sufficiently influenced in current AMCW ToF cameras since their modulation frequencies are relatively low (less than 100 MHz). By analysis, the standard deviation of the phase error (typical) from the stochastic model was evident that the random jitter is starting to influence range measurements when $f = 400\,\text{MHz}$ with random jitter of 140 ps has a measurable effect on ranging for the phase error of $10^{-4}$ rad and this RMS amount is further reduced to 115 ps when $f = 600\,\text{MHz}$. As examples, when $T = 1\,\text{ms}$ and the RMS $\sigma_{\text{RJ}} > 800\,\text{ps}$, the typical phase errors are in the range of (0.6–4) mrad for the modulation frequencies 400 and 600 MHz that are relatively large values with respect to the corresponding $\overline{\sigma_{\phi_{\text{err}}}}$. However, the potentially measurable amounts of the range error when the integration period $T = 1\,\text{ms}$ are within the region of $\sigma_{d_{\text{err}}} \approx [4 - 42]\,\text{mm}$ with the spanning of the RMS $\sigma_{\text{RJ}} = [600 - 1000]\,\text{ps}$ and the frequencies $f = [400 - 1000]\,\text{MHz}$. When $T = 1\,\mu\text{s}$, the potential measurable range errors are spanned $\sigma_{d_{\text{err}}} \approx [2 - 53]\,\text{mm}$ within the range of the RMS $\sigma_{\text{RJ}} = [400{-}1000]\,\text{ps}$ among all frequencies considered, i.e., $f = [10{-}1000]\,\text{MHz}$.

## 11.2 Future Investigations

During the last decade, the ToF range imaging cameras have significantly improved and have been used in various applications from agricultural sectors to engineering industries. With the improvement of the technology to build the depth sensor, novel applications of ToF range imaging will increase. At present, more researchers are mainly focused on the noise investigation in these cameras. However, from this book, we found that the jitter significantly influences range measurements; therefore, it is worth further investigating jitter in ToF range imaging systems. Several investigations could be extended to further perform the ToF range imaging systems due to the jitter influence on measurements as follows.

The proposed methodology in Chaps. 5 and 7 could be extended to correct range measurements after extracting and measuring the RJ and PJ in the light signal of the camera. This will be very useful since the available jitter can be compensated before the measurement. Since for a particular configuration of the AMCW camera (e.g., with default modulation frequency and integration time), the periodic and random jitter in the illumination source are fixed amounts that can be computed with the proposed methodology. Then, considering the benchmark numerical results in Chaps. 9 and 10 for the influence of the periodic and random jitter on range measurements, respectively, can be applied for compensation of the measurement. In addition, the proposed algorithm may be applied to extract the jitter in other types of ToF range imaging cameras such as pulse-based and pseudo-noise based modulation cameras. This methodology can be used for jitter measurement in any kind of application where the reference clock signal cannot be accessed.

A cheaper SDR USB dongle with the proposed algorithm for jitter extraction in ToF cameras at lower frequencies in the RF signal was investigated in Chap. 8. The high sample rates and high quality of SDRs can be used to explore the relatively large range of frequencies of the PJ and RJ in the light source of the AMCW ToF cameras by setting a suitable local oscillator frequency to get a large intermediate frequency (less than the Nyquist frequency) in the dongle. Also, this concept may be used to find the jitter in the light source of the aforementioned two modulation types of ToF cameras as well as in any other relevant applications. In addition, it may be possible to modify the DSP engine in the dongle that is a software based controller in order to get maximum benefit, but this may not be a straightforward process.

In Chap. 9, the influence of the periodic jitter on range measurements was investigated. Can we compensate the PJ amount in the first phase step (by determining and correcting it) before it proceeds to the second phase step and so on, in the correlation function within the integration period? With the influence of the PJ in each phase step is a constant (a fixed amount), then it can be further analyzed with the crosstalking between phase steps, during the integration period. But if it is not a constant (i.e., random behavior), then it will be difficult to compensate for the influence of the PJ on range measurements. Thus, this is a more challenging task for future investigation that arises from Chap. 9. On the other hand, here the periodic jitter at a single frequency was considered. But in practice, the ToF cameras may have periodic jitter at multiple frequencies. Then, it is very interesting to know how the analytical model will change accordingly. Will the influence be linear or non-linear and will it be partially or fully cancel on range measurements? So, some interesting questions arise from this future investigation.

A finite number of evaluations of the Monte Carlo simulation were used to investigate the influence of random jitter on range measurements in Chap. 10. In order to understand the impact of uncertainty of the results obtained, it is necessary to analyze the stochastic model for the full range of possible outcomes, to get much more accurate results. However, according to the central limit theorem, the Gaussian distribution was assumed for random jitter in the correlation function. But is it true, maybe another known or unknown distribution? This study can be further investigated by choosing a different standard distribution or prediction one for the random jitter.

Another interesting extension of this study is what kind of behavior occurs when both periodic and random jitter present in the correlation function. Will it be possible to obtain a comprehensive analytical model to investigate their influence on range measurements?

# Appendix A
# Mathematical Derivations

Here, we briefly introduce the Bessel function, error function, and monotonic property of invertible functions that were mainly used in Chap. 9.

## A.1 Bessel Function

Let $i$ be the imaginary unit for complex $z = x + iy$, where $x, y \in \mathbb{R}$. One of the solutions of the differential equation

$$z^2 \frac{d^2\omega}{dz^2} + z\frac{d\omega}{dz} + \left(z^2 - \nu^2\right)\omega = 0 \tag{A.1}$$

is $J_{\pm\nu}(z)$, the Bessel function of the first kind with indices $\pm\nu$. This is a regular function of $z$ throughout the $z$-plane cut along the negative real axis and for fixed $z(\neq 0)$ is an entire function of $\nu$. When $\nu$ is an integer (i.e., $\pm m$), then $\nu$ has no branch point and is an entire function of $z$. One of the integral representations of the Bessel function of the first kind of order $m$ is given by [1]

$$J_m(x) = \frac{1}{2\pi}\int_{-\pi}^{\pi} e^{i(x\sin\theta - m\theta)}\, d\theta, \tag{A.2}$$

where $\theta$ is any angle in unit of radians by convention.

 
G. Anthonys, *Timing Jitter in Time-of-Flight Range Imaging Cameras*,
https://doi.org/10.1007/978-3-030-94159-8

## A.2 Error Function

The error function is encountered in integrating the normal distribution, which is a normalized form of the Gaussian function. The error function erf(z) of the argument $z$ (real or complex) with respect to $\eta$ is defined by [1]

$$\mathrm{erf(z)} \equiv \frac{2}{\sqrt{\pi}} \int_0^z e^{-\eta^2}\, d\eta. \tag{A.3}$$

Some useful properties of error function [2–5] are

$$\mathrm{erf}(0) = 0, \quad \mathrm{erf}(\infty) = 1, \quad \mathrm{erf}(-z) = -\mathrm{erf}(z). \tag{A.4}$$

## A.3 Monotonic Functions

An invertible function $y = g(x)$ on an interval $(a, b)$ is:

1. Decreasing if $\forall x_1, x_2 \in (a, b) : x_1 < x_2 \implies g(x_1) \geq g(x_2)$.
2. Strictly decreasing if $\forall x_1, x_2 \in (a, b) : x_1 < x_2 \implies g(x_1) > g(x_2)$.
3. Increasing if $\forall x_1, x_2 \in (a, b) : x_1 < x_2 \implies g(x_1) \leq g(x_2)$.
4. Strictly increasing if $\forall x_1, x_2 \in (a, b) : x_1 < x_2 \implies g(x_1) < g(x_2)$.

If a function $y = g(x)$ is invertible on the interval $(a, b)$ and belongs to one of the above types, then this function is called monotonic on the given interval.

## References


1. Milton Abramowitz and Irene A Stegun. *Handbook of mathematical functions: with formulas, graphs, and mathematical tables*, volume 55. Courier Corporation, 1965.
2. Murray Geller and Edward W Ng. A table of integrals of the exponential integral. *Journal of Research of the National Bureau of Standards B*, 73B(3):1–20, 1969.
3. Murray Geller and EW Ng. A table of integrals of the error function. ii. additions and corrections. *J. Res. Natl. Bur. Stand.*, 75:149–163, 1971.
4. Otto Neall Strand. A method for the computation of the error function of a complex variable. *Mathematics of computation*, 19(89):127–129, 1965.
5. HE Salzer. Formulas for calculating the error function of a complex variable. *Mathematical Tables and Other Aids to Computation*, 5(34):67–70, 1951.

# Appendix B
# Statistical Theorems

In this appendix, we point out some definitions and proofs of the standard statistical main theorems used in Chap. 10.

## B.1 Definitions

This includes the *law of the unconscious statistician* theorem and the definition of the *cumulative distribution function* of a continuous random variable.

### *B.1.1 Law of the Unconscious Statistician*

When the probability distribution of a continuos random variable $X$ is known but the same is not known for a function $g(X)$, then we can use the law of the unconscious statistician (LOTUS) theorem to find the expected value of $g(X)$. Thus, the expected value of any function $g : \mathbb{R} \rightarrow \mathbb{R}$ of the variable $X$ is given by [1]

$$\mathbb{E}[g(X)] = \int_{-\infty}^{\infty} g(x) f_X(x)\, dx. \tag{B.1}$$

### *B.1.2 Cumulative Distribution Function of a Random Variable*

The cumulative distribution function (CDF) of a continuous random variable $X$ is a method to describe the distribution of random variables and is defined as [1]



G. Anthonys, *Timing Jitter in Time-of-Flight Range Imaging Cameras*,
https://doi.org/10.1007/978-3-030-94159-8

$$F_X(x) = \mathrm{P}(X \leq x), \qquad \text{for} \quad \forall x \in \mathbb{R}, \tag{B.2}$$

where $\mathrm{P}(X \leq x)$ is the probability that $X$ is less than or equal to an element $x$.

### *B.1.3 Gaussian (Normal) Distribution*

The normal distribution is a probability distribution that associates the random variable $X$ with a cumulative probability and is defined by

$$f_X(x) = \frac{1}{\sigma\sqrt{2\pi}} \exp\left\{ -\frac{1}{2}\left(\frac{x-\mu}{\sigma}\right)^2 \right\}, \tag{B.3}$$

where $x \in X$ and $\mu$ and $\sigma^2$ are the mean and variance of the variable $X$, respectively.

## B.2 Proofs

This section focuses on the proof of some theorems, namely the LOTUS theorem, the Gaussian integral, and the means and variances of some statistical parameters.

### *B.2.1 LOTUS Theorem*

If $X$ is a continuous random variable with probability density function (PDF) $f_X(x)$, then the mean of any function $g : \mathbb{R} \to \mathbb{R}$ is given by

$$\mathbb{E}[g(X)] = \int_{-\infty}^{\infty} g(x) f_X(x)\, dx, \tag{B.4}$$

where $y = g(x)$ for $\forall x \in \mathbb{R}$. Let $Y = g(X)$ be a strictly monotonic function (i.e., $g(x)$ is differentiable while its inverse $g^{-1}(y)$ is also monotonic) and the derivative is given by

$$\frac{d}{dy}\left(g^{-1}(y)\right) = \frac{1}{g'(g^{-1}(y))}. \tag{B.5}$$

By using the change of variables method, the right side of (B.1) can be formulated as

$$\int_{-\infty}^{\infty} g(x) f_X(x)\, dx = \int_{-\infty}^{\infty} y\, f_X(g^{-1}(y)) \frac{1}{g'(g^{-1}(y))}\, dy. \tag{B.6}$$

Also, we have

$$x = g^{-1}(y) \quad \rightarrow \quad dx = \frac{1}{g'(g^{-1}(y))} \qquad (\because \text{(B.5)}). \tag{B.7}$$

On the other hand, from the cumulative distribution function, we have

$$F_Y(y) = \mathrm{P}(Y \leq y) = \mathrm{P}(g(X) \leq y) = \mathrm{P}(X \leq g^{-1}(y)) = F_X(g^{-1}(y)). \tag{B.8}$$

Then, by the chain rule, the PDF of $Y$ is given by

$$f_Y(y) = f_X(g^{-1}(y)) \frac{1}{g'(g^{-1}(y))}. \tag{B.9}$$

Then from (B.6) or (B.9), we have

$$\begin{aligned}
\int_{-\infty}^{\infty} g(x) f_X(x)\, dx &= \int_{-\infty}^{\infty} y\, f_Y(y)\, dy \\
&= \mathbb{E}[Y] \qquad (\because \text{by definition})
\end{aligned}$$

$$\therefore \int_{-\infty}^{\infty} g(x) f_X(x)\, dx = \mathbb{E}[g(X)]. \tag{B.10}$$

### *B.2.2 Gaussian Integral*

The Gaussian integral (also the probability integral) is related to the erf(.), which is the integral of one-dimensional Gaussian function over $(-\infty, \infty)$ as

$$I = \int_{-\infty}^{\infty} e^{-x^2} dx. \tag{B.11}$$

It can be computed by combining two one-dimensional Gaussian variables. Then

$$I^2 = \int_{-\infty}^{\infty} e^{-x^2} dx \int_{-\infty}^{\infty} e^{-y^2} dy = \int_{-\infty}^{\infty} \int_{-\infty}^{\infty} e^{-(x^2+y^2)} dx\, dy.$$

Transforming to polar coordinates, $x^2 + y^2 = r^2$ and $dx\, dy = r\, dr\, d\theta$, we have

$$I = \sqrt{\int_0^{2\pi} \int_0^{\infty} e^{-r^2} r\, dr\, d\theta} = \sqrt{2\pi \left. \frac{e^{-r^2}}{-2} \right|_0^{\infty}} = \sqrt{\pi}. \tag{B.12}$$

Note that this integral can be used to find the corresponding area of a normal distribution for the given interval such as $\mu \pm \sigma$, $\mu \pm 2\sigma$, and $\mu \pm 3\sigma$ with the result approximately 68.27%, 95.45%, and 99.73% of the distribution, respectively.

## B.2.3 *Means and Variances of Some Parameters*

This section proves the means and variances of: a continuous random variable, a linear combination of random variables, and the Gaussian distribution.

### B.2.3.1 A Continuous Random Variable

Let $X$ be a continuous random variable and then the expected value of $X$ is defined by [1]

$$\mathbb{E}[X] = \int_{-\infty}^{\infty} x f_X(x)\, dx, \tag{B.13}$$

and the variance gives

$$\begin{aligned}
\mathbb{V}[X] &= \mathbb{E}\left[(X - \mathbb{E}[X])^2\right] \\
&= \int_{-\infty}^{\infty} (X - \mathbb{E}[X])^2 f_X(x)\, dx \quad (\because \text{(B.1)}) \\
&= \int_{-\infty}^{\infty} x^2 f_X(x)\, dx - 2\,\mathbb{E}[X] \int_{-\infty}^{\infty} x f_X(x)\, dx + (\mathbb{E}[X])^2 \int_{-\infty}^{\infty} f_X(x)\, dx \\
&= \mathbb{E}[X^2] - 2\,(\mathbb{E}[X])^2 + (\mathbb{E}[X])^2 \qquad (\because \text{(B.1)})
\end{aligned}$$

$$\therefore\ \mathbb{V}[X] = \mathbb{E}[X^2] - (\mathbb{E}[X])^2, \tag{B.14}$$

where $\mathbb{E}[X^2]$ is the expectation of the variable $X^2$. Then, from the LOTUS theorem, we have

$$\mathbb{E}[X^2] = \int_{-\infty}^{\infty} x^2 f_X(x)\, dx. \tag{B.15}$$

#### B.2.3.2 A Linear Combination of Random Variables

Let $Y = \sum_{k=1}^{K} \alpha_k X_k$ be the linear combination of $K \in \mathbb{Z}^+$ independent random variables $X_k$ with $\alpha_k$ real constants. Let $\mu_k$ and $\sigma_k^2$ be the mean and variance of the random variable $X_k$, respectively. Then, the mean $\mu_Y$ and variance $\sigma_Y^2$ of $Y$ can be found by

$$\begin{aligned}
\mu_Y = \mathbb{E}[Y] &= \mathbb{E}\left[\sum_{k=1}^{K} \alpha_k X_k\right] \\
&= \mathbb{E}\left[\alpha_1 X_1\right] + \mathbb{E}\left[\alpha_2 X_2\right] + \cdots + \mathbb{E}\left[\alpha_K X_K\right] \\
&= \alpha_1 \mathbb{E}\left[X_1\right] + \alpha_2 \mathbb{E}\left[X_2\right] + \cdots + \alpha_K \mathbb{E}\left[X_K\right] \\
&= \sum_{k=1}^{K} \alpha_k \mathbb{E}\left[X_k\right] = \sum_{k=1}^{K} \alpha_k \mu_k,
\end{aligned} \tag{B.16a}$$

and from the definition of variance in Appendix B.2.3.1, we have

$$\begin{aligned}
\sigma_Y^2 = \mathbb{V}[Y] = \mathbb{E}\left[(Y - \mathbb{E}[Y])^2\right] \\
&= \mathbb{E}\left[\left(\sum_{k=1}^{K} \alpha_k X_k - \sum_{k=1}^{K} \alpha_k \mu_k\right)^2\right] \\
&= \mathbb{E}\left[\left(\sum_{k=1}^{K} \alpha_k \left(X_k - \mu_k\right)\right)^2\right] \\
&= \mathbb{E}\left[\left(\sum_{k=1}^{K} \alpha_k (X_k - \mu_k)\right)\left(\sum_{l=1}^{K} \alpha_l (X_l - \mu_l)\right)\right] \\
&= \sum_{k=1}^{K} \sum_{l=1}^{K} \alpha_k \alpha_l \mathbb{E}[(X_k - \mu_k)(X_l - \mu_l)] \qquad (\because \text{linear property}) \\
&= \begin{cases} \sum_{k=1}^{K} \alpha_k^2 \, \mathbb{E}[(X_k - \mu_k)^2], & l = k \\ 0, & l \neq k \end{cases}
\end{aligned}$$

$$= \begin{cases} \sum_{k=1}^{K} \alpha_k^2 \mathbb{V}[X_k], & l = k \\ 0, & l \neq k \end{cases}$$

$$\therefore \mathbb{V}[Y] = \sum_{k=1}^{K} \alpha_k^2 \sigma_k^2. \tag{B.16b}$$

#### B.2.3.3 The Gaussian Distribution

Consider the definition of the mean given in (B.13). Then, substituting (B.3) into (B.13) gives

$$\mathbb{E}[X] = \int_{-\infty}^{\infty} \frac{x}{\sigma\sqrt{2\pi}} \exp\left\{ -\frac{1}{2}\left(\frac{x-\mu}{\sigma}\right)^2 \right\} dx \tag{B.17}$$

Substituting $t = (x-\mu)/\sqrt{2}\sigma$ into (B.17) gives

$$\begin{aligned} \mathbb{E}[X] &= \frac{1}{\sqrt{\pi}} \int_{-\infty}^{\infty} \left(\sqrt{2}\sigma t + \mu\right) e^{-t^2}\, dt \\ &= \frac{1}{\sqrt{\pi}} \left( \sqrt{2}\sigma \int_{-\infty}^{\infty} t\, e^{-t^2} dt + \mu \int_{-\infty}^{\infty} e^{-t^2}\, dt \right) \\ &= \frac{1}{\sqrt{\pi}} \left( \sqrt{2}\sigma \left. \frac{e^{-t^2}}{-2} \right|_{-\infty}^{\infty} + \mu\sqrt{\pi} \right) \qquad (\because \text{(B.12)}) \end{aligned}$$

$$\therefore \mathbb{E}[X] = \mu. \tag{B.18}$$

Now for the variance, substituting (B.3) or (B.18) into the definition of the variance (B.14) gives

$$\mathbb{V}[X] = \int_{-\infty}^{\infty} x^2 \frac{1}{\sigma\sqrt{2\pi}} \exp\left\{ -\frac{1}{2}\left(\frac{x-\mu}{\sigma}\right)^2 \right\} dx - \mu^2 \tag{B.19}$$

Substituting $t = (x-\mu)/\sqrt{2}\sigma$ into (B.19) gives

$$\begin{aligned} \mathbb{V}[X] &= \frac{1}{\sqrt{\pi}} \int_{-\infty}^{\infty} \left(\sqrt{2}\sigma t + \mu\right)^2 e^{-t^2}\, dt - \mu^2 \\ &= \frac{1}{\sqrt{\pi}} \left( 2\sigma^2 \int_{-\infty}^{\infty} t^2 e^{-t^2} dt + 2\sqrt{2}\mu\sigma \int_{-\infty}^{\infty} t e^{-t^2} dt + \mu^2 \int_{-\infty}^{\infty} e^{-t^2} dt \right) - \mu^2 \\ &= \frac{1}{\sqrt{\pi}} \left( 2\sigma^2 \int_{-\infty}^{\infty} t^2 e^{-t^2} dt + 0 + \mu^2\sqrt{\pi} \right) - \mu^2 \qquad (\because \text{(B.12)}) \end{aligned}$$

$$= \frac{2\sigma^2}{\sqrt{\pi}} \left( \frac{t\, e^{-t^2}}{-2} \bigg|_{-\infty}^{\infty} + \frac{1}{2} \int_{-\infty}^{\infty} e^{-t^2} dt \right) \qquad (\because \text{Integration by parts})$$

$$= \frac{2\sigma^2}{\sqrt{\pi}} \left( 0 + \frac{1}{2}\sqrt{\pi} \right) = \sigma^2. \qquad (\because \text{(B.12)}) \tag{B.20}$$

## Reference

1. Joseph K Blitzstein and Jessica Hwang. *Introduction to probability*. Chapman and Hall/CRC, 2014.

# Appendix C
# Specific Matlab Scripts

In this appendix, some Matlab scripts used in Chaps. 6, 7, 9, and 10 are included.

## C.1 Jitter Measurements on Simulated Data

This section corresponds to Chap. 6 that tested the proposed algorithm described in Sect. 5.2. This includes two files: one for generating the simulation data and the other for extracting the jitter in those simulated signals.

### *C.1.1 Generating Simulation Data: genSG10to50.m*

```
%% Generating simulation data for check the propsed algorithm:
%% File name: genSG10to50.m ---> includes for RJ, PJ and RJ + PJ;
%% Prepared by Gehan Anthonys - on September, 2017;

%% Parameters as for oscilloscope:
sampRate = 8e9;          % 8 GSa/s
Ts = 1/sampRate;         % 125 ps
tt = (0:1:2^14-1)*Ts;    % 65536 = 2^16 data points
lenTT = length(tt);
amplSig = 1;
modFreq = (10:10:50)*1e+6;
%% 1 ---> RJ only, 2 ---> PJ only, 3 ---> both.
boolPjRjBoth = 3;
switch boolPjRjBoth
    case 1
        foldName = 'withRJ\20sets\';
    case 2
        foldName = 'withPJ\20sets\';
    case 3
        foldName = 'withRJ+PJ\20sets\';
end
% for RJ
if boolPjRjBoth == 1 || boolPjRjBoth == 3
    lenPts = lenTT/16 + 1; % no of random numbers: 4097 for ../16
    rmsRJ = 5e-12; % rms: 5, 10, 20, 30, 40, 50 ps
    valRJ = rmsRJ*randn(1, lenPts); % with 5ps (rms) RJ
    for k = 1:lenPts - 1
        coeff = polyfit([tt(16*(k-1)+1), tt(16*k)],[valRJ(k), valRJ(k+1)], 1);
        interPts{k} = coeff(1)*tt(16*(k-1)+1:16*k) + coeff(2);
    end
```

 
G. Anthonys, *Timing Jitter in Time-of-Flight Range Imaging Cameras*,
https://doi.org/10.1007/978-3-030-94159-8

```
    horzIntRJ = interPts{1};
    for k = 2:length(interPts)
        horzIntRJ = horzcat(horzIntRJ, interPts{k});
    end
    intpolRJ = horzIntRJ;   % rms 5ps of RJ
    %%     To check the RJ:
    figure; plot(tt, intpolRJ, '.-'); grid on;
    % XOR —
    consRJ = repmat(rmsRJ, 1, lenTT);   % const rms 5ps of RJ, but the jitter
        % will be cancelled out even the 1st element set to 0,
    consRJ(1) = 0;  % since ideal and data signals aligned with 1st element,
        % THEN, nearly ALL POINTS ARE EXACTLY ALINGED. Therfore, this is USELESS.
    % XOR —
    halfConsRJ = horzcat(zeros(1, lenTT/2), repmat(rmsRJ, 1, lenTT/2));
        % half zero + half const rms 5ps of RJ
end

% for PJs
if boolPjRjBoth == 2 || boolPjRjBoth == 3
    amplPJ1 = 5e-12; % ampl: 5, 10, 20, 30, 40, 50 ps
    freqPJ1 = 4.37e+6;
    amplPJ2 = 0;
    freqPJ2 = 10e+6;
end

nSet = 20;      % no of data sets
for nn = 1:nSet
    for ii = 3:3
        switch boolPjRjBoth
            case 1      % with RJ only
                vol = amplSig*sin(2*pi*modFreq(ii)*(tt + intpolRJ));
                        % with rms value of RJ, XOR —
                % vol = amplSig*sin(2*pi*modFreq(ii)*(tt + consRJ));
                        % with const rms 5ps of RJ, XOR —
                % vol = amplSig*sin(2*pi*modFreq(ii)*(tt + halfConsRJ));
                        % with half const rms 5ps of RJ
            case 2      % with PJs only
                vol = amplSig*sin(2*pi*modFreq(ii)*(tt + amplPJ1*cos(2*pi*freqPJ1*tt) + amplPJ2*cos(2*pi*
                    freqPJ2*tt)));
            case 3      % with (PJ + RJ)
                vol = amplSig*sin(2*pi*modFreq(ii)*(tt + amplPJ1*cos(2*pi*freqPJ1*tt) + intpolRJ));
            otherwise   % WITHOUT jitter
                vol = amplSig*sin(2*pi*modFreq(ii)*tt);
        end
        %%     To check the RJ:
        figure; plot(tt, vol, '.-'); grid on;
        data = [tt', vol'];
        pathDir = strcat('E:\GA\Matlab\JitterAnalysis\wavesSR\jitsCalculation\SR4000\sSG8\', foldName); % sSG6
            , sSG7, sSG8
        cd(pathDir);
        filename = strcat('sSG', num2str(10*ii), '_8GSa', num2str(nn),'.csv');
        csvwrite(filename, data, 0, 0);
    end
end
pathDir = 'E:\GA\Matlab\JitterAnalysis\wavesSR\jitsCalculation\SR4000\';
cd(pathDir);
%% End of the file: genSG10to50.m————
```

## C.1.2 Extracting the Jitter: rstSG10to50.m

```
%% Extracting the jitter which the simulated data by genSG10to50.m
%% File name: rstSG10to50.m —> includes for RJ, PJ and RJ + PJ;
%% Prepared by Gehan Anthonys — on September, 2017;
%%
%% DATA SAVED in .xlsx are NOT in t—domain, they are in f—domain abs(jitter) values, i.e. mean jitter of FFT
        spectra

clear all;
folTOF = 8;
dataDir = strcat('E:\GA\Matlab\JitterAnalysis\wavesSR\jitsCalculation\SR4000\sSG', num2str(folTOF), '\');
cd(dataDir);
spltDataDir = strsplit(dataDir, '\');
simData = 1;     % set 1 for simulation (generated) data and 0 for SigGen data
nSet = 20;
sigFrq = (10:10:50)*1e+6;

ifSmth = 0;      %% 1 for smoothing, 0 for NOT —> specially for simulated data only
ifWndw = 0;      %% 1 for windowing, 0 for NOT, before fft of jitter

if ifWndw == 1
    strWndw = ', with windowing';
    crrctFctrEny = sqrt(8/3);   % for Hann, energy
    crrctFctrAmp = 2;           % for Hann, amplitude
else
```

```
        strWndw = '. without windowing';
        crrctFctrEny = 1;
        crrctFctrAmp = 1;             % for Hann, amplitude
end
for i = 1:5
    if simData == 1                % simulation (generated) data
        sheet = strcat('sSG', num2str(i*10));
        if ifSmth == 1
            if ifWndw == 1
                dataSG = abs(xlsread('rstWndwSimSG10to50.xlsx', sheet, 'B:B'));
                    % in order to generalise take the absolute
            else
                dataSG = abs(xlsread('rstNoWndwSimSG10to50.xlsx', sheet, 'B:B'));
            end
        else
            if ifWndw == 1
                dataSG = abs(xlsread('rstWndwSimSG10to50wo.xlsx', sheet, 'B:B'));
            else
                dataSG = abs(xlsread('rstNoWndwSimSG10to50wo.xlsx', sheet, 'B:B'));
            end
        end
    else
        sheet = strcat('SG', num2str(i*10));          % Sig Gen data
        if ifWndw == 1
            dataSG = abs(xlsread('rstWndwSG10to50.xlsx', sheet, 'B:B'));         % Sig Gen data: fft
        else
            dataSG = abs(xlsread('rstNoWndwSG10to50.xlsx', sheet, 'B:B'));       % Sig Gen data: fft
        end
    end
    valFfSG{i} = dataSG;
    lenFfSG(i) = length(dataSG);
    meanFfSG(i) = mean(dataSG);
    stdFfSG(i) = std(dataSG);
    rmsFfSG(i) = rms(dataSG);

    trshldLvlDataSG = meanFfSG(i);

    jFrq = sigFrq(i)*(0 : lenFfSG(i) - 1)/lenFfSG(i);
    jFrqHlf = sigFrq(i)*(0 : lenFfSG(i)/2 - 1)/lenFfSG(i);

    maxFfSG(i) = max(dataSG);

    % % To find the area under the curve (total, RJ, PJ) by using fft
    % To CHECK: whether have the PJ or not, in the signals
    figure;
    plot(jFrqHlf/1e+6, dataSG(1:end/2)*1e+12, 'o-', jFrqHlf/1e+6, repmat(trshldLvlDataSG, 1, lenFfSG(i)/2)*1e
          +12, '.-');
    grid on;

    sheet = 'SET_A';     %'SET_A - for RJ, 'SET_B - for PJ, 'SET_C - for RJ+PJ
    if strcmp(upper(sheet), 'SET_A') == 1
        thrshld = 3;
    else
        thrshld = 1;
    end

    indMaxFfSG(i) = find(dataSG == maxFfSG(i), 1, 'first'); % peak index,
                                                             % (i.e. @ PJ)
    if indMaxFfSG(i) < length(jFrqHlf) && fix(maxFfSG(i)./meanFfSG(i)) > thrshld
        % if more than 3, ASSUMED exist PJ. 'fix: Round toward zero'
        existPJ = 1;
        freqMaxSG(i) = jFrqHlf(indMaxFfSG(i));
        indPrevMax(i) = indMaxFfSG(i) - 1;
        indNextMax(i) = indMaxFfSG(i) + 1;

        indsLvlPrevMax = find(dataSG(1:indMaxFfSG(i)) > trshldLvlDataSG);
        indsLvlNextMax = find(dataSG(indMaxFfSG(i):end/2) > trshldLvlDataSG);
        % to find prev index wrt max
        if isempty(find(diff(indsLvlPrevMax) == 1, 1))
            indPrevMax(i) = indsLvlPrevMax(end) - 1;
        else
            for aa = length(indsLvlPrevMax):-1:1
                if dataSG(indsLvlPrevMax(aa) - 1) > trshldLvlDataSG
                    indPrevMax(i) = indPrevMax(i) - 1;
                else
                    break;
                end
            end
        end
        % to find next index wrt max
        if isempty(find(diff(indsLvlNextMax) == 1, 1, 'last'))
            indNextMax(i) = indMaxFfSG(i) + 1;
        else
            for zz = 1:length(indsLvlNextMax)
                if dataSG(indMaxFfSG(i) + indsLvlNextMax(zz)) > trshldLvlDataSG
                    if dataSG(indMaxFfSG(i) + indsLvlNextMax(zz)) < dataSG(indMaxFfSG(i) + indsLvlNextMax(zz)
                        - 1)
                        indNextMax(i) = indNextMax(i) + 1;
                    else
```

```
                        indNextMax(i) = indNextMax(i) - 1;
                        break;
                    end
                else
                    break;
                end
            end
        end
    else
        existPJ = 0;
        freqMaxSG(i) = 0;
        indPrevMax(i) = 1;
        indNextMax(i) = floor(lenFfSG(i)/2);
        dataLvlPJ_RJ = 0;
    end
    %% To find the PSD of partial RJ,
    %% we need the intermediate points inside the PJ bar -
    %%—— by using linear interpolation
    intpolPartRJ{i} = interp1([jFrqHlf(indPrevMax(i)) jFrqHlf(indNextMax(i))], ...
                               [dataSG(indPrevMax(i)) dataSG(indNextMax(i))], ...
                               jFrqHlf(indPrevMax(i):indNextMax(i)));

    %% Calculating areas & power of the signal
    %% Half: total area under the curve (ie. PJ + RJ)
        areaFfSG(i) = 1*trapz(dataSG(1:end/2));
        %% Total power spectrum
        psFfSG(i) = sum(dataSG(1:end/2).*dataSG(1:end/2));        % by half
        psFfSG_A(i) = sum(dataSG(1:end/2));      % by amplitude, by half
        psFfSG_byFull(i) = 1*sum(dataSG.*dataSG);         % by full,
            % BOTH ARE SAME when twice of half as, sqrt(2*...)

    if existPJ == 1
        figure;
        plot(jFrqHlf/1e+6, dataSG(1:end/2)*1e+12, 'o-', 'color', 'blue'); grid on;
        if indNextMax(i) <= length(jFrqHlf)
            hold on; plot(jFrqHlf(indPrevMax(i):indNextMax(i))/1e6, intpolPartRJ{i}*1e12, '+-');
            xlabel('Frequecy (MHz)'); ylabel('Amplitude (ps)');
            legend(strcat(num2str(sigFrq(i)/1e6),' MHz'), 'interpolated level');
        end
        hold on;
        fill(jFrqHlf(indPrevMax(i):indNextMax(i))/1e6, dataSG(indPrevMax(i):indNextMax(i))*1e12, 'b');

        % OR, direct calculation from plot (half side)
        lenPJpart = length(dataSG(indPrevMax(i):indNextMax(i)));
        psPJFfSG(i) = sum(dataSG(indPrevMax(i):indNextMax(i)).*dataSG(indPrevMax(i):indNextMax(i)) - ...
                                                intpolPartRJ{i}'.*intpolPartRJ{i}');
        psPJFfSG_A(i) = sum(dataSG(indPrevMax(i):indNextMax(i)) - intpolPartRJ{i}');    % by amplitude
    else
        figure;
        plot(jFrqHlf/1e+6, dataSG(1:end/2)*1e+12, 'o-', 'color', 'blue'); grid on;
        xlabel('Frequecy (MHz)'); ylabel('Amplitude (ps)');
        legend(strcat(num2str(sigFrq(i)/1e6), ' MHz'));

        lenPJpart = 0;
        psPJFfSG(i) = 0;
        psPJFfSG_A(i) = 0;
    end
    psdRJFfSG(i) = psFfSG(i) - psPJFfSG(i);
    psdRJFfSG_A(i) = psFfSG_A(i) - psPJFfSG_A(i);
end
meanFfSG;
stdFfSG;

%% Total jitter, from spectrum
rmsFfSgnl = rmsFfSG.*sqrt(lenFfSG);
    % from spectrum, to check the accuracy of total jitter,
ampFfSgnl = sqrt(2)*rmsFfSgnl;
    % This should be EQUAL to 'calAmpFfSG' given by 'sqrt(signal's POWER)'
%% with correction factor
rmsFfSgnl_cF = rmsFfSG.*sqrt(crrctFctrEny).*sqrt(lenFfSG);
    % corrected on Nov 17, 2018 —> factor should in sqrt()
ampFfSgnl_cF = sqrt(2)*rmsFfSgnl_cF
%%
%% Calculating jitter parameters:
% NOTE: FROM THE SPECTRUM IT GIVES "RMS value of each freq of time signal",
% therefore, to get the amplitude we need to MULTIPLE by sqrt(2) for PJ and
% for Total J. But for RJ it can be get as it is, since RJ measured in RMS
%%——— on Jan 31, 2018.

% Total jitter
if strcmp(upper(sheet), 'SET.A') == 1
        %'SET.A - for RJ, 'SET.B - for PJ, 'SET.C - for RJ+PJ
    calRmsFfSG = sqrt(2*psFfSG);        % since two sided
    calRmsFfSG_cF = sqrt(2*psFfSG)*sqrt(crrctFctrEny)
        % corrected on Nov 17, 2018 —> factor should in sqrt()
else                        % 'SET.B' or 'SET.C'
    calAmpFfSG = sqrt(2)*sqrt(2*psFfSG);        % since two sided
    calAmpFfSG_cF = sqrt(2)*sqrt(2*psFfSG)*sqrt(crrctFctrEny)
        % corrected on Nov 17, 2018 —> factor should in sqrt()
```

```
end

%% for PJ --> [ampl (or rms), freq], by (in f and PSD - domains)
freqPJSG = freqMaxSG
% by PSD is given amplitude, but spectrum is given the rms amounts
calAmplPJSG = sqrt(2)*sqrt(2*psPJFfSG);           % since two sided
calAmplPJSG_cF = sqrt(2)*sqrt(2*psPJFfSG)*sqrt(crrctFctrEny)
        % corrected on Nov 17, 2018 --> factor should in sqrt()

%% for RJ --> [rms], by (in f and PSD - domains)
if strcmp(upper(sheet), 'SET B') == 1        % if PJ only --> SET B
    calRmsRJSG = 0;
    calRmsRJSG_cF = 0
elseif strcmp(upper(sheet), 'SET A') == 1          % if RJ only --> SET B
        calRmsRJSG = sqrt(2)*sqrt(2*psdRJFfSG)       % since two sided
    calRmsRJSG_cF = sqrt(2)*sqrt(2*psdRJFfSG)*crrctFctrEny
    else              % if PJ + RJ  --> SET C
    calRmsRJSG = sqrt(2*psdRJFfSG)      % since two sided
    calRmsRJSG_cF = sqrt(2*psdRJFfSG)*crrctFctrEny
        % corrected on Nov 17, 2018 --> factor should in sqrt()
end
%%
%%  for ploting
for i = 1:5
    padFfSG{i} = num2cell(padarray(valFfSG{i}(1:end/2), (max(lenFfSG)-lenFfSG(i))/2, 0, 'post'));
end
dataSG10 = cell2mat(padFfSG{1});
dataSG20 = cell2mat(padFfSG{2});
dataSG30 = cell2mat(padFfSG{3});
dataSG40 = cell2mat(padFfSG{4});
dataSG50 = cell2mat(padFfSG{5});

jFs = 30e+6;
jffN =  max(lenFfSG);
jFreq = jFs*(0 : jffN-1)/jffN;
jFreqHalf = jFs*(0 : jffN/2-1)/jffN;

if simData == 1       % simulation (generated) data
    deviceSG = 'Simulation data : ';
    measureDev = 'using computer';
else                % for Sig Gen data
    deviceSG = 'Sig/Gen Agilent 8648B : ';
    measureDev = ' by using oscilloscope: Hp infiniium 2.25 GHz, 8 GSa/s';
end
textStr = num2str(calRmsRJSG);

%% FFT spectrum for all together, THIS is better
figure;
plot(jFreqHalf/1e+6, dataSG10*1e+12, 'o-', 'color', 'blue'); hold on;
        %text(1, 13, num2str(calRmsRJSG(1)), 'FontSize', 16);
plot(jFreqHalf/1e+6, dataSG20*1e+12, '+-', 'color', 'magenta'); hold on;
        %text(6,12, num2str(calRmsRJSG(2)), 'FontSize', 16);
plot(jFreqHalf/1e+6, dataSG30*1e+12, 's-', 'color', 'green'); hold on;
        %text(11,11, num2str(calRmsRJSG(3)), 'FontSize', 16);
plot(jFreqHalf/1e+6, dataSG40*1e+12, 'p-', 'color', 'red'); hold on;
        %text(16,10, num2str(calRmsRJSG(4)), 'FontSize', 16);
plot(jFreqHalf/1e+6, dataSG50*1e+12, 'x-', 'color', 'cyan'); grid on;
        %text(21,9, num2str(calRmsRJSG(5)), 'FontSize', 16);%ylim([0 5]*1)
title({[deviceSG, 'FFT spectrum (1st half) of the total jitter (mean) of data sets: ', num2str(nSet), ' @ ',
      cell2mat(spltDataDir(end-1))], ...
          [measureDev, ' for 10 MHz to 50 MHz', strWndw]}, 'FontSize', 12);
xlabel('Frequency (MHz)','FontSize', 24,'FontWeight','bold');
ylabel('Amplitude (ps)','FontSize', 24,'FontWeight','bold');
legend({'10 MHz', '20 MHz', '30 MHz', '40 MHz', '50 MHz'}, 'FontSize', 20);
set(gca,'FontSize',20);
%%
%% save the data
if simData == 1      % simulation (generated) data
    if ifWndw == 1
        xlsName = 'rstWndwSimAEF.xlsx';
            % with windowing before fft of jitter
    else
        xlsName = 'rstNoWndwSimAEF.xlsx';
    end
    [Num,Txt,Raw] = xlsread(xlsName, sheet);
    nRows = size(Raw,1);
    dataAtOut =  strcat('A', num2str(nRows+1));
    dataCal = [sigFrq; freqPJSG; calAmplPJSG; calRmsRJSG];

    xlswrite(xlsName, dataCal', sheet, dataAtOut);
end
dataDir = 'E:\GA\Matlab\JitterAnalysis\wavesSR\jitsCalculation\SR4000\';
cd(dataDir);
%% End of the file: rstSG10to50.m-----------
```

## C.2 Capturing Multiple Sets of Data from Two Oscilloscopes

In order to control the uncertainty of the experimental results, multiple data needs to be captured. This section corresponds to Chap. 7 where multiple sets of data are captured from the oscilloscope and saved automatically to a computer. For medium speed oscilloscope (Keysight Infiniium S-Series DSOS604A High Definition [1]), the automation process can be set up on the device, while for low speed oscilloscope (HP Infiniium 54846B [2]) it is needed to connect via USB/GBIP interface with a Matlab script. The multiple sets of data from the two cameras and the signal generator are captured by the two oscilloscopes, separately.

Since the two cameras are getting warm due to running for a long period, the light signals of the cameras are not captured continuously. Instead, approximately 30 min of slots are used for data capturing in each camera. Cameras are switched off for the cooling process between slots of data capturing.

### *C.2.1 DSOS604A High Definition, 6 GHz and 20 GSa/s*

This medium speed oscilloscope can be automated to capture the signals continuously. The following sequence is used to capture and store the data. Select:

1. The "Trigger" tab
2. On "Trigger Action ..."
3. The box next to "Perform Multipurpose On Trigger"
4. The number of times that needs to save on a trigger under "Max No Of Actions"
5. The "Setup ..." in the "Multipurpose On Trigger" section
6. The "QuickWaveform" under the "Multipurpose" drop down list
7. The file format and the location to save the files

The files are saved in the comma-separated value (*.csv) format.

### *C.2.2 HP Infiniium 54846B, 2.25 GHz and 8 GSa/s*

This is a low speed oscilloscope and can connect via USB/GBIP interface to capture the multiple signals continuously with the following Matlab script.

```
% 03 May, 2017 ——————————————
% To collect the data from HP infinnium 54846B oscilloscope via USB/GBIP interface:
% Capturing 65536 points for SR4000/ DS325/ SigGen 8648B using the oscilloscope
%% ——————————————————

clc      %Clear the MATLAB command window
clear    %Clear the MATLAB variables

% Setup:
instrumentConnected = true;
% Ignore the instrument when doing a test:
if instrumentConnected
    % Instrument
    visaObjOscope = visa('agilent','GPIB0::7::INSTR');
```

```
    % Set the buffer size
    visaObjOscope.InputBufferSize = 100000;
    % Set the timeout value
    visaObjOscope.Timeout = 10;
    % Set the Byte order
    visaObjOscope.ByteOrder = 'littleEndian';
    % Open the connection
    fopen(visaObjOscope);
end

for k = 501:1050        % Use 500 acquisition slots (taken ~30 mins), since the camera is bit heating.
    if instrumentConnected
        fprintf(visaObjOscope, ':DISK:STOR CHAN1, "F:\\SR400\\SR30_8GSa%d.csv", TEXT, XYP, ON', k);
    end
    pause(4);            % About 3.3 sec needs to store each file for SR4000 & DS325
end

if instrumentConnected
    % Check for any instrument errors
    instrumentError = query(visaObjOscope, ':SYST:ERR?')

    % Close the instrument
    fclose(visaObjOscope);
end
```

## C.3 Romberg Integration Algorithm

This script is used in Chap. 9 that evaluates the influence of the periodic jitter in range measurements by using one of the numerical approaches used. This Matlab function computes the integral using the Romberg integration method [3].

```
function [R] = rombf(f,a,b,N)
% 18 June, 2019 ----------------
% Romberg method is used for computing the integral of function f in the range [a, b] using
% at most k Richardson extrapolations. Stop when q extrapolations have been performed.
% Inputs:
%         f - function.
%         a,b - lower and upper bound of the integral
%         N - Romberg table N by N with 2^(N-1) panels.
% Output:
%         R - Romberg table with R[N, N] is the best estimate of the value of the integral.
%
% An example:
%         [R] = rombf(corrFun, -tau/2, tau/2, rmbgN); with number of panels = 2^(rmbgN-1).
%
% Thus, if rmbgN = 18 then number of panels is 2^(17) = 131072.
% %----------------------

h = b - a; np = 1;

% First term R(1,1)
R(1,1) = h*(feval(f,a) + feval(f,b))/2;
for i = 2:N
    h = h/2;
    np = 2*np;
    sum = 0;              % Compute midpoint sum
    for k = 1:2:np-1
        sum = sum + feval(f, a + k*h);
    end
    % First column entry R(i,1)
    R(i,1) = R(i-1,1)/2 + h*sum;
    m = 1;
    % Other columns R(i,2), ... ,R(i,i)
    for k = 2:i           % Repeated Richardson extrapolation
        m = 4*m;
        R(i,k) = R(i,k-1) + (R(i,k-1) - R(i-1,k-1))/(m - 1);
    end
end
```

## C.4 Monte Carlo Simulation

This section corresponds to Chap. 10, which evaluates the influence of the random jitter in range measurements. The section includes two files: one for generating data with the Monte Carlo method [4] and for obtaining the PDF using non-parametric estimations. The second file is for generating the results of the phase error and analyzing the uncertainty of the results.

### *C.4.1 Generating Data with MCS: genMCsRJ.m*

```
%% July 04, 2019
%%— Related to the chapter 8 of thesis —> 'genMCsRJ.m'
%% This is for analyse the effect of the random jitter (RJ)
%% (i.e., epsS & epsL (random variates)) on range measurements by frequency domain.
%% Light function is funL = cos(2*pi*frqL*(t + epsL) - phiToF)
%% Shutter function is funS = cos(2*pi*frqS*(t + epsS) + theta)

clear all;

nTrpz = 60e6;                % [3 5 20 ...]*1e6; % number of steps in 'trapz' integration
nEval = 500;                 % The number of function evaluations
nSet = 30;                   % Repeating the number of set for uncertainty calculation
setNo = 1:nSet;              % for x axis plot

T = 1e-3;                    % integration period in sec (FIXED)
phiToF = (0:6)*pi/6;         % ToF phase shift in rad
rad2deg(phiToF)
frqL = 600e6;                % mod frequency of light signal —> [30 50 200 ... 1000] MHz
frqS = frqL;                 % mod frequency of shutter signal —> [30 50 200 ... 1000] MHz

% Random jitter parameters
muL = 0e-12;                 % mean in sec
sgmL = (0:50:1000)*1e-12;                  %[0 50 100 200 400 ... 1000]*1e-12;     % std dev in sec
muS = 0e-12;                 % mean in sec
sgmS = [0 sgmL(2:end) + 1e-12]; %[0 51 101 201 401 ... 1001]*1e-12;     % std dev in sec

tt = linspace(-T/2, T/2, nTrpz);

%—— Run the correlation function
ifRJ = 1;   % 1 for RJ
for ii = 2:3%length(phiToF)                % 0 30 60 ... degrees
    tic;               % starting the time
    for nL = 2:2%length(sgmL)              % 0 50 100 ... ps
        for nS = nL %2:2:4% length(sgmS)
            for nSt = 1:nSet
                parfor nEvl = 1:nEval
                    esL = muL + randn(nTrpz,1)*sgmL(nL);
                    esS = muS + randn(nTrpz,1)*sgmS(nS);
                    if ifRJ == 0
                        funI = @(tt) exp(2i*pi*frqS*tt).*cos(2*pi*frqL*tt - phiToF(ii));
                    else
                        funI = exp(2i*pi*frqS*(tt + esS')).*cos(2*pi*frqL*(tt + esL') - phiToF(ii));
                    end
                    angIone = angle(pi*trapz(tt, funI'));
                    errAngPhi(nEvl) = abs(angIone) - abs(phiToF(ii));
                end
                stdErrAngPhi(nSt) = std(errAngPhi);
%% Plotting the density estimations
                %{
%% for histogram
                optBWhist = 2/nEval^(1/3)*iqr(errAngPhi);
                if sgmL(nL) == 0
                    noBinsHist = 1;
                else
                    noBinsHist = floor((max(errAngPhi) - min(errAngPhi))/optBWhist);
                end
%% for kde
                optBWkde = 1.06/nEval^(1/5)*min(stdErrAngPhi(nSt), iqr(errAngPhi)/1.34);
                xPos = min(errAngPhi); %min
                yPosH = max(hist(errAngPhi, noBinsHist));
                yPosK = max(ksdensity(errAngPhi));

                figure;
```

```
                subplot(1,2,1); hh = histfit(errAngPhi, noBinsHist); % hist(errAnglone, noBinsHist, 'Normalization','pdf');
                hh(1).FaceColor = 'b';
                text(xPos, yPosH, strcat('\sigma_{\phi_{err}} = ', num2str(stdErrAngPhi)), 'FontSize', 14);
                title(['Histogram for the PDF of phase error \phi_{err} for \phi = ', num2str(phiToF(ii)), ' rad']);
                xlabel('Phase error \phi_{err} (rad)'); ylabel('Frequency count');
                set(gca,'FontSize',20);
                [yVal, xVal] = ksdensity(errAngPhi, 'width', optBWkde);
                subplot(1,2,2); plot(xVal, yVal, 'LineWidth', 1.5, 'Color', 'b');
                text(xPos, yPosK, strcat('\sigma_{\phi_{err}} = ', num2str(stdErrAngPhi)), 'FontSize', 14);
                title(['KDE for the PDF of phase error \phi_{err} for \phi = ', num2str(phiToF(ii)), ' rad']);
                xlabel('Phase error \phi_{err} (rad)'); ylabel('Estimated density function for \phi_{err}');
                    set(gca,'FontSize',20);
                %}
%% In order to calculate the uncertainty we need "(mean +/- std err)" of the standard deviation of phase error
                muErrAngPhi(nSt) = mean(errAngPhi);
                varErrAngPhi(nSt) = var(errAngPhi);
            end
        end
%% meanStdErr... are same for both methods, but for uncStdErr... this method is not given correct answer:
%%    Since, "sqrt(mean(sum of var))" not EQUAL "mean(sum of std)" --> see Chapter 8
        meanStdErrAngPhi1(nL) = mean(stdErrAngPhi);
        uncStdErrAngPhi1(nL) = std(stdErrAngPhi)/sqrt(length(stdErrAngPhi));
%%          OR
        meanStdErrAngPhi(nL) = sqrt(sum(varErrAngPhi)/length(varErrAngPhi));
        uncStdErrAngPhi(nL) = sqrt(sum((stdErrAngPhi - sqrt(sum(varErrAngPhi)/length(varErrAngPhi))).^2))/length(varErrAngPhi);

        %% Plotting std. dev. of the phase error by nSet wise (i.e., repetition wise)
        %{
                figure;
                plot(setNo, stdErrAngPhi{nL}, '-'); grid on;
                title(strcat('Standard deviation of the phase error \phi_{err} for the number of sets : ', ...
        num2str(nSet), ' for \phi = ', num2str(phiToF(ii)), ' rad when f_{mod} = ', num2str(f1/1e6), ...
                    ' MHz, \sigma_l = ', num2str(sgmL(nL)*1e12), ' ps and \sigma_s =', num2str(sgmS(nS)*1e12), ' ps'));
                xlabel('Set number'); ylabel('Standard deviation of phase error \sigma_{\phi_{err}}');
                text(max(setNo)/2, max(stdErrAngPhi{nL}), ...
        strcat('Standard deviation = ', num2str(meanStdErrAngPhi(nL)), ' \pm ', num2str(uncStdErrAngPhi(nL))));
        %}
    end
    endT = toc;                          % completing time
        strT = strcat(num2str(endT/60), ' mins');
    valPhiToF = phiToF(ii);
%% Plotting uncertainty of the std. dev. of the phase error by sigL wise -->
    %%          as STD_\phi_err = \mu_{std of \phi_err} \pm \unc_{std of \phi_err}
    %{
    figure;
    plot(sgmL*1e12, meanStdErrAngPhi, 'o', 'LineWidth', 1, 'MarkerSize', 10, 'color', 'b'); grid on;
    title(strcat('Uncertainty of the standard deviation of the phase error \phi_{err} by sigL wise with sets : ', ...
                num2str(nSet), ' for \phi = ', num2str(valPhiToF), ' rad when f_{mod} = ', num2str(frqL/1e6), ' MHz'));
    xlabel('RMS of the injected random jitter \sigma_{RJ} (ps)');
    ylabel('Typical phase error $\overline{\sigma_{\phi_{err}}}$ (rad)', 'interpreter', 'latex');
    set(gca,'FontSize',20); xlim([0 1050]);
    text(0, 0, strT);
    %}
    %% Save the work space as a .mat file
    %{
    filename = strcat('phaErrRJ', num2str(rad2deg(valPhiToF)), 'deg', num2str(frqL/1e6), 'MHz', num2str(nSet), 'sets.mat');
    dataDir = 'E:\GA\Matlab\JitterAnalysis\wavesSR\jitsCalculation\SR4000\mcsRJ\'; cd(dataDir);
    save(filename, 'nSet', 'frqL', 'sgmL', 'sgmS', 'valPhiToF', 'stdErrAngPhi', 'meanStdErrAngPhi', 'uncStdErrAngPhi', 'strT', '-v7.3');
    %}
end

dataDir = 'E:\GA\Matlab\JitterAnalysis\wavesSR\jitsCalculation\SR4000\'; cd(dataDir);
```

## C.4.2 Results with Uncertainty: rstMCsRJ.m

```
%% Modified on September 12, 2019
%%-- Related to the chapter 8 of thesis --> 'rstMCsRJ.m'
%% This is for -->
%% (1) to plot the phase error for all frequencies on same graph and
%% (2) to analyse the uncertainty of the resulted phase individually

clear all; close all;

phaToF = [30 60];                        % ToF phase shift
```

```
modFrq = [30 50 100 200 400 600 800 1000];                  % Modulation frequencies in MHz
dataDir = 'E:\GA\Matlab\JitterAnalysis\wavesSR\jitsCalculation\SR4000\mcsRJ\'; cd(dataDir);
symMrk = {'*-c', '+-k', '^-m', 'd-g', 's-b', 'o-y', '.-r', 'x-k'};
symMrkUnc = {'*c', '+k', '^m', 'dg', 'sy', 'ob', '.r', 'xk'};
for pp = 1:length(phaToF)
    matName = strcat('phaErrRJ', num2str(phaToF(pp)), '*30sets.mat');
    matFiles = dir(matName);
    %% (1) to plot the phase error ---
    figure;
    for qq = 1:length(matFiles)
        matNameFrqWise = strcat('phaErrRJ', num2str(phaToF(pp)), 'deg', num2str(modFrq(qq)), 'MHz30sets');
        load(matNameFrqWise);
        meanStdErr(qq, :) = meanStdErrAngPhi;
        uncStdErr(qq, :) = uncStdErrAngPhi;
        rmsRJ(qq, :) = sgmL;
        modFrq(qq) = frqL;
%       plot(rmsRJ(qq, :)*1e12, meanStdErr(qq, :), symMrk{qq}, 'LineWidth', 1, 'MarkerSize', 10, 'color',
    symMrk{qq}(end));
        semilogy(rmsRJ(qq, :)*1e12, meanStdErr(qq, :), symMrk{qq}, 'LineWidth', 1, 'MarkerSize', 10, 'color',
            symMrk{qq}(end));
%       errorbar(rmsRJ(qq, :)*1e12, meanStdErr(qq, :), uncStdErr(qq, :), symMrk{qq}, ...
%                'LineWidth', 1, 'MarkerSize', 10, 'color', symMrk{qq}(end));
        strLegNum{qq} = num2str(frqL/1e6); hold on;
    end
    hanLegNum = legend(strLegNum, 'Location', 'northwest', 'FontSize', 20, 'interpreter', 'latex'); legNumTtl
        = get(hanLegNum, 'Title');
    set(legNumTtl, 'String', '{\it f} (MHz)', 'FontWeight', 'normal', 'FontSize', 20, 'interpreter', 'latex');
    xlabel('RMS of the injected random jitter $\sigma_{RJ}$ (ps)', 'interpreter', 'latex');
    ylabel('Typical phase error $\overline{\sigma_{\phi_{err}}}$ (rad)', 'interpreter', 'latex');
    set(gca, 'FontSize',20); xlim([0 1050]); hold off;

    %% (2) to analyse the uncertainty
%%      Individual plot
    %{
    for qq = 1:length(matFiles)
        figure;
        rmsVals = [rmsRJ(qq, :)*1e12, rmsRJ(qq, :)*1e12];
        uncVals = [uncStdErr(qq, :), -uncStdErr(qq, :)];
%% As stems
        stem(rmsVals, uncVals, symMrk{qq}, 'LineWidth', 1, 'MarkerSize', 10, 'color', symMrk{qq}(end));
%% XOR
        semilogy(rmsVals, uncVals, symMrk{qq}, 'LineWidth', 1, 'MarkerSize', 10, 'color', symMrk{qq}(end));
%% XOR
%% As error bars
        errorbar(rmsRJ(qq, :)*1e12, meanStdErr(qq, :), uncStdErr(qq, :), symMrk{qq}, ...
                'LineWidth', 1, 'MarkerSize', 10, 'color', symMrk{qq}(end)); grid on;
        xlabel('RMS of the injected random jitter $\sigma_{RJ}$ (ps)', 'interpreter', 'latex');
        ylabel('Error bars for phase error $\overline{\sigma_{\phi_{err}}}$ (rad)', 'interpreter', 'latex');
        set(gca, 'FontSize',25); xlim([0 1050]); %ylim([-5e-4 5e-4]);
        title(strcat('Uncertainty of the mean of standard deviation of the phase error \overline{\sigma_{\phi_
            {err}}} (rad) ', ...
            num2str(nSet), 'sets for \phi = ', num2str(valPhiToF), ' rad when f_{mod} = ', strLegNum{qq}, '
                MHz'), 'FontSize', 10);
    end
    hold off;
    %}
        %%      Plot the corresponding 0.1 mrad phase error for RJ against Mod freq.
    %{
    modFrq = [30 50 100 200 400 600 800 1000];        %in MHz
    sgmRJ = [530,410,288,203,142,115,97,87]; % in ps
    figure;
    plot(modFrq, sgmRJ, 'ok', 'MarkerSize', 10); hold on;
    legend('RMS \sigma_{RJ}', 'Fitted curve');
    xlabel('Modulation frequency $f$ (MHz)', 'interpreter', 'latex');
    ylabel('Random jitter of RMS $\sigma_{RJ}$ (ps)', 'interpreter', 'latex');
    set(gca, 'FontSize',25, 'FontName', 'Times New Roman'); xlim([0 1050]); ylim([0 550]);
    %}
%%%%%%%%%%%%%%%%%%%%%%%%
    %% (2) to analyse the uncertainty
%%      Individual plot
    %{
    for qq = 1:length(matFiles)
        figure;
        rmsVals = [rmsRJ(qq, :)*1e12, rmsRJ(qq, :)*1e12];
        uncVals = [uncStdErr(qq, :), -uncStdErr(qq, :)];
%% As stems
        stem(rmsVals, uncVals, symMrk{qq}, 'LineWidth', 1, 'MarkerSize', 10, 'color', symMrk{qq}(end));
                set(gca, 'yscal', 'log');
%% XOR
%% As error bars
%           errorbar(rmsRJ(qq, :)*1e12, meanStdErr(qq, :), uncStdErr(qq, :), symMrk{qq}, ...
%                   'LineWidth', 1, 'MarkerSize', 10, 'color', symMrk{qq}(end));
        grid on;
        xlabel('RMS of the injected random jitter $\sigma_{RJ}$ (ps)', 'interpreter', 'latex');
        ylabel('Error bars for phase error $\overline{\sigma_{\phi_{err}}}$ (rad)', 'interpreter', 'latex');
        set(gca, 'FontSize',25, 'FontName', 'Times New Roman'); xlim([0 1050]);
        title(strcat('Uncertainty of the mean of standard deviation of the phase error \overline{\sigma_{\phi_
            {err}}} (rad) ', ...
```

```
                num2str(nSet), 'sets for \phi = ', num2str(valPhiToF), ' rad when f_{mod} = ', strLegNum{qq}, '
                    MHz'), 'FontSize', 10);
    end
    hold off;
    %}
%%      All plots in same figure
    %{
    figure;
    for qq = 1:length(matFiles)
        rmsVals = [rmsRJ(qq, :)*1e12, rmsRJ(qq, :)*1e12];
        uncVals = [uncStdErr(qq, :), -uncStdErr(qq, :)];
        plot(rmsVals, uncVals, symMrk{qq}, 'LineWidth', 1, 'MarkerSize', 10, 'color', symMrk{qq}(end));
        set(gca,'yscal','log'); hold on;
        strLegUnc{qq} = strLegNum{qq};
    end
    hanLegUnc = legend(strLegUnc, 'Location', 'northwest', 'FontSize', 20, 'interpreter', 'latex');
    legUncTtl = get(hanLegUnc,'Title');
    set(legUncTtl,'String', '{\it f} (MHz)', 'FontWeight', 'normal', 'FontSize', 20, 'interpreter', 'latex');
    xlabel('RMS of the injected random jitter $\sigma_{RJ}$ (ps)', 'interpreter', 'latex');
    ylabel('Standard error for phase error $\overline{\sigma_{\phi_{err}}}$ (rad)', 'interpreter', 'latex');
    set(gca,'FontSize',20); xlim([0 1050]); %ylim([-5e-4 5e-4]);
    title(strcat('Uncertainty of the mean of standard deviation of the phase error \overline{\sigma_{\phi_{err
        }}} (rad) ', ...
            num2str(nSet), 'sets for \phi = ', num2str(valPhiToF), ' rad when f_{mod} = ', strLegUnc{1}, ' to
                ', ...
            strLegUnc{end}, ' MHz'), 'FontSize', 10); hold off;
    %}
end

%% Calculating corresponding range errors when sigmaRJ = [200 400 600 800 1000] ps
fMod = 1e6*[30; 50; 200; 400; 600; 800; 1000];          % Modulation frequency in Hz
pErr = [3.777e-5 7.555e-5 1.140e-4 1.517e-4 1.910e-4;
        4.890e-5 9.767e-5 1.482e-4 1.963e-4 2.476e-4;
        6.914e-5 1.387e-4 2.131e-4 2.947e-4 3.820e-4;
        9.864e-5 2.081e-4 3.501e-4 5.897e-4 1.076e-3;
        1.462e-4 4.184e-4 1.536e-3 9.044e-3 8.917e-2;
        2.019e-4 1.252e-3 2.149e-2 7.719e-1 9.047e-1;
        2.988e-4 6.477e-3 7.406e-1 9.128e-1 9.080e-1;
        4.846e-4 5.640e-2 9.099e-1 9.063e-1 9.093e-1];       % Phase error in radian

c = 299792458;                                   % Speed of light ms^(-1)

dErr_mm = (c*pErr./4/pi./fMod)*1e3               % range error in mm
dErr_Rmm = round(dErr_mm, 2)                     % range error round to 2 decimal places (mm)

dataDir = 'E:\GA\Matlab\JitterAnalysis\wavesSR\jitsCalculation\SR4000\'; cd(dataDir);
```

# References


1. *Infiniium S-Series High-Definition Oscilloscopes – The New Standard for Superior Measurements*. Keysight Technologies, August 2018. URL https://literature.cdn.keysight.com/litweb/pdf/5991-3904EN.pdf?id=2447379.
2. *Agilent Technologies - Infiniium 54800 Series Oscilloscopes – Now with Mixed-Signal Oscilloscope, Models and MegaZoom, Deep Memory*. Agilent Technologies, November 2002. URL https://www.keysight.com/en/pd-1000004491%3Aepsg%3Apro-pn-54846B/infiniium-dso?cc=NZ&lc=eng#.
3. Alejandro L Garcia. *Numerical methods for physics*. Prentice Hall Englewood Cliffs, NJ, 2 edition, 2000.
4. Christopher Z Mooney. *Monte Carlo simulation*. Number 116. Sage, 1997.

# Appendix D
# Maple Computations

In this appendix, some expressions obtained in Chap. 9 are verified by using the mathematical tool, Maple ver. 16.02 by Maplesoft, and are included.

## D.1 Correlation Model with Periodic Jitter

This corresponds to Sect. 9.3. For convenience, both the reflect and shutter signals are assumed to have a unity amplitude. The expression for the background portion is named as *nonDCpart*.



G. Anthonys, *Timing Jitter in Time-of-Flight Range Imaging Cameras*,
https://doi.org/10.1007/978-3-030-94159-8

$restart;$
$Parameters\left(f_l, f_s, \Theta, \phi, T, A_{pj}, f_{pj}, \beta_l, \beta_s, n, m\right):$

$with(inttrans):$
$u := \text{Heaviside} : rect := t \rightarrow u\left(t+\frac{T}{2}\right) - u\left(t-\frac{T}{2}\right):$

$T := 1:$
$plot(rect(t), t=-1..1, thickness=2)$

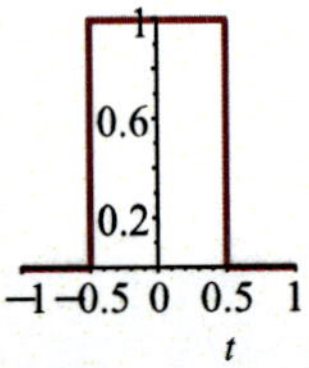


$lt := \cos\left(2\cdot\pi\cdot f_l\cdot\left(t+A_{pj}\cdot\sin\left(2\cdot\pi\cdot f_{pj}\cdot t\right)\right)-\phi\right)$

$$lt := \cos\left(-2\,\pi f_l\left(t+A_{pj}\sin\left(2\,\pi f_{pj}\,t\right)\right)+\phi\right) \quad (1)$$

$eTht := \exp\left(I\cdot f_s\cdot 2\cdot\pi\cdot\Theta\cdot t\right)$

$$eTht := \mathrm{e}^{2 I f_s \pi \Theta t} \quad (2)$$

$eJt := \exp\left(I\cdot 2\cdot\pi\cdot f_s\cdot\Theta\cdot A_{pj}\cdot\sin\left(2\cdot\pi\cdot f_{pj}\cdot t+\phi^s_{pj}\right)\right)$

$$eJt := \mathrm{e}^{2 I \pi f_s \Theta A_{pj} \sin\left(2\pi f_{pj} t+\phi^s_{pj}\right)} \quad (3)$$

$nonDCpart := \int_{-\frac{T}{2}}^{\frac{T}{2}} lt\cdot eJt\cdot eTht\,\mathrm{d}t$

$$nonDCpart := \int_{-\frac{1}{2}}^{\frac{1}{2}} \cos\left(-2\,\pi f_l\left(t+A_{pj}\sin\left(2\,\pi f_{pj}\,t\right)\right)+\phi\right)\mathrm{e}^{2 I \pi f_s \Theta A_{pj} \sin\left(2\pi f_{pj} t+\phi^s_{pj}\right)}\,\mathrm{e}^{2 I f_s \pi \Theta t}\,\mathrm{d}t \quad (4)$$

$FTnonDCpart := fourier\left(rect(t)\cdot\cos\left(2\cdot\pi\cdot\left(f_l+n\cdot f_{pj}\right)\cdot t-\phi\right),\ t,\ 2\cdot\pi\cdot\left(f_s\cdot\Theta+m\cdot f_{pj}\right)\right)$

$$FTnonDCpart := \frac{\sin\left(-\pi\,n\,f_{pj}-\pi f_l+\pi\left(f_s\Theta+m f_{pj}\right)\right)\mathrm{e}^{-I\phi}}{2\,\pi\,n\,f_{pj}+2\,\pi f_l-2\,\pi\left(f_s\Theta+m f_{pj}\right)} + \frac{\sin\left(\pi\,n\,f_{pj}+\pi f_l+\pi\left(f_s\Theta+m f_{pj}\right)\right)\mathrm{e}^{I\phi}}{2\,\pi\,n\,f_{pj}+2\,\pi f_l+2\,\pi\left(f_s\Theta+m f_{pj}\right)} \quad (5)$$

## D.2 Correlation Model without Periodic Jitter

This corresponds to Sect. 9.4. As before, both the reflect and shutter signals are assumed to have a unity amplitude. The expressions for the amplitude and background portions are named as *DCpart* and *nonDCpart*, respectively.

*restart*
*Parameters*$(f_l, f_s, \Theta, \phi, T)$ :
*with*(*inttrans*) :
$u :=$ Heaviside : $rect := t \rightarrow \left(u\left(t+\frac{T}{2}\right) - u\left(t-\frac{T}{2}\right)\right)$ :
$T := 1$ :
*plot*(*rect*(*t*), *t* = −1 ..1, *thickness* = 2)

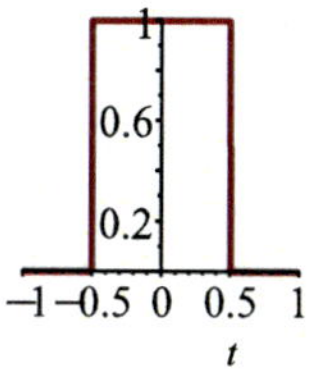


*with*(*DynamicSystems*) :

$$nonDCpart := \int_{-\frac{T}{2}}^{\frac{T}{2}} \cos(2\cdot\pi\cdot f_l\cdot t - \phi)\cdot\exp(I\cdot 2\cdot\pi\cdot f_s\cdot\Theta\cdot t)\,\mathrm{d}t$$

$$nonDCpart := \frac{1}{2\,\pi\left(\Theta^2 f_s^2 - f_l^2\right)}\Big(\Big(\mathrm{I}\,\mathrm{e}^{2\,\mathrm{I}\,\pi f_s\Theta}\cos(\pi f_l)\cos(\phi)\,\Theta f_s + \mathrm{I}\,\mathrm{e}^{2\,\mathrm{I}\,\pi f_s\Theta}\sin(\pi f_l)\sin(\phi)\,\Theta f_s \tag{1}$$
$$- \mathrm{e}^{2\,\mathrm{I}\,\pi f_s\Theta}\cos(\pi f_l)\sin(\phi)\,f_l + \mathrm{e}^{2\,\mathrm{I}\,\pi f_s\Theta}\cos(\phi)\sin(\pi f_l)\,f_l - \mathrm{I} f_s\Theta\cos(\pi f_l)\cos(\phi)$$
$$+ \mathrm{I} f_s\Theta\sin(\pi f_l)\sin(\phi) + f_l\cos(\pi f_l)\sin(\phi) + f_l\sin(\pi f_l)\cos(\phi)\Big)\,\mathrm{e}^{-\mathrm{I}\,\pi f_s\Theta}\Big)$$

$FTnonDCpart := fourier(rect(t)\cdot\cos(2\cdot\pi\cdot f_l\cdot t - \phi),\ t,\ 2\cdot\pi\cdot f_s\cdot\Theta)$

$$FTnonDCpart := \frac{\sin(\pi\,\Theta f_s + \pi f_l)\,\mathrm{e}^{\mathrm{I}\phi}}{2\,\pi\,\Theta f_s + 2\,\pi f_l} + \frac{\mathrm{e}^{-\mathrm{I}\phi}\sin(\pi\,\Theta f_s - \pi f_l)}{2\,\pi\,\Theta f_s - 2\,\pi f_l} \tag{2}$$

$$DCpart := \int_{-\frac{T}{2}}^{\frac{T}{2}} \exp(I\cdot 2\cdot\pi\cdot f_s\cdot\Theta\cdot t)\,\mathrm{d}t$$

$$DCpart := \frac{-\frac{1}{2}\left(\mathrm{e}^{2\,\mathrm{I}\,\pi f_s\Theta} - 1\right)\mathrm{e}^{-\mathrm{I}\,\pi f_s\Theta}}{\pi\,\Theta f_s} \tag{3}$$

$FTDCpart := fourier(\mathrm{rect}(t), t, 2\cdot\pi\cdot f_s\cdot\Theta)$

$$FTDCpart := \frac{\sin(\pi\,\Theta f_s)}{\pi\,\Theta f_s} \tag{4}$$

## D.3 Verification of the Analytical Models

In order to verify the models obtained in Sects. 9.3 and 9.4 with the Maple computations, it is needed to use the sifting property, which was defined by [1]

$$\int_{-\infty}^{\infty} \delta(t-a)\,g(t)\,dt = g(a), \tag{D.1}$$

where $g(t)$ is continuous function and finite at $g(a)$. Then, by using (D.1), Eqs (5) and (2) in Sects. D.1 and D.2, respectively, can be shown as

$$
\begin{aligned}
&\frac{e^{-i\phi}}{2}\operatorname{sinc}(\pi(f_s\Theta - f_l + (m-n)f_{\mathrm{PJ}})) + \frac{e^{i\phi}}{2}\operatorname{sinc}(\pi(f_s\Theta + f_l + (m+n)f_{\mathrm{PJ}}))\\
&= \frac{1}{2}e^{-i\left(\frac{f_s\Theta + m f_{\mathrm{PJ}}}{f_l + n f_{\mathrm{PJ}}}\right)\phi}\Big[\operatorname{sinc}(\pi(f_s\Theta + m f_{\mathrm{PJ}} - f_l - n f_{\mathrm{PJ}}))\\
&\quad + \operatorname{sinc}(\pi(f_s\Theta + m f_{\mathrm{PJ}} + f_l + n f_{\mathrm{PJ}}))\Big] \qquad \text{(D.2)}
\end{aligned}
$$

and

$$
\begin{aligned}
&\frac{e^{i\phi}}{2}\operatorname{sinc}(\pi(\Theta f_s + f_l)) + \frac{e^{-i\phi}}{2}\operatorname{sinc}(\pi(\Theta f_s - f_l))\\
&= \frac{e^{i\left(\frac{\Theta f_s}{-f_l}\right)\phi}}{2}\operatorname{sinc}(\pi(\Theta f_s + f_l)) + \frac{e^{-i\left(\frac{\Theta f_s}{f_l}\right)\phi}}{2}\operatorname{sinc}(\pi(\Theta f_s - f_l))\\
&= \frac{e^{-i\left(\frac{\Theta f_s}{f_l}\right)\phi}}{2}\big[\operatorname{sinc}(\pi(\Theta f_s + f_l)) + \operatorname{sinc}(\pi(\Theta f_s - f_l))\big]. \qquad \text{(D.3)}
\end{aligned}
$$

## Reference

1. Alan V. Oppenheim, Ronald W. Schafer, and John R. Buck. *Discrete-time Signal Processing (2nd Ed.)*. Prentice-Hall, Inc., Upper Saddle River, NJ, USA, 1999. ISBN 0-13-754920-2.

# Appendix E
# Units and Uncertainty

This appendix point outs the units of measurement and some useful formulas for calculating the uncertainty of the measurements used in this book.

## E.1 Units of Measurement

All measurements consist of a unit that discloses what is measured and a number that expresses how many units are measured (Tables E.1 and E.2) [1].

**Table E.1** Units of the measurements used in this book

| Property | Main unit | Abbreviation |
|---|---|---|
| Angle | radian or degree | rad or ° |
| Angular frequency | per radian | $rad^{-1}$ |
| Current | ampere | A |
| Distance | meter | m |
| Frequency | hertz | Hz |
| Time | second | s |
| Velocity | meter per second | m/s |
| Voltage | volt | V |

### *Examples*

5 GHz = 5 gigahertz = 5000000 hertz
1 kHz = 1 kilohertz = 1000 hertz
2 ms = 2 milliseconds = 0.002 seconds



G. Anthonys, *Timing Jitter in Time-of-Flight Range Imaging Cameras*,
https://doi.org/10.1007/978-3-030-94159-8

**Table E.2** Prefixes of the units of measurements

| Prefix | Abbreviation | Factor w.r.t. the main unit |
|---|---|---|
| giga | G | $10^9$ |
| mega | M | $10^6$ |
| kilo | k | $10^3$ |
| centi | c | $10^{-2}$ |
| milli | m | $10^{-3}$ |
| micro | $\mu$ | $10^{-6}$ |
| nano | n | $10^{-9}$ |
| pico | p | $10^{-12}$ |
| femto | f | $10^{-15}$ |

1 ns = 1 nanoseconds = 0.000000001 seconds
3 fs = 3 femtoseconds = 0.000000000000003 seconds

## E.2 Uncertainty of Measurement

For each measurement, the errors are quantified by associating an uncertainty. The repeating measurements allow to know a better idea of the actual value and also enable to characterize the uncertainty of the measurement.

Let $\{x_1, x_2, x_3, \dots x_N : N \in \mathbb{Z}^+\}$ be the measurements from an experiment by repeating $N$ times. Table E.3 shows some useful expressions for calculating uncertainty of measurements [2, 3].

**Table E.3** Used expressions for uncertainty calculation

| (Typical) The average of all measurements $x_{\text{avg}} = \frac{1}{N}\sum_{i=1}^{N} x_i$ | | |
|---|---|---|
| *for* | $N \leq 10$ | $N > 10$ |
| *where* | | |
| *Uncertainty* | | |
| A measurement $\delta x$ when $x = \{x_1, x_2, \dots x_N\}$ | $\delta x = \frac{\max(x)-\min(x)}{2}$ | $\sigma = \sqrt{\frac{\sum_{i=1}^{N}(x_i - x_{\text{avg}})^2}{N}}$ |
| the mean $\delta x_{\text{avg}}$ | $\frac{\delta x}{\sqrt{N}}$ | $\frac{\sigma}{\sqrt{N}}$ |
| With confidence level of 68.27%, the measured value $x_m = x_{\text{avg}} \pm \delta x_{\text{avg}}$ | | |
| The average of the measured values with uncertainty $x_{m_{\text{avg}}}$ $x_{m_{\text{avg}}} = \frac{1}{M}\sum_{j=1}^{M}\left(x_{\text{avg}_j} \pm \delta x_{\text{avg}_j}\right)$ | | |

# References

1. Russ Rowlett, *How many?: a dictionary of units of measurement*. University of North Carolina, 2005.
2. John R Taylor. *An Introduction to Error Analysis*. University Science Books, Sausalito, California, 1997.
3. Philip R Bevington and D Keith Robinson. *Data reduction and error analysis for the physical sciences*. McGraw-Hill, 3 edition, 2003.

# Index





G. Anthonys, *Timing Jitter in Time-of-Flight Range Imaging Cameras*,
https://doi.org/10.1007/978-3-030-94159-8